Lecture Notes in Economics and Mathematical Systems

T5-ARJ-298

505

Springer
Berlin
Heidelberg
New York
Barcelona
Hong Kong
London
Milan
Paris
Singapore
Tokyo

Stefan Voß
Joachim R. Daduna (Eds.)

Computer-Aided Scheduling of Public Transport

 Springer

Editors

Prof. Dr. Stefan Voß
University of Technology Braunschweig
Department of Business Administration,
Information Systems and Information Management
Abt-Jerusalem-Straße 7
38106 Braunschweig, Germany

Prof. Dr. Joachim R. Daduna
University of Applied Business Administration Berlin
Badensche Straße 50–51
10825 Berlin, Germany

Cataloging-in-Publication data applied for

Die Deutsche Bibliothek - CIP-Einheitsaufnahme

Computer aided scheduling of public transport / Stefan Voß ; Joachim R.
Daduna (ed.). - Berlin ; Heidelberg ; New York ; Barcelona ; Hong Kong ;
London ; Milan ; Paris ; Singapore ; Tokyo : Springer, 2001
 (Lecture notes in economics and mathematical systems ; 505)
 ISBN 3-540-42243-9

ISSN 0075-8450
ISBN 3-540-42243-9 Springer-Verlag Berlin Heidelberg New York

Springer-Verlag Berlin Heidelberg New York
a member of BertelsmannSpringer Science+Business Media GmbH

http://www.springer.de

© Springer-Verlag Berlin Heidelberg 2001
Printed in Germany

Typesetting: Camera ready by the editors
Printed on acid-free paper   SPIN: 10796653   55/3142/du   5 4 3 2 1 0

Preface

This proceedings volume consists of selected papers presented at the Eighth International Conference on *Computer-Aided Scheduling of Public Transport* (CASPT 2000), which was held at the conference center of the Konrad-Adenauer-Foundation in Berlin, Germany, from June 21st to 23rd, 2000. The CASPT 2000 is the continuation of a series of international workshops and conferences presenting recent research and progress in computer-aided scheduling in public transport. Previous workshops and conferences were held in

- Chicago (1975),
- Leeds (1980),
- Montréal (1983 and 1990),
- Hamburg (1987),
- Lisbon (1993) and
- Cambridge, Mass. (1997).[1]

With CASPT 2000, our series of workshops and conferences celebrated its 25th anniversary. Starting with a *Workshop on Automated Techniques for Scheduling of Vehicle Operators for Urban Public Transportation Services* in 1975 the scope and purpose has broadened since and still continues to do so. The previous workshops and conferences were focused on public mass transit, and while this remained the primary focus of the 2000 conference, it included also computer-aided scheduling methods being developed and applied in related means of passenger transport systems. Commonalities regarding operations research techniques such as, e.g., column generation techniques and

---

[1] While there were no formal proceedings for the first workshop but only a pre-printed copy of all papers issued to participants on arrival, the subsequent ones are well documented as follows:

Wren, A. (Ed.) (1981). *Computer Scheduling of Public Transport.* North-Holland, Amsterdam.

Rousseau, J.-M. (Ed.) (1985). *Computer Scheduling of Public Transport, 2.* North-Holland, Amsterdam.

Daduna, J.R. and A. Wren (Eds.) (1988). *Computer-Aided Transit Scheduling, Lecture Notes in Economics and Mathematical Systems,* 308. Springer, Berlin.

Desrochers, M. and J.-M. Rousseau (Eds.) (1992). *Computer-Aided Transit Scheduling, Lecture Notes in Economics and Mathematical Systems,* 386. Springer, Berlin.

Daduna, J.R., I. Branco, and J.M.P. Paixão (Eds.) (1995). *Computer-Aided Transit Scheduling, Lecture Notes in Economics and Mathematical Systems,* 430. Springer, Berlin.

Wilson, N.H.M. (Ed.) (1999). *Computer-Aided Transit Scheduling, Lecture Notes in Economics and Mathematical Systems,* 471. Springer, Berlin.

meta-heuristics have arisen in public transit as well as railways and air traffic. Moreover, the increased availability of information technology allows the effective use of transactional data and will also enhance planning functionality and its integration into efficient information systems. Successful developments in research and applications are reported.

The conference was addressed to individuals from transport operators, academic institutions and consulting firms who are involved in research, development and application of computer-aided scheduling in public transport. During the conference a total of 58 papers were presented from various fields of research and application. As in previous conferences, a parallel exhibition of software demonstrations accompanied the conference to demonstrate state-of-the-art systems for computer-aided planning in public transport. Out of the 42 submitted full papers we strived to select a high quality subset to be included in this proceedings volume.

The papers are organized in five parts, in accordance with the main topics of the CASPT 2000. Following the tradition, we have put together the first two parts dealing with vehicle and crew scheduling. While it is yet open whether integrated planning is possible, it is believed that advances are seen on both sides, the methodical one as well as the application oriented one. Advances in operations research techniques allow a more detailed modeling while modeling issues ask for better techniques – an ongoing challenge.

The third part deals with advanced transit service including integrated transit and demand responsive service. The fourth part is devoted to monitoring and control, important issues in public transit, for railways as well as for air traffic. Finally, strategic aspects such as investments into railway networks and revenue management strategies are considered in the fifth part.

While we believe that these proceedings provide insights into the state-of-the-art of the field, we also hope and know that the story is never-ending. That is, new advances on different levels are expected, taking into consideration algorithmic innovations, computing devices as well as information and communication technology that allow us to incorporate more and more planning functionality into our systems. Therefore, we not only look back at the present conference and its contributions but also at future conferences reporting on further advances. Following a good custom, the conferences are alternating between North America and Europe. While this may change in the future with offers from other continents we follow this tradition in the sense that the relay goes to America – and we break the tradition as it goes to South America for the first time, i.e., it is intended to be held in Brazil organized by Nicolau D.F. Gualda, Sao Paulo.

Acknowledgements

Of course such an ambitious project like organizing the conference and publishing a high quality proceedings to reflect its outcome would not have been possible without the most valuable input of a large number of people. The sci-

entific program was put together with the help of the international program
committee, consisting of the following members:

- Joachim R. Daduna, University of Applied Business Administration Berlin, Berlin, Germany (*General Co-Chair*)
- Raymond S.K. Kwan, University of Leeds, Leeds, United Kingdom
- Avi Ceder, Technion-Israel Institute of Technology, Haifa, Israel
- Uwe Pape, University of Technology Berlin, Berlin, Germany
- Jean-Marc Rousseau, GIRO Inc. and University of Montréal, Montréal, Canada
- Paolo Toth, University of Bologna, Bologna, Italy
- Stefan Voß, University of Technology Braunschweig, Braunschweig, Germany (*General Co-Chair*)
- Nigel H.M. Wilson, Massachusetts Institute of Technology, Cambridge, USA

Moreover, we wish to thank all the authors for their contributions, the referees (see Appendix 1 for an appropriate list) for their input and the attendees for their participation and fruitful discussion. As not all papers have been included in these proceedings, Appendix 2 provides a list of those presentations made at the conference without being documented here. Last but not least we acknowledge the help we achieved by some sponsors; a list of all sponsors and exhibitors is given in Appendix 3.

Running a conference and compiling the proceedings cannot be done by one or two general co-chairs alone and we received valuable help from various sides. Most of all we have to thank the whole Braunschweig team for their support and understanding. While we do not mention everybody individually, we emphasize and appreciate the help and support we obtained from Jürgen Böse. In particular, we are also indebted to Andreas Fink, Kai Gutenschwager, Torsten Reiners, Gabriele Schneidereit and Thomas Wiedemann for their enthusiasm and help when running the conference and throughout the project, and to Werner Müller at Springer-Verlag for his help and encouragement.

The Editors:

Stefan Voß, Braunschweig
Joachim R. Daduna, Berlin

April 2001

Contents

Part 1: Vehicle and Crew Scheduling – Methodical Advances

A New Model for the Mass Transit Crew Scheduling Problem   1
Mohamadreza Banihashemi, Ali Haghani

A Global Method for Crew Planning in Railway Applications.   17
Alberto Caprara, Michele Monaci, Paolo Toth

Efficient Timetabling and Vehicle Scheduling for Public
Transport.. 37
Avi Ceder

Rail Crew Scheduling and Rostering Optimization Algorithms   53
*Andreas Ernst, Houyuan Jiang, Mohan Krishnamoorthy, Helen Nott,
David Sier*

Applying an Integrated Approach to Vehicle and Crew
Scheduling in Practice 73
Richard Freling, Dennis Huisman, Albert P.M. Wagelmans

A Network Flow Approach to Crew Scheduling Based on an
Analogy to an Aircraft/Train Maintenance Routing Problem .   91
Taïeb Mellouli

Tabu Search for Driver Scheduling............................ 121
Yindong Shen, Raymond S.K. Kwan

Part 2: Vehicle and Crew Scheduling – Practical Issues

Experiences with a Flexible Driver Scheduler 137
Sarah Fores, Les Proll, Anthony Wren

Scheduling Train Crews: A Case Study for the Dutch
Railways... 153
Richard Freling, Ramon M. Lentink, Michiel A. Odijk

Evaluating a DSS for Operational Planning in Public
Transport Systems: Ten Years of Experience with the
GIST System 167
Teresa Galvão Dias, José Vasconcelos Ferreira, João Falcão e Cunha

Crew Scheduling for Netherlands Railways "Destination: Customer" 181
Leo Kroon, Matteo Fischetti

Selecting and Implementing a Computer Aided Scheduling System for a Large Bus Company 203
Michael Meilton

Days-off Scheduling in Public Transport Companies 215
Dulce Pedrosa, Miguel Constantino

Part 3: Advanced Transit Service and Vehicle Routing

Modeling Cost and Passenger Level of Service for Integrated Transit Service 233
Mark Hickman, Kelly Blume

Adaptive Memory Programming for a Class of Demand Responsive Transit Systems 253
Federico Malucelli, Maddalena Nonato, Teodor Gabriel Crainic, Francois Guertin

A Cycle Based Optimization Model for the Cyclic Railway Timetabling Problem 275
Leon Peeters, Leo Kroon

Minmax Vehicle Routing Problems: Application to School Transport in the Province of Burgos 297
Cristina R. Delgado Serna, Joaquín Pacheco Bonrostro

Part 4: Monitoring and Control

An Approach Towards the Integration of Bus Priority, Traffic Adaptive Signal Control, and Bus Information/ Scheduling Systems 319
Pitu Mirchandani, Anna Knyazyan, Larry Head, Wenji Wu

An Optimal Integrated Real-time Disruption Control Model for Rail Transit Systems 335
Su Shen, Nigel H.M. Wilson

Design of Customer-oriented Dispatching Support for Railways 365
Leena Suhl, Claus Biederbick, Natalia Kliewer

Determining Traffic Delays through Simulation 387
Penglin Zhu, Eckehard Schnieder

Optimization Approach to Support the Grouping and
Scheduling of Air Traffic Control Sectors 399
Ana Paula Barbosa-Póvoa, Paula Leal de Matos, Lúcio Rocha

Part 5: Strategic Decision Problems

New Revenue Management Strategies for Railway Network
Providers 415
*Imma Braun, Karl Albrecht Klinge, Martin Schroeder, Eckehard
Schnieder*

Impacts of Deregulation on Planning Processes and
Information Management Design in Public Transit 429
Joachim R. Daduna

Cost-benefit-analysis of Investments into Railway Networks
with Periodically Timed Schedules........................... 443
Michael Kolonko, Ophelia Engelhardt-Funke

Appendices

Appendix 1: Referees 461

Appendix 2: List of Presented Papers not Included in this
    Volume 463

Appendix 3: Exhibitors and Sponsors 465

A New Model for the Mass Transit Crew Scheduling Problem

Mohamadreza Banihashemi[1] and Ali Haghani[2]

[1] A/E Group, Inc.,
Geometric Design Lab, Turner-Fairbank Highway Research Center,
FHWA, McLean, VA 22101-2296 U.S.A.
mohamadreza.banihashemi@fhwa.dot.gov
[2] Department of Civil and Environmental Engineering,
University of Maryland College Park, MD 20742 U.S.A.
haghani@eng.umd.edu

Abstract. A new formulation for the "Mass Transit Crew Scheduling" (MTCS) problem is presented. The proposed model is a "task-based" multi-commodity network flow problem in which the variables are defined in conjunction with the tasks and the tasks compatibilities.

Based on the union contracts in the United States the calculations of the task compatibility costs usually cannot be finalized until we establish the workdays and solve the problem. In our approach, we start from an initial model, the relaxed MTCS problem, in which we consider minimum costs for these compatibilities. Then we propose to go through an iterative procedure for establishing the workdays and adjusting the compatibility costs if necessary. This would be accomplished by generating new variables corresponding to the established feasible workdays and a "soft" constraint associated with each new variable.

The relaxed model also lacks the constraints that prevent the construction of the workdays that are illegal based on the union agreements or other rules. For each infeasible workday we could establish a "hard" constraint to be added to the relaxed problem. An exact solution procedure for small instances of this problem could be a constraint and variable generation approach in which the workday variables as well as the soft and the hard constraints would be added to the problem in an iterative procedure.

1 Introduction

Crew scheduling is the last major step in the four-step transit planning process. These four major consecutive steps are network route design, setting frequencies and building timetables, vehicle scheduling, and crew scheduling. The crew scheduling step uses block information obtained from the vehicle scheduling step, the driver work rules (basically determined by union contracts) and workday cost structure (driver transfer costs) as input, and

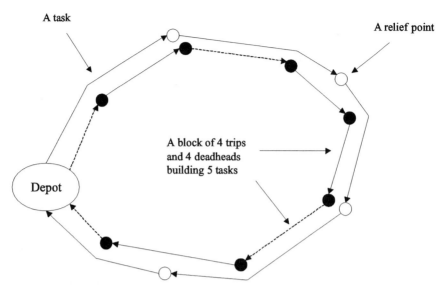

A task

A relief point

A block of 4 trips
and 4 deadheads
building 5 tasks

Depot

Figure 1. A block of trips and deadheads partitioned into tasks at relief points.

produces work schedules for the drivers (workdays). This step is also known as run cutting.

Some definitions related to the elements of the transit planning process used in this paper are as follows. Figure 1 clarifies the relations between trips and tasks.

- **Relief Point** is a stop along a route where one crew can go on break and a second crew can begin service on the same vehicle. Relief points partition the blocks into separate tasks.
- **Driver Transfer Time** is the time required for a driver to travel in his/her workday, without driving a vehicle, from the end of a task to the beginning of the next one.
- **Spread Time** is the period of time between the starting time and the end time of the workday. There is usually a maximum allowable spread time in transit agencies.
- **Straight Workday** is a workday for a full time driver with no unpaid break except one as meal breaks with the length greater or equal than a specific time (usually 30 minutes) and less or equal than another specific time (usually 60 minutes).
- **Swing Workday** is a workday for a full time driver with one or more unpaid breaks with the longest one longer than a specific time (usually 60 minutes).
- **Tripper** or part-time workday is a short workday assigned to part-time drivers.
- **Spread Time Penalty** is the penalty for swing workdays if the spread time is more than a specific time.

- **Hold-over Time** is a paid break between two assignments.
- **Preparatory Time** is the time required for a driver for pull out. There are usually two different preparatory times. The "first preparatory time" is the one that belongs to the first pull out for the driver during the day. "Regular preparatory time" is for any other pull out.
- **Storage Time** is the time required for a pull in. There are usually two different storage times. The "last storage time" is the one that belongs to the last pull in for the driver during the day. "Regular storage time" is for any other pull in.
- **Paid-time** is the total time during which each driver is paid.
- **Minimum Paid-time** is the minimum hours that the full time drivers should be paid.
- **Platform Time** is the total of the time that a driver spends on serving the tasks or on the preparation and storage.
- **Make-up Time** is the difference between the paid-time and the minimum paid-time if the former is less than the latter.
- **Maximum Make-up Time** is the maximum make-up time that is allowed for a legal workday. It is usually 30 minutes.
- **Break Time** is the time between serving two tasks but not part of the driver transfer time.
- **Lunch Break Time** is the longest break time among the breaks for a straight workday.
- **Minimum Lunch Break** is the minimum lunch break time for straight workdays. It is usually 30 minutes.
- **Maximum Lunch Break** is the maximum allowable lunch break time for straight workdays. It is usually 60 minutes.
- **Overhead Cost** is part of the overhead cost of the agency that is affected by the number of full time drivers. An example could be the driver's insurance costs. This cost should be stated in terms of the equivalent operational cost (time) of crew.

2 Review of the Literature

2.1 Set Covering Problem Formulation

Usually, the MTCS problem is formulated as a set covering problem (SCP). These are "workday-based" formulations in which the set of feasible workdays and the costs associated with these workdays are considered as the elements in the optimization process. A SCP is established to find a set of feasible workdays covering all tasks (or pieces) and minimizing the total cost of the operation. A general form of this formulation is as follows:

$$\text{Minimize} \sum_{j=1}^{n} c_j \, x_j \qquad (1)$$

$$\text{subject to} \quad \sum_{j=1}^{n} a_{ij}\, x_j \geq 1 \qquad \qquad \forall\, i = 1, \ldots, m \qquad (2)$$

$$\sum_{j=1}^{n} x_j \leq r \qquad \qquad (3)$$

$$x_j \in \{0,1\} \qquad \qquad \forall\, j = 1, \ldots, n \qquad (4)$$

with:

$x_j = \begin{cases} 1 & \text{if workday } j \text{ is chosen in schedule} \\ 0 & \text{otherwise} \end{cases}$

$a_{ij} = \begin{cases} 1 & \text{if task (or piece) } i \text{ is a part of workday } j \\ 0 & \text{otherwise} \end{cases}$ (a_{ij} are parameters)

c_j Cost associated with workday j

m Total number of driver tasks (or pieces)

n Total number of feasible driver workdays

r Maximum number of workdays

The objective function (1) in the above formulation minimizes the total cost associated with workdays and constraint set (2) ensures that all tasks (or pieces) are covered by the selected workdays. Constraint set (3) limits the number of workdays. The feasible workdays are established based on the compatibility of the tasks (or pieces) and also based on the rules embedded in union contracts. In a real-world problem there may be thousands of tasks (or pieces) and millions of feasible workdays and in practice it is impossible to consider all of the feasible workdays explicitly.

If the inequalities in constraint set (2) were equalities then the problem becomes a "set partitioning" problem in which all of the tasks (or pieces) should be covered exactly once. There is also another version of the formulation that is proposed by Mitra and Darby-Dowman (1985), and is called "generalized set partitioning" formulation. This formulation allows the algorithm to choose a solution that does not cover all of the tasks (or pieces) if there is no solution covering all the tasks (or pieces).

There are three different approaches for solving this problem. In all three approaches a limited number of feasible workdays are constructed in advance. In two of these approaches the process would be finding the best combination of the preconstructed workdays. The third approach starts from a set of feasible preconstructed workdays but it continues to produce other feasible workdays if they could improve the solution.

In the first approach, formulating the problem as an integer linear programming problem, the best combination of the feasible preconstructed workdays covering all of the tasks (or pieces) is found as the solution to the problem. Smith and Wren (1988) followed this approach but with preconstructed pieces in their formulation. They generated feasible workdays and based on some criteria reduced the size of the workday set by eliminating some of them. They used slack and surplus variables in the constraint set (2) by which their formulation had become equivalent to the formulation of Mitra and Darby-Dowman (1985). They used the branch-and-bound algorithm to get the integer solution for the constructed SCP. Their formulation is embedded into the IMPACS software used by many transit agencies. The largest problem they have reported in their paper is a problem with 379 pieces, 309 constraints, 4892 feasible workdays (variables) requiring 74 duties.

Paixão (1990) presented a computerized procedure for solving the MTCS problems on PCs. He used his own algorithm presented in the Paixão (1984), for obtaining bounds on the optimal solution. Then by formulating the SCP as a dynamic programming problem he solved this problem using state space relaxation (SSR). The largest problem solved and reported in the paper is a problem with 103 pieces and 3082 feasible workdays (variables).

Paias and Paixão (1993) continued working on SSR for the SCP. They showed that SSR provides a lower bound for the original SCP. They also built feasible solutions upon the SSR solution with a reasonable gap with the lower bound. The largest problems solved and reported are a problem with 103 pieces and 580 feasible workdays (variables) and a problem with 80 pieces and 1710 feasible workdays (variables). For all of the 18 instances the gap reported is less than 2%.

In the second approach a column generation approach is used but within a limited number of feasible workdays. The procedure starts from a subset of feasible workdays as a schedule and continues to replace some of the workdays that are in the subset with the workdays that are not in the subset but could improve the solution. The procedure finally finds the best combination of workdays within the set of the feasible preconstructed workdays.

Carraresi et al. (1995) used a combined Lagrangean relaxation and column generation approach to solve the MTCS problem. The preconstructed duties are built from pieces of work. Their mathematical model is embedded in the MTRAM package. The two test problems that they have reported to solve have 1323 and 1123 pieces of work. The number of preconstructed feasible workdays for these problems was 72260 and 9859, respectively. They achieved gaps of 3.48% and 1.32% in the solution of these two problems.

Fores et al. (1999) reported another implementation of the column generation approach. This was for problems with a limited but large number of workdays. Their approach is embedded into the TRACS II software. Solutions reported in the paper belong to problems with around 90 workdays that probably are the result of some hundred tasks.

In the third approach the solution procedure starts from the preconstructed feasible workdays and finds the best combination of these workdays covering all of the tasks (or pieces). The procedure then goes through generating new columns (workdays) by which the existing solution can be improved. If such an improvement is possible the new columns are generated and the optimization of the relaxed problem is repeated. Otherwise the existing solution would be optimal. In the following paragraphs we review some of the results obtained for solving the MTCS problem.

Desrochers and Soumis (1989) have implemented this exact solution procedure. It is embedded into the CREW-OPT software. In their approach they start from a subset of the feasible workdays and solve the SCP. They then establish a subproblem that generates other feasible workdays if such workdays are capable of improving the previous solution. With putting in the new workdays and eliminating some of the workdays from the subset the new SCP is built and solved. The process continues until no more workdays are built by the subproblem. The problems solved and reported in the paper are two problems, one from an American city and another from a British city with the number of tasks equal to 167 and 235, respectively.

2.2   Task-based Formulations

Beasley and Cao (1996, 1998) and Mingozzi et al. (1999) use this type of the formulation. The formulations are similar but the solution approaches are different. The models are very similar to the single-commodity formulation of the vehicle-scheduling problem. The formulations only consider a single type of workdays and, therefore, the corresponding compatibility costs associated with different types of workdays are not considered.

Beasley and Cao (1996) used Lagrangean relaxation to provide a lower bound for the problem. They improved this lower bound using sub-gradient optimization. As the last step they used a tree search algorithm to incorporate this lower bound to get to the optimal solution. Test problems with up to 500 tasks are reported to be solved to optimality. Beasley and Cao (1998) used a similar approach. This time they apply a dynamic programming algorithm to find the initial lower bound. Then they followed a similar procedure as reported in Beasley and Cao (1996) and solve the problem to optimality. The solution of the same test problems is reported in this paper.

Mingozzi et al. (1999) used a similar task-based formulation as well as a set partitioning formulation. Different heuristic solution procedures were presented and were applied to the same test problems used by Beasley and Cao (1996, 1998). In the first and the second procedures they used the task-based formulation. They created two relaxed problems whose feasible solutions were feasible to the original problem. The optimal solutions of these new problems were then considered as a heuristic solution of the original problem. The gaps with the optimal solutions were between 14%-20% for the first heuristic solution and between 5%-22% for the second one. In the third heuristic solution procedure they used the set partitioning formulation. The dual of the linear relaxation of this problem was used by a heuristic to obtain a lower bound. The number of variables of this set partitioning problem was then reduced using this lower bound, and the reduced-size problem was solved using branch-and-bound. All test problems were solved with a little gap to the optimality (gaps less than 0.12%).

There is a major characteristic of the test problems reported in the above three references that do not match the characteristics of the MTCS problems in the United States. The starting and the ending times of the tasks for these test problems are generated randomly with no specific ties between the ending time of the tasks and the starting times of some other tasks. In most of the transit agencies in the United States tasks are constructed by partitioning the vehicle blocks. This process results in generating a task with the starting time and location exactly the same as the ending time and location of another task (except the first and the last tasks of the blocks). This results in presenting very strong ties between pairs of consecutive tasks in the blocks. These strong ties are too many and too difficult to break. They make the MTCS problem much more difficult to solve than the multiple depot vehicle scheduling (MDVS) problem.

2.3 Genetic Algorithms

In the Genetic Algorithm (GA) presented by Clement and Wren (1995) they used pieces of work as basis and established a list of feasible workdays. Arrays of digits 0 and 1 are chosen as chromosomes, genes of 0 for workdays not in the schedule and genes of 1 for the ones in the schedule. Therefore, each chromosome represents a feasible crew schedule. Several greedy algorithms had been used for crossovers in which the duties are indexed into pieces of work. The strategy of randomly using these crossovers found to be the best one. A Problem with 97 duties as the solution was solved finding a solution just a few duties away from the optimal solution.

The most recent GA approach is by Kwan et al. (1999). They used a workday-based set covering formulation with workdays preconstructed by combining pieces of work. The main purpose of their research was to improve the process embedded into the TRACS II system. This system solves the LP relaxation of the SCP in the first stage and uses branch-and-bound in the second stage to find an integer solution. The idea was to replace the second stage of this process with a GA to get to integrality. They chose a similar definition for chromosomes as Clement and Wren (1995) except that the number of the genes was limited to the number of "preferred workdays" instead of all workdays. A heuristic was used to repair the chromosomes representing infeasible schedules and a single point crossover was adopted. Problems with about 150-500 pieces were solved. The potential number of workdays was between 10000 and 30000, with the preferred number of workdays between 113 and 392. The number of workdays in the solution was between 14 and 113. Gaps with the ILP solutions were all between 0% and 3.8%.

3 The New Formulation

Inspired by the multi-commodity formulation of the MDVS and the constraint generation approach presented in Banihashemi (1998), we develop a similar

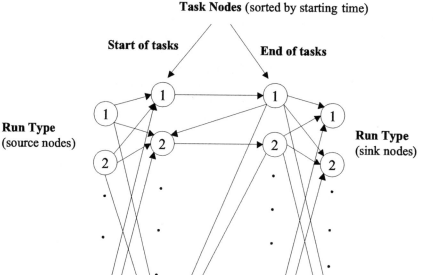

Figure 2. Network corresponding to the multi-commodity formulation.

formulation for the MTCS problem. We first present a relaxed model (R-MTCS) in which the constraints enforcing different regulations and the union contract agreements are relaxed. Then we present the way that we can handle the generation of the constraints associated with different regulations and union contract agreements.

3.1 A Relaxed Model

There are different types of workdays including straight workdays and swing workdays for full time drivers, and short workdays for part time drivers. Considering each workday as a chain of tasks, and considering different types of workdays as different types of commodities, the R-MTCS problem can be formulated as a multi-commodity network flow formulation. This new formulation is task-based. Figure 2 shows the network corresponding to this formulation.

The multi-commodity network flow formulation of the R-MTCS problem can be written as:

$$\text{Minimize} \quad \sum_{j,d} c_{0,j,d} X_{0,j,d} + \sum_{i,j,d} c_{i,j,d} X_{i,j,d} + \sum_{i,d} c_{i,0,d} X_{i,0,d} \quad (5)$$

subject to

$$\sum_j X_{0,j,d} \leq r_d \qquad \forall\, d \qquad (6)$$

$$X_{0,i,d} + \sum_j X_{j,i,d} - w_{i,d} = 0 \qquad \forall\, i,d \qquad (7)$$

$$X_{i,0,d} + \sum_j X_{i,j,d} - w_{i,d} = 0 \qquad \forall\, i,d \qquad (8)$$

$$\sum_j X_{i,0,d} \leq r_d \qquad \forall\, d \qquad (9)$$

$$\sum_d w_{i,d} = 1 \qquad \forall\, i \qquad (10)$$

$$f(X_{0,j,d}) \geq 0 \qquad (11)$$

$$\textit{All variables are binary} \qquad (12)$$

with:

$$X_{0,j,d} = \begin{cases} 1 & \text{if task } j \text{ is the first task of a run of type } d \\ 0 & \text{otherwise} \end{cases}$$

$$X_{i,j,d} = \begin{cases} 1 & \text{if compatible tasks } i \text{ and } j \text{ are consecutive tasks contained} \\ & \text{in a run of type } d \\ 0 & \text{otherwise} \end{cases}$$

$$X_{i,0,d} = \begin{cases} 1 & \text{if task } i \text{ is the last task of a run of type } d \\ 0 & \text{otherwise} \end{cases}$$

$$w_{i,d} = \begin{cases} 1 & \text{if task } i \text{ is in a run of type } d \\ 0 & \text{otherwise} \end{cases}$$

$f(\ )$ A linear function restricting the relations among the number of workdays with different workday types.

$c_{0,j,d}$ Driver transfer time between the depot and the start point of task j plus the extra preparatory time plus the overhead cost in term of time.

$c_{i,j,d}$ Driver transfer time between the end point of task i to the start point of task j plus the time of task i for compatible tasks.

$c_{i,0,d}$ The time of task i plus the extra storage time (if returning to the depot after the last task is necessary then driver transfer time to the depot is also added).

r_d Maximum number of workdays of type d

This formulation can be streamlined as follows:

$$\text{Minimize} \quad \sum_{j,d} c_{0,j,d}\, X_{0,j,d} + \sum_{i,j,d} c_{i,j,d}\, X_{i,j,d} + \sum_{i,d} c_{i,0,d}\, X_{i,0,d} \qquad (13)$$

$$\text{subject to} \quad \sum_j X_{0,j,d} \leq r_d \qquad\qquad\qquad\qquad \forall\, d \quad (14)$$

$$\sum_d X_{i,0,d} + \sum_{j,d} X_{i,j,d} = 1 \qquad\qquad\qquad \forall\, i \quad (15)$$

$$X_{0,i,d} + \sum_j X_{j,i,d} - X_{i,0,d} - \sum_j X_{i,j,d} = 0 \quad \forall\, i,d \quad (16)$$

$$f(X_{0,j,d}) \geq 0 \qquad\qquad\qquad\qquad\qquad\qquad\qquad (17)$$

All variables are binary $\qquad\qquad\qquad\qquad\qquad\qquad (18)$

We can easily prove that the multi-commodity formulation of the R-MTCS problem and the streamlined formulation are equivalent. In the remaining of this paper the R-MTCS formulation is referred to the streamlined formulation.

Function f is presenting the restrictions between the numbers of different types of runs. For example, for the Mass Transit Administration (MTA) of the Baltimore City the number of the straight runs should be at least 50% of the total number of the runs. If we consider three types of runs 1, 2, and 3, respectively, as straight runs, swing runs, and part time runs we would have the following constraint:

$$\sum_j X_{0,j,1} - \sum_j X_{0,j,2} - \sum_j X_{0,j,3} \geq 0$$

3.2 Considering the Union Contract Agreements and Other Constraints

Any constraints other than the ones stated in the presented formulation can be grouped into either hard constraints or soft constraints. The definitions of these two groups of constraints are given below:

Hard constraints: These constraints are the ones that restrict the building of specific workdays. We call these workdays "restricted workdays." These constraints are basically the result of limitations regarding

- spread time,
- maximum and minimum workday time for full time as well as workdays,
- limitation in the maximum and the minimum platform time for full time as well as workdays,
- transfer time,
- the length of the longest break time (lunch break) for straight workdays, these are usually a maximum and a minimum limitation,
- the length of the longest break time for swing workdays.

Soft constraints: These constraints are the ones that do not restrict the building of specific workdays but add a penalty to the objective function if these workdays are built in the solution. We call these workdays "penalized workdays." A common list of these penalties is as follows:

- Penalty associated with the straight workdays and swing workdays if the workday time is less than full paid time (full paid time is usually 8 hours).
- Penalty associated with the swing workdays if the spread time is more than a specific time. The penalty might be different for different ranges of spread time. As an example, for the MTA, if the spread time were more than 10.5 hours, half of the excess time would be added to the paid time. If the spread time were more than 12 hours, another half of the excess time (with respect to 12 hours) would be added to the paid time as well.

From any solution to the R-MTCS problem we can build a set of workdays. Each of these workdays is a chain of tasks feasible to be a workday one after another. Each task is only in one workday and all of the tasks are covered. These workdays are not necessarily satisfying the above hard constraints and the value of the objective function does not include the penalty associated with the penalized workdays. To be able to consider the hard and the soft constraints we need to add all possible penalty terms to the objective function and to add all possible hard and soft constraints to the list of the constraints of the R-MTCS problem. The general form of the penalty terms in the objective function is as follows:

$$+ \sum_{k,d} e_{k,d} \, Y_{k,d} \qquad (19)$$

And the hard and the soft constraints are as follows:

$$X_{0,t_1,d} + X_{t_1,t_2,d} + X_{t_2,t_3,d} + \cdots + X_{t_{p-1},t_p,d} + X_{t_p,0,d} \le p$$
$$\forall \, d, \forall \, \textit{restricted workdays} \quad (20)$$

$$X_{0,t_1,d} + X_{t_1,t_2,d} + X_{t_2,t_3,d} + \cdots + X_{t_{q-1},t_q,d} + X_{t_q,0,d} - Y_{k,d} \le q$$
$$\forall \, d, \forall \, \textit{penalized workdays} \quad (21)$$

with:

$Y_{k,d}$ workday variable for workday k with workday type d

$e_{k,d}$ penalty associated with workday variable $Y_{k,d}$

$X_{0,t_1,d}$ variable associated with task t_1 as the first task of a workday of type d

$X_{i,j,d}$ variable associated with the compatible tasks, i and j, of a workday of type d

$X_{t_p,0,d}$ variable associated with task t_p as the last task of a workday of type d

$X_{t_q,0,d}$ variable associated with task t_q as the last task of a workday of type d

p, q number of tasks of the restricted and penalized workdays, respectively

Constraint set (20) prevents the construction of the illegal workdays. Constraint set (21) enforces workday variable $Y_{k,d}$ be 1 if all of the X variables associated with workday k are 1. If any of the X variables were 0, $Y_{k,d}$ would be zero as well.

3.3 Advantages and Disadvantages of the New Formulation vs. the Previous Formulations

Advantages:

1. The formulation can consider all tasks compatibilities. However, we can limit the number of compatibilities while we are building the R-MTCS problem to eliminate the ones that have less possibility to be in the optimal solution and to reduce the problem size.
2. A single computer code can take care of analyzing the R-MTCS solution and evaluating this solution to build the workday variables (19), hard constraints (20), and soft constraints (21) regardless of the complexity of the union agreements or other rules.
3. A general MIP optimization code and an analysis code for analyzing the solution of the R-MTCS problem are the only tools we need for dealing with this problem. However, we have come to the conclusion that an exclusive optimization code would probably be more efficient and we are in the process of developing such an application.
4. The feasible workdays are built in some optimization process and they are more reliable than the feasible workdays that are built in other processes such as greedy adding processes.

Disadvantages:

1. The original R-MTCS problem and also the updated problems are larger than the problems established using previous formulations.
2. As we go through adding the generated variables and constraints to the problem, it becomes much more complicated.

3.4 Exact Solution Procedure

The exact solution procedure of the MTCS problem has the following steps:

1. Solve the R-MTCS problem to optimality and analyze the solution. Start building a list of feasible workdays based on all restrictions and from the feasible workdays in the first solution to the R-MTCS problem. Build a workday variable and two soft constraints for any feasible workday. Also build a hard constraint for any infeasible workday.

2. Add any new workday variable (19) to the objective function and any new hard constraint (20) and/or new soft constraint (21) to the problem.
3. Solve the newly updated R-MTCS problem and analyze the solution. If there is any infeasible workday or any new feasible workday in the solution, go to step 4, otherwise stop, the solution is optimal.
4. Build a new workday variable and two new soft constraints for any new feasible workday. Also build a new hard constraint for any new infeasible workday and go to step 2.

This procedure is a row-column-generation approach. The procedure ends when there is no infeasible workday built in the solution. The final solution would be the optimal solution to the problem considering all task compatibilities.

As we expected we could only apply this procedure to very small problems. Problems with up to 30 tasks were solved to optimality. After a few iterations the solution time increases exponentially. Currently we are in the process of developing an exclusive optimization code for this problem.

3.5 Some Thoughts on a Heuristic Solution Procedure

We can always establish workdays from the solution of the R-MTCS problem regardless of the integrality of the solution. If there are non-integer solutions in the LP relaxation of the R-MTCS problem the workdays that we can build would have a weight of less than 1 but we still can analyze the workdays and the sub-workdays to build hard constraints, feasible workdays and corresponding variables and constraints. We are in the process of finalizing and testing a heuristic solution procedure with the following steps:

1. Solve the LP relaxation of the R-MTCS problem and analyze the solution.
2. Add the feasible workdays to a list and add the workday variables to the objective function and the hard and the soft constraints to the problem.
3. For each infeasible workday appearing in the solution consider the weight of the workday in the solution and build a constraint to prevent the same workday with the same weight being constructed in the next iteration. In this manner we force the optimization process to build more workdays from which we would extract more feasible workdays. If a certain number of iterations is reached, go to step 4, return to step 1.
4. Review the list of the feasible workdays and choose a subset of mutually independent feasible workdays and fix them in the problem by fixing the corresponding variables to 1. Stop if all of the tasks are assigned to workdays, otherwise return to step 1.

3.6 Lower Bound

If the number of available drivers (number of different workdays for different workday types in the formulation) is large enough and if we set the overhead cost equal to zero the first integer solution to the R-MTCS problem would be exactly chains of tasks in the same order that they appear in the blocks without any driver transfer time in the solution. The solution to this problem is obviously a lower bound to the optimum solution to the MTCS problem.

As we put a non-zero overhead and limit the number of available drivers we would get a larger value as the solution to the R-MTCS problem. At each iteration of the exact solution procedure and with solving the new R-MTCS problem with some hard and soft constraints and some workday variables we come up with a larger lower bound to the MTCS problem. In the process of any iterative heuristic procedure we can also follow the exact solution procedure to the required extent to get a reasonable lower bound with a reasonable gap between that lower bound and the last feasible solution found. Our study on this part is also continuing.

3.7   Test Problem Generation

One of the issues observed in the literature is that the characteristics of the test problems used by some researchers do not match the characteristics of the MTCS problems in the United States. The main issue is the strong ties between pairs of consecutive tasks in the blocks that exist in the real-world problems. We have developed a procedure for creating test problems. We are using the trip patterns of the MTA of Baltimore City to produce MDVS problems. Solving these problems to optimality we build the blocks and establish the tasks. The created tasks have the required characteristics of the tasks of the MTCS problems.

4   Conclusion

In this paper we have considered a new model for the mass transit crew scheduling problem. This is an on-going research. On one hand, the new formulation has added more accuracy to the problem and on the other hand it has added more complexity to it. Considering the rate of the increase in computer capabilities we are very optimistic about this direction.

We still need to work on the exact solution procedure to make it more efficient. This is not just for getting the exact solution, that seems too ambitious for medium and large problems. Depending on available computer capabilities and also depending on the time we want to spend we can get different lower bounds by applying the required number of iterations of the exact solution procedure. The heuristic solution procedure also needs to be improved. Before we work on any other exact or heuristic solution procedure we need to complete the development of an exclusive optimization application that we are developing for our vehicle and crew scheduling models.

Bibliography

Banihashemi, M. (1998). *Multiple Depot Transit Scheduling Problem Considering Time Restriction Constraints*. Ph.D. thesis, Civil Engineering Department, University of Maryland, College Park, USA.

Beasley, J.E. and E.B. Cao (1996). A tree search algorithm for the crew scheduling problem. *European Journal of Operational Research 94*, 517–526.

Beasley, J.E. and E.B. Cao (1998). A dynamic programming based algorithm for the crew scheduling problem. *Computers & Operations Research 25*, 567–582.

Carraresi, P., M. Nonato, and L. Girardi (1995). Network models, Lagrangean relaxation and subgradients bundle approach in crew scheduling problems. In J.R. Daduna, I. Branco, and J.M.P. Paixão (Eds.), *Computer-Aided Transit Scheduling, Lecture Notes in Economics and Mathematical Systems*, 430, Springer, Berlin, 188–212.

Clement, R. and A. Wren (1995). Greedy genetic algorithms, optimizing mutations and bus driver scheduling. In J.R. Daduna, I. Branco, and J.M.P. Paixão (Eds.), *Computer-Aided Transit Scheduling, Lecture Notes in Economics and Mathematical Systems*, 430, Springer, Berlin, 213–235.

Desrochers, M. and F. Soumis (1989). A column generation approach to the urban transit crew scheduling problem. *Transportation Science 23*, 1–13.

Fores, S., L. Proll, and A. Wren (1999). An improved ILP system for driver scheduling. In N.H.M. Wilson (Ed.), *Computer-Aided Transit Scheduling, Lecture Notes in Economics and Mathematical Systems*, 471, Springer, Berlin, 43–61.

Kwan, A.S.K., R.S.K. Kwan, and A. Wren (1999). Driver scheduling using genetic algorithms with embedded combinatorial traits. In N.H.M. Wilson (Ed.), *Computer-Aided Transit Scheduling, Lecture Notes in Economics and Mathematical Systems*, 471, Springer, Berlin, 81–102.

Mingozzi, A., M. A. Boschetti, S. Ricciardelli, and L. Bianco (1999). A set partitioning approach to the crew scheduling problem. *Operations Research 47*, 873–888.

Mitra, G. and K. Darby-Dowman (1985). CRU-SCHED: A computer based bus crew scheduling system using integer programming. In J.M. Rousseau (Ed.), *Computer Scheduling of Public Transport 2*, North-Holland, Amsterdam, 223–232.

Paias, A. and J.P. Paixão (1993). State space relaxation for set-covering problems related to bus driver scheduling. *European Journal of Operational Research 71*, 303–316.

Paixão, J.P. (1984). *Algorithms for Large Scale Set-covering Problems*. Ph.D. thesis, Department of Management Science, Imperial College, London.

Paixão, J.P. (1990). Transit crew scheduling on a personal workstation (MS/DOS). In H. Bradley (Ed.), *Operational Research'90*, Pergamon Press, Oxford, 421–432.

Smith, B.M. and A. Wren (1988). A bus crew scheduling system using a set covering formulation. *Transportation Research A 22*, 97–108.

A Global Method for Crew Planning in Railway Applications

Alberto Caprara, Michele Monaci, and Paolo Toth

DEIS, University of Bologna,
Viale Risorgimento 2, I-40136 Bologna, Italy
{acaprara,mmonaci,ptoth}@deis.unibo.it

Abstract. Crew planning is a typical problem arising in the management of large transit systems (such as railway and airline companies). Given a set of train services to be performed every day, the problem calls for a set of crew rosters covering the train services at minimum cost. Although the cost may depend on several factors, the main objective is to minimize the number of crews needed to perform the rosters. The process of constructing the rosters from the train services has been historically subdivided into three independent phases, called pairing generation, pairing optimization, and rostering optimization, as is the case for the approach presented in Caprara et al. (1999a) for the Italian railways. In that paper, it is suggested that a feedback between the last two phases may significantly improve the quality of the final solution. In this paper, we illustrate the implementation of a new crew planning system within the EU Project TRIO. In particular, we describe the design of a new module for pairing generation, as well as an effective technique for integrating the pairing and rostering optimization phases into a unique one. The improvements over the previous approach are shown through computational results on real-world instances.

1   Introduction

Crew planning is a typical problem arising in the management of large transit systems (such as railway and airline companies). In this paper, we focus on the problem arising in railway applications, and in particular at Ferrovie dello Stato SpA (FS for short), the Italian railway company.

As input, one is given a set D of depots, where the available crews are located, and a planned timetable for the *train services* (i.e., both the actual journeys for passengers or freight, and the transfers of empty trains or equipment between different stations) to be performed every day of a certain time period. Each train service is first split into a sequence of *trips*, defined as segments of train journeys which must be serviced by the same crew without rest. Each trip is characterized by a departure time, a departure station, an arrival time, an arrival station, a travel distance, and possibly by additional attributes. All the times are expressed in minutes from 1 to 1440 (i.e., 24

hours). During the given time period each crew can perform a *roster*, defined
as a cyclic sequence of trips whose operational cost and feasibility depend
on several rules laid down by union contracts and company regulations. To
ensure that each daily occurrence of a trip is performed by one crew, for each
roster whose length is, say, k days, k crews are needed. In fact, in a given calendar day each of these crews performs the activities of a different day of the
roster. Moreover, in consecutive calendar days each crew performs the activities of (cyclically) consecutive days of the roster. The crew planning problem
then consists of finding a set of rosters for the crews located in the depot set
D, covering once every trip, so as to satisfy all the operational constraints at
minimum cost. Although the cost may depend on several factors, the main
objective is to minimize the global number of crews needed to perform all
the rosters, i.e., required to cover all the timetabled trips.

Analogous crew planning problems arising in different transportation areas are examined in Wren (1981); Bodin et al. (1983); Carraresi and Gallo
(1984); Rousseau (1985); Daduna and Wren (1988); Desrochers and Rousseau
(1992); Desrosiers et al. (1995); Barnhart et al. (1998); Wilson (1999); Desaulniers et al. (1998); Barnhart et al. (1999) and Klabjan (1999), where
several references for specific problems are provided.

Railway crew planning represents a very complex and challenging problem
due to both the size of the instances to be solved and the type and number
of operational constraints. In practice, the overall crew management problem
is approached in three phases, according to the following scheme:

1. **Pairing generation:** Starting from the given timetabled trips, a very
   large number of feasible *pairings* (also called *duties* or *shifts*) is generated.
   Each pairing starts and ends at the same depot and represents a sequence
   of trips to be covered by a single crew belonging to that depot within a
   given time period overlapping few consecutive days (typically 1 or 2 days).
   A pairing has various characteristics (e.g., associated depot, starting and
   ending time instants, working time duration, with/without intermediate
   rest periods, night working periods, etc.) which determine its cost.
2. **Pairing optimization:** A selection is made of the best subset of the
   pairings generated in phase 1, so as to guarantee that all the timetabled
   trips are covered at minimum cost. This phase follows quite a general
   approach, based on the solution of set-covering or set-partitioning problems, possibly with additional constraints (e.g., lower and upper bounds
   on the number of pairings selected for each depot or having given characteristics).
3. **Rostering optimization:** The pairings selected in phase 2 are sequenced
   into rosters, defining a periodic duty assignment to each crew which guarantees that all the pairings are covered for a certain number of consecutive days (e.g., a month). Rosters are generated separately for each depot
   $d \in D$. In this phase, trips are no longer taken into account explicitly,
   but determine the *attributes* of the associated pairings which are relevant
   for the roster feasibility and cost.

Phases 1 and 2 are often referred to as the *Crew Scheduling Problem*. In railway applications, a crew can travel as a passenger on some trip at no extra cost, a main difference with respect to airline applications. Accordingly, phase 2 is typically solved as a set covering rather than a set partitioning problem. In addition, only inclusion-maximal feasible pairings, among those with the same cost, need to be considered in the pairing generation.

Decomposition is motivated by several reasons. First of all, as previously mentioned, each crew is located in a given depot $d \in D$, which represents the starting and ending point of its work segments. A natural constraint imposes that each crew must return to its home depot within few days, which leads to the concept of pairing as a short-term work segment starting and ending at the home depot and overlapping few consecutive days. Secondly, constraints affecting the short-term work segments are different in nature from those related to the overall crew rosters. Finally, decomposition constitutes the current practice adopted by most railway companies. This is the reason why, in the design of the ALPI system, the computer-based crew planning system developed by FS in the mid 90's (see, e.g., Caprara et al. (1999a)), FS required that each phase was approached without taking into account the subsequent ones.

However, in the conclusions of Caprara et al. (1999a), the authors suggested that a feedback between phases 2 and 3 could lead to better results. Trying to exploit this issue, we integrated phases 2 and 3 together within the EU Project TRIO, where we were asked for taking care of the crew planning part. Moreover, whereas in the ALPI system FS took care of the implementation of phase 1, in the TRIO project we designed our own pairing generator, taking into account what is done in the following phases.

In this paper, we describe the implementation of our pairing generator and the integration of the pairing and rostering optimizers, showing how the quality of the final solution is improved with respect to the ALPI approach on real-world instances from FS.

2 Development of the Pairing Generator

In principle, the pairing generation phase calls for the determination of the complete set of feasible pairings from the given timetabled trips. However, in railway applications, the complete set of feasible pairings may be too large to be practically handled by the following phases of the optimization process, even for medium-size instances involving a few hundred trips. Hence, one has to fix some rules to reduce the number of the pairings generated, trying to keep the "best ones" according to some scoring criteria.

In this section, we first illustrate the rules that a pairing must satisfy in order to be feasible. Then, we describe how one could in principle generate the complete list of the feasible pairings, and finally we propose some heuristic criteria to reduce the number of pairings generated.

2.1 Pairing Feasibility Rules

The illustration of the rules is similar to the one given in Caprara et al.
(1999a). However, we report the complete list of constraints both in order
to make the paper self-contained and because some of the union contract
rules were changed with respect to Caprara et al. (1999a). Moreover, within
the description we point out the implications of these constraints on the
subsequent optimization phases. The numerical values given in the following
correspond to the rules currently applied at FS.

Each pairing is a trip sequence starting and ending at the same depot $d \in$
D, and satisfying the sequencing rules and operational constraints described
below.

Sequencing rules require that each pair of consecutive trips i and j in a
pairing is *compatible*, namely:

- The arrival station of i coincides with the departure station of j.
- If trips i and j belong to different train services, the time interval between
  the arrival of i and the departure of j is at least equal to a *technical
  transfer time*, depending on i and j, and whose order of magnitude is
  about 30 minutes. This includes the times possibly required to change
  trains, to perform maneuver and other technical operations in the station.

Sometimes one is allowed to directly sequence trips i and j also if the first
condition is violated. In particular, this is possible when the arrival station
of trip i and the departure station of trip j are located in the same *node*, i.e.,
a city where many stations are present. In this case, the technical transfer
time must include transfers within the node, and so it is typically larger than
50 minutes.

As mentioned in the introduction, a crew may travel on some trip as a
passenger. Besides allowing a trip to be covered by more than one pairing in
the pairing optimization phase, this allows us to construct a pairing where
the crew is a passenger on some trips, i.e., these trips, called *passenger trips*
in the pairing, are not covered by the pairing. The advantage is that the
technical transfer times are smaller when dealing with passenger trips. In
particular, the technical transfer time between trips i and j is decreased by
about 10 minutes if either i or j is a passenger trip in the current pairing, and
about 20 minutes if both are passenger trips. The trips covered by a pairing
will be called *driver trips*.

An *external rest* of a pairing is a rest interval between two compatible consecutive trips i and j, ending and starting, respectively, at a station where a
rest service is offered, which exceeds the technical transfer time by at least 7
hours. A pairing can contain at most one external rest. The following characteristics are associated with each pairing:

- *start time*, defined as the departure time of its first trip i minus a *technical
  departure time* depending on i, equal to about 10 minutes if i is a driver
  trip and equal to 0 otherwise;

- *end time*, defined as the arrival time of its last trip i plus a *technical arrival time* depending on i, equal to about 10 minutes if i is a driver trip and equal to 0 otherwise;
- *paid time*, defined as the time elapsing between the start time and the end time of the pairing;
- *driving time*, defined as the sum of the driver trip durations plus all rest periods between consecutive driver trips on the same train service shorter than 30 minutes;
- *working period*, defined as the time interval between the start time and the end time of the pairing, minus the external rest, if any;
- *working time*, defined as the duration of the working period;
- *driving distance*, defined as the sum of the travel distances of the driver trips in the pairing.

A pairing is defined to be *overnight* if its working period overlaps the time interval between midnight and 5 am. The *operational constraints* require that for each pairing:

- The *paid time* must be smaller than 24 hours.
- If the pairing does not contain an external rest, the working time must be at least equal to 45 minutes, without exceeding 7 hours and 5 minutes if it is overnight and 8 hours and 50 minutes otherwise. If the pairing contains an external rest, the working time must be at least equal to 45 minutes without exceeding 7 hours and 5 minutes for each part before and after the rest.
- If the pairing does not contain an external rest, the driving time must be strictly positive without exceeding 4 hours and 30 minutes. Otherwise, the driving time must be strictly positive without exceeding 4 hours and 30 minutes for each part before and after the external rest.
- In the intervals between 11 am and 3 pm, and between 6 pm and 10 pm, a *meal rest* of at least 30 minutes is required.
- If the pairing does not contain an external rest, the driving distance must not exceed 600 kilometers. If the pairing contains an external rest, the driving distance must not exceed 600 kilometers for each part before and after the rest.

In the ALPI system, the cost of a pairing is set to 1 if the pairing does not contain an external rest and to 2 otherwise. As mentioned also in Caprara et al. (1999a), a more careful definition of the costs, taking into account the binding constraints in the rostering optimization phase, leads to better results. However, this information is not known *a priori*. Therefore, the costs of the pairings are initially set to the values above and later adjusted within the pairing optimization phase, as discussed in Section 3.

2.2 Enumeration of the Feasible Pairings

We have implemented this phase according to a depth-first branch-and-bound technique. For each depot $d \in D$ separately, we enumerate the pairings associated with d.

At the root node of the branching tree, we branch by deciding the first driver trip to be inserted in the (initially empty) current pairing. More generally, at each level k of the branching tree we branch by selecting the k-th driver trip in the current pairing p. Each time we consider the insertion of a new driver trip in p, we check whether this insertion is feasible with respect to the sequencing rules and the operational constraints.

At each node of the branching tree, we try to complete the current pairing with the available trips, used as passenger trips, by going from the arrival station of its last trip h, to depot d, and minimizing the corresponding time. More precisely, letting τ_h be the arrival time of trip h plus the technical arrival time of h, we determine which is the feasible sequence of passenger trips that guarantees the earliest arrival at depot d by departing from the arrival station of h not earlier than τ_h. Such a sequence will be referred to as the *shortest path* from trip h to depot d, and its duration denoted by $\delta(h, d)$. (Below, we will discuss how to compute these shortest paths efficiently.) If the pairing obtained in this way is feasible, we store it. Note that the constraint on the minimum driving time after the possible external rest forbids the insertion of an external rest within the sequence of passenger trips used to return to the depot.

Whenever no additional trip can feasibly be added to the current pairing p, we backtrack by removing the last driver trip inserted. Furthermore, simple lower bounds on the working and paid time needed to return to the depot are used to detect infeasibilities and backtrack as soon as possible. More precisely, let h be the last driver trip inserted in pairing p and $\delta(h, d)$ be defined as above. In addition, let W denote the maximum working time for each part of a pairing with external rest. Let a_p and w_p be the paid time and the working time of pairing p so far (including h), respectively. If $a_p + \delta(h, d)$ exceeds the maximum paid time, or $w_p + \delta(h, d)$ exceeds $2W$, we backtrack. If p does not contain an external rest before h, and $w_p = a_p$ exceeds W, the current pairing p is not allowed to contain an external rest. Hence, if $w_p + \delta(h, d)$ is larger than the maximum working time for a pairing without external rest, we backtrack. Similarly, if p contains an external rest before h, denoting by w_p the working time of the second part of p, if $w_p + \delta(h, d) > W$, we backtrack. It should be noted that the pairings are required to have minimum working and driving times, so a procedure must be called after the construction of each pairing in order to check these constraints. While infeasibilities associated with an excess of the working and paid times can be detected immediately (and so we can stop the construction of the current pairing), the above kind of constraints can be checked only at the end of the construction, and so it is impossible to detect earlier these infeasibilities.

Besides computing the shortest path from each trip to each depot, as mentioned above, we compute the shortest path from each depot to each trip to consider the possible presence of passenger trips in the beginning of the pairing before the first driver trip. Moreover, the possible insertion of passenger trips between two consecutive driver trips within the current pairing requires the determination of the shortest path from every trip to every other trip. All the above mentioned shortest paths can be obtained by applying a preprocessing procedure which computes the shortest path from every trip h to every station s, and from every station s to every trip h.

Actually, the following considerations show that the pure shortest path computation is not sufficient for our purposes. Consider two driver trips i and j within a pairing, such that the arrival station of i, say e_i, and the departure station of j, say s_j, do not coincide (nor belong to the same node). If a meal break may be assigned between i and j, it may be the case that such a break cannot be assigned if the shortest path from i to station s_j is used, whereas the break can be assigned (in e_i or in an intermediate station) if another path (arriving later at station s_j) is used. Note that in this case there is still a "best" path between i and j, i.e., any path giving a meal break, if possible, and arriving at station s_j before the departure of j (including the technical time).

Furthermore, we also consider the presence of an external rest between two driver trips i and j. In this case, there may not be a "best" path between i and j. Indeed, we may have two alternative paths, one for which the external rest ends later, and one for which the external rest is longer. Suppose the set of trips to complete the current pairing (after trip j) is fixed. By choosing the first path, the working time of the second part of the pairing will be smaller than by choosing the second path, and, therefore, the pairing could be feasible only in the first case. On the other hand, if the pairing is feasible by choosing the second path, its global working time will be smaller, and hence the pairing will be better for the subsequent phases (see the next section). Moreover, different paths may or may not assign a meal break between the external rest and trip j. Accordingly, given two paths P_1 and P_2 between i and j with an external rest, we will say that P_1 is *dominated* by P_2 if

- the external rest in P_1 is not longer than that in P_2,
- the external rest in P_1 ends not later than in P_2,
- if P_1 contains a meal break after the end of the external rest then also P_2 does.

For this reason, we have to consider more than one possibility when driver trips i and j are consecutive in a pairing and have an external rest in between, i.e., the branching process after the insertion of trip i has to generate more than one node corresponding to the selection of trip j, each node being associated with a nondominated external rest.

The computation of the "best" paths mentioned above is carried out considering each driver trip i as the starting one, and defining an acyclic directed graph $G = (V, A)$, where V contains driver trip i as well as the set of passenger trips that have no overlap with i. (Note that the trips that overlap i cannot be contained in a same pairing as i as the maximum paid time is 24 hours.) We associate with each trip a *value* equal to the difference (modulo 24 hours) between its starting time and the starting time of i. The vertices in V are numbered from 1 to $|V|$ according to increasing values of the corresponding trips, breaking ties arbitrarily. Let t_j denote the trip associated with vertex $j \in V$ (noting that $t_1 = i$). For $j, k \in V$, we have an arc $(j, k) \in A$ if $j < k$ and trip t_k can follow trip t_j within a pairing according to the sequencing rules described in the previous subsection. Each arc $(j, k) \in A$ is assigned the binary flags $l_{(j,k)}$ and $d_{(j,k)}$ equal to 1 if a lunch break and a dinner break can be assigned between trips t_j and t_k, respectively. Note that the graph defined in this way is *acyclic* and that each sequence of passenger trips following i in a feasible pairing corresponds to a path in G.

By using a standard labeling procedure for directed acyclic graphs, we determine, for each passenger trip corresponding to a vertex in $V \setminus \{1\}$, if the trip can be reached by a path starting from i, and if such a path can include a lunch break, a dinner break, or both. Since the interesting information is the one associated with the "best" path from trip i to each station s, we do not store all the paths found. Instead, for each station s, we consider all passenger trips, corresponding to vertices of G, arriving at station s and store

- the shortest path from i to s (if any),
- the shortest path from i to s containing a lunch break (if any),
- the shortest path from i to s containing a dinner break (if any).

In a similar way, for each trip i and station s, we compute and store the "best" path from s to i.

Having computed the information above, when we branch considering the insertion of driver trip j just after driver trip i in the current pairing, if no external rest is assigned between i and j, we know the "best" path between i and j, as mentioned above. Otherwise, we may have several possibilities for the intermediate external rest. We consider all possible stations for an external rest, i.e., stations s such that the shortest path from i to s and the shortest path from s to j allow the presence of an external rest. Among these stations, we consider for the branching those for which the external rest is not dominated, as defined above. Note that, for a same station s, we may have to consider two branches, one corresponding to a meal break assigned between the end of the external rest and driver trip j, and the other without meal break.

For large size instances, the computation of the paths above in all graphs G, one for each trip, may be very time consuming. Moreover, for these instances, in order to reduce the number of pairings generated, we typically forbid the presence of passenger trips within the pairings. Nevertheless, the information about the shortest paths from trips to stations (but not vice versa) is still very useful for the infeasibility detection and backtracking within the branch-and-bound procedure described above. In order to obtain this information, instead of considering one graph for each trip, we consider a directed graph $H = (W, F)$ for each depot $d \in D$, defined as follows. Vertex set W contains a dummy vertex w associated with depot d, as well as one vertex j associated with each driver trip j. For $j \in W \setminus \{w\}$, there is an arc (j, w) if the arrival station s of trip j coincides with depot d or belongs to the same node. In the first case, the cost of the arc is set to 0, whereas in the second case it is equal to the transfer time associated with the node of s and d, as defined in the previous subsection. Moreover, for $i, j \in W \setminus \{w\}$, there is an arc (i, j) if driver trip j can follow driver trip i in a pairing according to the sequencing rules. In this case, the cost of arc (i, j) is set to the difference (modulo 24 hours) between the arrival time of trip j and the arrival time of trip i. The computation of the backward shortest paths from depot d to all the driver trips from which d can be reached within 24 hours is performed through a standard shortest path procedure (e.g., the Dijkstra algorithm).

2.3   Heuristic Reduction of the Number of Pairings

As mentioned above, the complete set of feasible pairings may be too large to be handled explicitly, as it would take too long to generate it and it would be very unconvenient for the subsequent phases to handle such a huge set. Hence, we have developed the following criteria to reduce the set of pairings during the generation.

A first considerable reduction of the number of feasible trips is achieved by avoiding the insertion of passenger trips in the pairings. In our implementation we have a binary flag called *start_end*, which, if set to 1, forbids the presence of passenger trips at the beginning of the pairing (before the first driver trip) and at the end of the pairing (after the last driver trip). Furthermore, we have a binary flag called *middle*, which, if set to 1, forbids the presence of passenger trips between driver trips in the pairing.

In addition, we have also defined some rules to distinguish between "good" and "bad" pairings. These rules consider the fraction of working time which is not spent driving a train. In particular, based on two parameters *max_diff* and *min_ratio*, we say that a pairing is "bad" if the difference between its working and driving times is at least *max_diff* or if the ratio between its driving and working times is smaller than *min_ratio*. If the pairing contains an external rest, the two conditions apply separately to each part of the pairing. These conditions are tested during the construction of the pairing so as to backtrack as soon as the current pairing is observed to be bad. Note that the second condition, concerning the ratio between the driving and the working times, should be checked only when the pairing is completed. Instead, we check the condition on the current (incomplete) pairing if the working time (of the current part, in case an external rest has been assigned) of the pairing is at least equal to *min_time*, a third parameter in our implementation.

3   Integration of Pairing and Rostering Optimization

As mentioned in the introduction, in the previous approaches to the crew planning problem the pairing generation phase is followed by two distinct phases: a pairing optimization phase, aimed at selecting a "good" subset of pairings covering all the trips, and a rostering optimization phase in which the selected pairings are arranged into rosters. In this section, we illustrate how these two phases may be joined together in an iterative way in order to obtain a better final solution. In particular, we will keep both the pairing optimization phase and the rostering optimization phase, with the difference that the selection of the pairings in the former is driven by the objective function of the latter, and that, each time a new candidate set of pairings is found by the first phase, the second one is called to check whether this pairing set leads to a set of rosters better than the incumbent one.

Before going into details, we point out that, in the implementation illustrated in this paper, the pairings generated in the first phase are not changed by the subsequent two phases, i.e., there is no feedback between these phases and the pairing generation phase. This issue is currently under investigation.

This section is organized as follows. First, we illustrate the crew rostering problem solved by the rostering optimization phase, and whose objective function coincides with the one of the overall crew planning problem. Both the definition of the crew rostering problem and the solution method used are unchanged with respect to the ALPI system, see Caprara et al. (1998) and Caprara et al. (1999a). Then, we discuss the changes made to the pairing optimization phase. In particular, this phase has a new objective function and calls several times the rostering optimization phase, as mentioned above.

3.1   Rostering Optimization

In this phase, we are given a set of pairings to be covered by a set of crew rosters. Each roster is a sequence of pairings associated with the same depot, hence, for each depot $d \in D$, a separate crew rostering problem is solved considering only the corresponding pairings.

The algorithm is executed with a given global computing time limit. This time limit is subdivided among the depots, assigning to each depot $d \in D$ a time limit proportional to the number of pairings associated with it. In practice, using the simple lower bound described below and a more sophisticated lower bound proposed in Caprara et al. (1998), the method often finds for the current depot a solution which is provably optimal within a short time, in which case the construction of the rosters for that depot is clearly terminated. As a consequence, the algorithm often terminates much earlier than the time limit elapses.

A roster contains a subset of pairings and spans a cyclic sequence of groups of six consecutive days, conventionally called *weeks*. Hence the number of days in a roster is an integer multiple of 6. The length of a roster is typically 30 days (5 weeks) and does not exceed 60 days (10 weeks), although these requirements are not explicitly imposed as constraints.

The crew rostering problem consists of finding, for each depot $d \in D$, a feasible set of rosters covering all the pairings associated with the depot and spanning a minimum number of weeks. As already discussed in the introduction, the global number of crews required to cover all the pairings every day is equal to 6 times the total number of weeks in the solution. Thus, the minimization of the number of weeks implies the minimization of the global number of crews required.

Among the pairing characteristics mentioned in Section 2, those to be taken into account in this phase are the *start time*, the *end time*, the *paid time* and the *working time*. In addition, it is necessary to know which are the *pairings with external rest*, the *overnight pairings*, and the *long pairings* (i.e., the pairings that do not include an external rest and whose working time is longer than 8 hours and 5 minutes).

The complete list of sequencing rules and operational constraints that a roster must satisfy in order to be feasible is given in Caprara et al. (1999a). Here we give only the constraints that are used to drive the new objective function of the pairing optimization phase. First of all, every week can contain at most 1 long pairing, 2 overnight pairings, and, on average, no more than 1.4 pairings with external rest. Moreover, for each week, the overall working time must not exceed 36 hours and, on average, the overall paid time must not exceed 34 hours. The *rest* time that must elapse between the end of a pairing and the start of the next pairing in a roster is at least 18 hours if the two pairings are contained in the same week, and, on average, at least 58 hours if the two pairings are contained in different weeks (the rest between two consecutive weeks is called *weekly rest*). Accordingly, given a set S of pairings associated with the same depot $d \in D$, the following values represent simple lower bounds on the number of weeks corresponding to the rosters associated with depot d:

$$L^d_{ext} = \text{(number of pairings in } S \text{ with external rest)}/1.4$$
$$L^d_{long} = \text{number of long pairings in } S$$
$$L^d_{night} = \text{(number of overnight pairings in } S)/2$$
$$L^d_{paid} = \sum_{j \in S} a_j/(34 \cdot 60)$$
$$L^d_{work} = \sum_{j \in S} w_j/(36 \cdot 60)$$

where, for each pairing $j \in S$, a_j and w_j denote, respectively, the paid time and the working time of the pairing expressed in minutes. Note that, with the values above, L^d_{work} will always be smaller than L^d_{paid}. Nevertheless, we use also the information given by L^d_{work} to update the pairing costs, as illustrated in § 3.3. Another simple lower bound, indicated by L^d_{seq}, is based on the minimum time that must elapse between two consecutive pairings in a roster. The bound can be iteratively computed, starting from another valid integer lower bound L on the number of weeks required by depot d, through the following formula:

$$L^d_{seq} = \frac{\sum_{j \in S} a_j + (|S| - L) \cdot (18 \cdot 60) + L \cdot (58 \cdot 60)}{6 \cdot 24 \cdot 60} \tag{1}$$

which takes into account the fact that the rosters must contain each pairing $j \in S$, whose duration is a_j, as well as at least L weekly rests, whose average duration must be at least 58 hours, and $|S| - L$ rests within the same week, whose minimum duration is 18 hours. If the integer rounded-up value of L^d_{seq}

given by (1) turns out to be larger than L, one can recompute a new value of L^d_{seq}, by defining $L = \lceil L^d_{seq} \rceil$, terminating when no lower bound improvement occurs. We denote by $L^d_{glo} = \max_{\gamma \in \Gamma} L^d_{\gamma}$ the global (real valued) lower bound for depot d, where $\Gamma = \{ext, long, night, paid, work, seq\}$.

For a detailed description of the heuristic method used to solve the crew rostering problem, we refer the reader to Caprara et al. (1998).

3.2   Pairing Optimization within the ALPI System

Within the ALPI system, the pairing selection is done by solving a set covering problem with additional base constraints, as illustrated in Caprara et al. (1999a). These base constraints are aimed at guiding the selection of the pairings, so as to avoid an excessive percentage of pairings with unconvenient characteristics (e.g., pairings with external rest or with a large paid time) in the crew rostering phase.

The heuristic method used in this phase is the one proposed by Caprara et al. (1999b) for the pure set covering problem, adapted so as to take into account the additional constraints. For the purposes of the current paper, we briefly illustrate the method as it applies to the pure set covering problem.

Let $N = \{1, \ldots, n\}$ be the given collection of pairings, determined in the pairing generation phase, associated with the given trip set $M = \{1, \ldots, m\}$. For each pairing $j \in N$, let c_j be the associated cost and I_j the associated set of driver trips. For each trip $i \in M$, let $J_i = \{j \in N : i \in I_j\}$ be the set of pairings covering it.

The *Set Covering Problem (SCP)* calls for

$$v(\text{SCP}) = \text{Minimize} \sum_{j \in N} c_j x_j \tag{2}$$

subject to

$$\sum_{j \in J_i} x_j \geq 1, \qquad i \in M \tag{3}$$

$$x_j \in \{0, 1\}, \qquad j \in N \tag{4}$$

where $x_j = 1$ if pairing j is selected in the optimal solution, $x_j = 0$ otherwise.

The heuristic method for SCP proposed by Caprara et al. (1999b) is based on dual information associated with a Lagrangian relaxation of model (2)-(4). For every vector $u \in R^m_+$ of Lagrangian multipliers associated with constraints (3), the Lagrangian subproblem reads:

$$L(u) = \min \left\{ \sum_{j \in N} c_j(u) x_j + \sum_{i \in M} u_i : x_j \in \{0, 1\}, j \in N \right\} \tag{5}$$

where $c_j(u) = c_j - \sum_{i \in I_j} u_i$ is the *Lagrangian cost* associated with pairing $j \in N$. The optimal solution of the Lagrangian subproblem is given by $x_j = 1$

if $c_j(u) \leq 0$, and $x_j = 0$ otherwise ($j \in N$). The Lagrangian dual problem associated with (5) consists of finding a Lagrangian multiplier vector $u^* \in R_+^m$ which maximizes the lower bound $L(u)$, and is solved through subgradient optimization; see, e.g., Held and Karp (1971).

The approach consists of three main steps. The first one is referred to as the *subgradient step*, aimed at quickly finding a near-optimal Lagrangian multiplier vector.

The second one is the *heuristic step*, in which a sequence of near-optimal Lagrangian vectors is determined. For each vector, the *scores* associated with the corresponding Lagrangian costs, are given as input to a fast greedy heuristic procedure to possibly update the incumbent solution. Letting u be the current Lagrangian multiplier vector, the greedy algorithm works as follows:

procedure greedy
begin
1.　Initialize $M^* := M$ to be the set of the currently uncovered trips, and $S := \emptyset$ to be the set of the currently selected pairings;
　　repeat
2.　　Compute the score $\sigma_j := score(j, u, M^*)$ for each $j \in N \setminus S$, and let $j^* \in N \setminus S$ be a pairing with minimum score;
3.　　$S := S \cup \{j^*\}$; $M^* := M^* \setminus I_{j^*}$
　　until $M^* = \emptyset$
end.

The key step of the procedure is step 2, in which the pairing scores σ_j are defined through the function $score(j, u, M^*)$, which takes into account the current Lagrangian cost of pairing j and the number of uncovered trips covered by j. An extensive computational analysis showed that the solutions produced with the function σ_j given below widely outperformed the ones produced by the scores proposed in the literature and other new scores:

$$\sigma_j := \begin{cases} \gamma_j/\mu_j, & \text{if } \gamma_j \geq 0, \\ \gamma_j \cdot \mu_j, & \text{if } \gamma_j < 0, \end{cases} \tag{6}$$

where $\mu_j = |I_j \cap M^*|$ and $\gamma_j = c_j - \sum_{i \in I_j \cap M^*} u_i$. The solution S returned by procedure greedy may contain redundant pairings, whose removal leads to another set covering problem. If the number of pairings in this problem is less then 10 (as is typically the case), the problem is solved exactly by complete enumeration, otherwise it is solved heuristically.

In the third step, called *fixing*, one selects a subset of pairings having a high probability of being in an optimal solution, and fixes to 1 the corresponding variables. In this way one obtains an SCP instance with a reduced number of pairings (and trips), on which the three-step procedure is iterated.

After each application of the three-step procedure, an effective *refining procedure* is used to produce improved solutions.

Since the SCP instances in crew planning typically contain a very large number of pairings (as discussed in Section 2), the computing time of the heuristic method is considerably reduced by defining a *core problem* containing a suitable small set of pairings, chosen among those having the lowest Lagrangian costs. The definition of the core problem is often very critical, since an optimal solution typically contains some pairings that, although individually worse than others, must be selected in order to produce a good overall solution. Hence it is better not to "freeze" the core problem, this is done by using a *variable pricing* scheme to update the core problem iteratively in a vein similar to that used for solving large scale linear programs.

Within few minutes of computing time, several thousands of different heuristic solutions are found by the method above, even for large size instances with hundreds or thousands of trips and millions of pairings. The cost of each solution is given by the sum of the costs of the pairings selected, and the best solution found within a given time limit is stored in a file and constitutes the input for the rostering optimization phase.

The main drawbacks of this approach are evident:

- The construction of the heuristic solutions takes into account only the pairing costs, and not directly the real objective function, i.e., the minimization of the global number of weeks for all the depots of set D.
- The pairing costs only partly reflect the constraints of the rostering optimization phase. In particular, it is difficult to find out, *a priori* and separately for each depot, which are the constraints that will make the construction of the rosters for this depot difficult.
- Only one solution is kept among those found, whereas the crew rostering phase could produce much better rosters starting from the pairings set selected in some other solution which was not stored because its pairing cost was not the best one.

These observations inspired the design of a new pairing optimization module, which calls the rostering optimization phase several times, as illustrated below.

3.3   The New Pairing Optimization Method

This section presents the main modifications made to the heuristic algorithm used in the pairing optimization phase with respect to the ALPI system. All these modifications take into account the constraints on the rostering optimization problem previously mentioned. In particular, we will use the simple lower bounds on the number of weeks required for each depot, as described in § 3.1.

The additional base constraints imposed on the pairing optimization phase in the ALPI system were aimed at selecting the pairings taking into account in an implicit way the rostering optimization constraints. Since the latter are explicitly considered by the new system, within this system the base constraints have been removed.

The first modification concerns the possible updating of the best solution found so far after every call to the greedy procedure. For each solution produced by this procedure, given by pairing set S, we compute, for each depot

$d \in D$, the simple bounds of § 3.1: Let L^d be the integer rounded-up value of the corresponding global lower bound L^d_{glo}, and $L = \sum_{d \in D} L^d$ be the overall lower bound for the original crew planning problem, i.e., the lower bound on the total number of weeks required to cover the complete set of timetabled trips. Moreover, let z be the value of the best crew planning solution found so far (initially, $z = \infty$), expressed in weeks. If $L < z$, we call the rostering optimization module, as the set S of pairings may lead to a better crew planning solution. Let z^H be the value of the solution found by the rostering optimization phase, which is executed with a given time limit (100 seconds in our implementation), even if typically the time required is shorter, as explained in § 3.1. If $z^H < z$, we update the best crew planning solution so far.

The second modification concerns the high-level organization into steps of the pairing optimization method. The *fixing* and *refining* steps have currently been removed, and at the end of the *heuristic* step, we redefine the pairing costs, as described in the following, and restart the method from the beginning with the new costs. In the description below, we will call *iteration* the execution of the *subgradient* and *heuristic* steps for a given cost vector.

Initially, the cost of a pairing is equal to 1 if it does not contain an external rest and 2 otherwise. In each iteration, the new costs take into account the binding rostering lower bounds for the solutions found with the previous costs. More precisely, let c^k_j denote the cost of pairing j in the generic iteration k, and let d be the depot associated with j. Let $\bar{L}^d_{glo}$ be the average value of the global lower bound L^d_{glo} computed over all "good" greedy solutions found in iteration k. By "good" we mean that the overall (i.e., over all depots) crew planning lower bound L was within 2% of the best overall crew planning lower bound found so far. Moreover, for each $\gamma \in \Gamma = \{ext, long, night, paid, work, seq\}$, let $\bar{L}^d_\gamma$ be the average value of lower bound L^d_γ for depot d, computed over all "good" greedy solutions found in iteration k.

The new cost c^{k+1}_j of pairing j is defined by

$$c^{k+1}_j := c^k_j + 0.1 \cdot \sum_{\gamma \in \Gamma} g(j, \gamma) \cdot f(\bar{L}^d_\gamma / \bar{L}^d_{glo}).$$

Here, the function $g(j, \gamma)$ is defined in the interval $[0, 1]$, and takes into account the value of characteristic γ for pairing j. In particular, for $\gamma \in \{ext, long, night\}$, $g(j, \gamma)$ is a binary flag equal to 1 iff pairing j is with external rest, long and overnight, respectively. Finally,

$$g(j, paid) = a_j / (24 \cdot 60)$$
$$g(j, work) = w_j / (14 \cdot 60 + 10)$$
$$g(j, seq) = a_j / (24 \cdot 60 + 18 \cdot 60)$$

noting that 24 hours and 14 hours plus 10 minutes are upper bounds on the paid time and on the working time of a pairing, respectively, and that $(24 + 18)$ hours is an upper bound on the contribution of a pairing to lower bound L_{seq}^d. The weight function $f(\alpha)$ is continuous and monotone in the interval $[0,1]$ with $f(0) = 0$, $f(1) = 1$, and increasing more steeply for values of α close to 1. Accordingly, for each depot $d \in D$, the closer is the lower bound $\bar{L}_\gamma^d$ to the global one, the larger is the corresponding weight, and hence the larger is the contribution of characteristic γ to the value of cost c_j^{k+1}.

The last modification concerns the procedure greedy used in the *heuristic step* and described in the previous subsection. In particular, let S be the current set of pairings selected by the procedure. In step 2, for each depot $d \in D$, we compute the global lower bound L_{glo}^d corresponding to the pairings in S associated with depot d. Recalling that L_{glo}^d represents a (real) number of weeks, we also compute the equivalent (integer) number of *days*, denoted by $\ell^d = \lceil L_{glo}^d \cdot 6 \rceil$. For each pairing $j \in N \setminus S$ associated with depot d, we compute the new global lower bound $L_{glo}^d(j)$ corresponding to the addition of pairing j to S, and denote by $\ell^d(j)$ the equivalent (integer) number of days. The score σ_j is initially defined by (6). Then, we penalize the pairing j if lower bound $\ell^d(j)$ is greater then ℓ^d. More precisely, if $\sigma_j > 0$ we define

$$
\sigma_j := \begin{cases}
2 \cdot \sigma_j & \text{if } \lceil L_{glo}^d(j) \rceil > \lceil L_{glo}^d \rceil \\
1.5 \cdot \sigma_j & \text{if } \lceil L_{glo}^d(j) \rceil = \lceil L_{glo}^d \rceil \text{ and } \ell^d(j) > \ell^d \text{ and } \ell^d(j) \mod 6 = 0 \\
1.1 \cdot \sigma_j & \text{if } \lceil L_{glo}^d(j) \rceil = \lceil L_{glo}^d \rceil \text{ and } \ell^d(j) > \ell^d \text{ and } \ell^d(j) \mod 6 = 5 \\
\sigma_j & \text{otherwise}
\end{cases}
$$

That is, we double the score if the integer rounded-up lower bound in weeks is increased by the addition of j, and also penalize the score if the lower bound in days after the addition of j is increased, and is equal, respectively, to an integer multiple of 6 or to an integer multiple of 6 minus 1, as these situations generally lead to crew rostering instances for which it is difficult to find a solution whose value equals the lower bound. Of course, if $\sigma_j < 0$, we divide σ_j by 2, 1.5, 1.1, respectively.

Moreover, in the procedure illustrated in the previous subsection, the removal of the redundant pairings in S was performed only when a feasible solution had been obtained. In the current context, the presence of redundant pairings may produce unreliable values for the global lower bounds used in the score computation. For this reason, the removal of possible redundant pairings is performed, in step 3, after *each* addition of a new pairing to S, as follows. For a set R of pairings, let $C(R) = \bigcup_{j \in R} I_j$ denote the set of trips covered by R. For each pairing $j \in N$, we have a flag f_j equal to 1 if the pairing has already been removed from the current solution S during the current execution of procedure greedy and equal to 0 otherwise. We store the set $S' = S$ of pairings just after the addition of the new pairing to S, and consider the pairings in S' according to the order in which they have been selected. For each pairing $j \in S$, if $C(S \setminus \{j\}) = C(S)$, we remove j from the current set S. Otherwise, if $C(S' \setminus \{j\}) = C(S')$, i.e., pairing j is redundant in S' but not in the current solution S, we remove j from S if $f_j = 0$. Note that in this way the number of currently covered trips may decrease during the removal phase, but procedure greedy is guaranteed to converge since each pairing j is removed at most once from S when $C(S \setminus \{j\}) \neq C(S)$.

4 Experimental Results

In this section, we illustrate the results of the old and new crew planning systems on a set of real-world instances provided by FS. All our programs were run on a Digital Ultimate Workstation. The pairing generator was coded in C, whereas the pairing and rostering optimization phases, both in the ALPI system and in the new integrated version, are implemented in FORTRAN.

In Table 1 we report the main characteristics of each instance, namely the number of trips, the number of depots, the number of pairings generated and the time required by the generation, expressed in seconds. We chose the pairing generator parameters described in § 2.3 so as to get (roughly) between half a million and one million pairings for each instance. The results with different parameters (as long as the number of pairings generated is tractable) are analogous to those presented in the sequel.

Instance	# trips	# depots	# pairings	time
MESTRE	121	1	1,024,448	166
MILAN	502	1	874,416	629
VERONA	86	1	484,139	94
BZ_TS_UD	118	3	457,021	109
A_CH_F_D_I_1	91	13	796,771	136
A_CH_F_D_I_2	309	15	497,847	370

Table 1. Characteristics of the instances considered.

The first three instances refer to trains which have to be covered by crews from a single depot, namely Mestre, Milan and Verona, respectively. The fourth instance is associated with trains to be covered by crews from the depots of Bolzano, Trieste and Udine, whereas the last two instances refer to international trains connecting Austria, France, Germany, Italy and Switzerland, where the various depots are located. Even if, at the moment, crews are handled separately for each country, these last instances simulate what would happen if all crews were handled by a unique European railway company.

Given the pairings generated as above, we ran the ALPI pairing optimization and rostering optimization modules, with time limits of 9,000 and 1,000 seconds, respectively. The time limit for the first module is much larger than the time limit for the second one since the pairing optimization phase, having to deal with several hundred thousand pairings for these instances, is typically much more time consuming than the rostering phase, which has to deal with a few hundred pairings selected in the previous phase, often subdivided among different depots. On the same instances, we also ran the integrated pairing and rostering optimization module. The overall time limit was 10,000 seconds, and the time limit for each internal execution of the rostering optimization phase was set to 100 seconds.

The results are reported in Table 2. For each instance and each system, we give the number of weeks in the final crew planning solution (# week), along with the associated number of selected pairings (# pair). Moreover, for the old system, we report a lower bound (LB) on the optimal value of the crew planning solution computed with respect to the pairings selected by the pairing optimization phase. This bound is possibly better than that corresponding to the simple lower bounds illustrated in the previous section as, within the rostering optimization phase, we also consider, for each depot, a more sophisticated lower bounding procedure, see Caprara et al. (1998). For the new system, we report the best overall crew planning lower bound (LB) computed over all set covering solutions found during the pairing optimization phase – note that this bound may not correspond to the set of pairings yielding the best crew planning solution. Finally, for both systems we report the CPU seconds required to obtain the best solution (time). For the old system, this time is equal to the sum of the times spent to find the best solutions required by the pairing optimizer and, for each depot, by the rostering optimizer.

Instance	Old System				New System			
	# week	LB	# pair	time	# week	LB	# pair	time
MESTRE	11	11	37	1118	11	11	21	148
MILAN	50	50	145	6082	48	47	152	7786
VERONA	8	8	20	102	7	7	22	344
BZ_TS_UD	11	11	30	103	10	9	28	138
A_CH_F_D_I_1	12	12	28	812	11	11	24	3434
A_CH_F_D_I_2	49	49	127	1208	41	40	94	633

Table 2. Solutions found with the old and new systems.

The table shows the considerable improvement, i.e., achieved by the integration of the pairing and rostering optimization phases, leading to an average percentage saving (in the number of weeks, and hence in the number of crews) of about 9.5%. Note that the time required to find the best solution is similar for the two systems, and that the number of pairings in the best solution is in some cases much smaller for the new system (for instances MESTRE and A_CH_F_D_I_2). We also observe that the improvement is particularly significant for instance A_CH_F_D_I_2, probably because the new system is able to subdivide the selected pairings among the depots in a much more effective way than the old one.

5 Conclusions and Future Work

In this paper we have shown the considerable improvements that can be achieved by integrating the pairing and rostering optimization phases into a unique one, even if the main structure of the two modules is essentially unchanged. (In particular, the rostering optimization module is identical to the one of the ALPI system.)

The main issue that we are currently investigating is how to construct a "good" set of candidate pairings in the pairing generation phase, since, with the current rules, the explicit generation of all the feasible pairings is completely impractical even for medium size instances with a few hundred trips. Quite likely, this will require some interaction between the pairing optimization and the pairing generation phases, possibly by using the Lagrangian costs to drive the generation of new pairings.

Moreover, it would be very interesting to derive some overall lower bounds on the optimal solution of the original crew planning problem, directly from the given trips, in order to evaluate the quality of the solutions found and possibly stop the process if optimality is proved.

Acknowledgements

This work was supported by E.U. Project EuROPE TRIO through Consorzio Padova Ricerche, Padova. We thank Pier Luigi Guida from FS, the project coordinator, for his support. Moreover, Francesco Bibbò, Vito Sante Achille, Vincenzo Autiero, and Francesco Olimipieri from FS provided us with the instances mentioned in this paper and with a detailed description of the crew planning constraints. Finally, we are grateful to Andrea Tabanelli for his help in carrying over some preliminary computational tests. The computational experiments were carried over at the LAB.O.R., the Laboratory of Operations Research of the University of Bologna. Finally, we are grateful to two anonymous referees for their helpful comments.

Bibliography

Barnhart, C., E.L. Johnson, G.L. Nemhauser, M.W.P. Savelsbergh, and P.H. Vance (1998). Branch-and-price: Column generation for solving huge integer programs. *Operations Research 46*, 316–329.

Barnhart, C., E.L. Johnson, G.L. Nemhauser, and P.H. Vance (1999). Crew scheduling. In R.W. Hall (Ed.), *Handbook of Transportation Science*, Kluwer, Boston, 493–521.

Bodin, L., B. Golden, A. Assad, and M. Ball (1983). Routing and scheduling of vehicles and crews: The state of the art. *Computers & Operations Research 10*, 63–211.

Caprara, A., M. Fischetti, P.L. Guida, P. Toth, and D. Vigo (1999a). Solution of large-scale railway crew planning problems: The Italian experience. In N.H.M. Wilson (Ed.), *Computer-Aided Transit Scheduling, Lecture Notes in Economics and Mathematical Systems*, 471, Springer, Berlin, 1–18.

Caprara, A., M. Fischetti, and P. Toth (1999b). A heuristic method for the set covering problem. *Operations Research 47*, 730–743.

Caprara, A., M. Fischetti, P. Toth, and D. Vigo (1998). Modeling and solving the crew rostering problem. *Operations Research 46*, 820–830.

Carraresi, P. and G. Gallo (1984). Network models for vehicle and crew scheduling. *European Journal of Operational Research 16*, 139–151.

Daduna, J.R. and A. Wren (Eds.) (1988). *Computer-Aided Transit Scheduling, Lecture Notes in Economics and Mathematical Systems*, 308. Springer, Berlin.

Desaulniers, G., J. Desrosiers, M. Gamache, and F. Soumis (1998). Crew scheduling in air transportation. In T.G. Crainic and G. Laporte (Eds.), *Fleet Management and Logistics*, Kluwer, Boston, 169–185.

Desrochers, M. and J.-M. Rousseau (Eds.) (1992). *Computer-Aided Transit Scheduling, Lecture Notes in Economics and Mathematical Systems*, 386. Springer, Berlin.

Desrosiers, J., Y. Dumas, M.M. Solomon, and F. Soumis (1995). Time constrained routing and scheduling. In M.O. Ball, T.L. Magnanti, C.L. Monma, and G.L. Nemhauser (Eds.), *Network Routing, Handbooks in Operations Research and Management Science*, 8, Elsevier, Amsterdam, 35–139.

Held, M. and R.M. Karp (1971). The traveling salesman problem and minimum spanning trees: Part II. *Mathematical Programming 1*, 6–25.

Klabjan, D. (1999). *Topics in Airline Crew Scheduling and Large Scale Optimization*. Ph.D. thesis, Georgia Institute of Technology, Atlanta, USA.

Rousseau, J.-M. (Ed.) (1985). *Computer Scheduling of Public Transport 2*. North-Holland, Amsterdam.

Wilson, N.H.M. (Ed.) (1999). *Computer-Aided Transit Scheduling, Lecture Notes in Economics and Mathematical Systems*, 471. Springer, Berlin.

Wren, A. (Ed.) (1981). *Computer Scheduling of Public Transport*. North-Holland, Amsterdam.

Efficient Timetabling and Vehicle Scheduling for Public Transport

Avi Ceder

Transportation Research Institute, Civil Engineering Faculty
Technion-Israel Institute of Technology, Haifa, Israel 32000
ceder@tx.technion.ac.il

Abstract. This work attempts to combine the creation of public transport timetables and vehicle scheduling so as to improve the correspondence of vehicle departure times with passenger demand while minimizing the resources (the fleet size required). The methods presented for handling the two components simultaneously can be applied for both single and interlining transit routes, and can be carried out in an automated manner. With the growing problems of transit reliability, and advances in the technology of passenger information systems, the importance of even and clock headways is reduced. This allows for the possibility to create more efficient schedules from both the passenger and operator perspectives. The procedures presented are accompanied by examples and graphical explanations. It is emphasized that the public timetable is one of the predominant bridges between the operator (and community) and the passengers.

1 Introduction

In general terms, the public transport operational planning process includes four basic components performed in sequence: *(1)* network route design, *(2)* setting timetables, *(3)* scheduling vehicles to trips, and *(4)* assignment of drivers (crew). It is desirable for all four components to be planned simultaneously to exploit the system's capability to the greatest extent and maximize the system's productivity and efficiency. However, this planning process is extremely cumbersome and complex, and, therefore, seems to require separate treatment of each component, with the outcome of one fed as an input to the next component. In the last 25 years, a considerable amount of effort has been invested in the computerization of the four components mentioned above, in order to provide more efficient controllable and responsive schedules. The best summary as well as the accumulative knowledge of this effort was presented in the books edited by Wren (1981); Rousseau (1985); Daduna and Wren (1988); Desrochers and Rousseau (1992); Daduna et al. (1995); Wilson (1999).

This work attempts to combine the two components of creating timetables and vehicle scheduling so as to improve the correspondence of vehicle departure times with passenger demand while minimizing the resources (the fleet size required). While the vehicle scheduling problem is treated extensively in the books mentioned above, only little attention is given to the problem of efficiently constructing vehicle frequencies and timetables.

Mathematical programming methods for determining frequencies and timetables have been proposed by several authors as follows. The objective in Furth and Wilson (1981) is to maximize the net social benefit, consisting of ridership benefit and wait time saving, subject to constraints on total subsidy, fleet size and passenger loading levels. Koutsopoulos et al. (1985) extended this formulation by incorporating crowding discomfort costs in the objective function and treating the time dependent character of transit demand and performance. Their initial problem comprises a non-linear optimization program relaxed by linear approximations. Ceder and Stern (1984) addressed the problem with an integer programming formulation and heuristic person-computer interactive procedure. The latter approach focuses on reconstructing timetables when the available vehicle fleet is restricted. Finally, Ceder and Tal (1999) used mixed integer programming and heuristic procedures for constructing timetables with maximum synchronization. That is maximization of the number of simultaneous arrivals of vehicles to connection stops.

Other methods for frequency and timetable determination are related to the type and adequacy of the input passenger count data. These methods aimed at practicability appear in Ceder (1984, 1987) and are briefly described in the following Section 2. In Section 3 the scope and framework of this study are outlined. In Section 4 a timetable construction procedure is integrated with the creation of chains of trips (blocks), and Sections 5 and 6 provide an example and concluding remarks.

2   Background

2.1   Optimal Timetables

Public transport timetable is commonly constructed for given sets of derived frequencies. The basic criteria for the determination of frequencies are: *(a)* to provide adequate vehicle's space to meet passenger demand, and *(b)* to assure a minimum frequency (maximum-policy headway) of service. Ceder (1984) described four different methods for calculating the frequencies. Two are based on point-check (counting the passengers on-board the transit vehicle at certain point(s)), and two on ride-check (counting the passengers along the entire transit route). In the point-check methods the frequency is the division between passenger load at the maximum (max) load point (either the one across the day or in each hour) and the desired occupancy or load factor. In the ride-check methods the frequency is the division between average or restricted-average passenger load and the desired occupancy. The average load is determined by the area under the load profile (in passenger-km) divided by the route length (km), and the restricted average is a higher value than the average one, in order to assure that in certain percentage of the route length the load does not exceed the desired occupancy. This desired occupancy (or load factor) is the desired level of passenger load on each vehicle, in each time period (e.g., number of seats).

In a follow-up study Ceder (1987) analyzed optional ways for generating public timetables. This analysis allows for establishing a spectrum of alternative timetables, based on three categories of options: *(a)* selection of type of headway, *(b)* selection of frequency determination method for each period, and *(c)* selection of special requests. In category (a) the headway (time interval between adjacent departures) can be equal or balanced. Equal headway refers to the case of evenly spaced headways and balanced headway – to the case of unevenly spaced headways but with even average passenger load at the hourly maximum load point. These cases are being extended in this work. In category (b) it is possible to select for each time period one of the four frequency determination methods mentioned above, or a given frequency by the scheduler. In category (c) it is possible to request clock headways (departure times that repeat themselves in each hour, easy-to-memorize) and/or certain number of departures (usually for cases with limited resources).

The outcome of these analyses is a set of optional timetables in terms of vehicle's departure times at all specified timepoints, using passenger load data. Each timetable is accompanied by two comparison measures which are used as an evaluation indicator in conjunction with resource saving. The first measure is the total required vehicle runs (departures) and the second is an estimate for the minimum required fleet size at the route level only.

2.2 Deficit Function

Following is a description of the Deficit Function (DF) approach described by Ceder and Stern (1981), for assigning the minimum number of vehicles to allocate for a given timetable. A DF is simply a step function that increases by one at the time of each trip departure and decreases by one at the time of each trip arrival. Such a function may be constructed for each terminal in a multiterminal transit system. To construct a set of DFs, the only information needed is a timetable of required trips. The main advantage of the DF is its visual nature. Let $d(k, t)$ denote the DF for the terminal k at the time t where its maximal value of $d(k, t, S')$ over the schedule horizon $[T_1, T_2]$ is designated $D(k, S)$. If the set of all terminals is denoted as T, the sum of $D(k)$ $\forall\, k \in T$ is equal to the minimum number of vehicles required to service T. This is known as the fleet size formula. Mathematically, for a given fixed schedule S:

$$D(S) = \sum_{k \in T} D(k) = \sum_{k \in T} \max_{t \in [T_1, T_2]} d(k, t) \tag{1}$$

where $D(S)$ is the minimum number of buses to service the set T, and T_1 and T_2 are the schedule horizon boundaries.

When deadheading (DH) trips are allowed, the fleet size may be reduced below the level described in Equation (1). Ceder and Stern (1981) described a procedure based on the construction of a unit reduction DH chain (UR-DHC), which, when inserted into the schedule, allows a unit reduction in the fleet size. The procedure continues inserting URDHCs until no more can be included or a lower boundary on the minimum fleet in reached. Initially, the lower bound was determined to be the maximum number of trips in a given timetable that are in simultaneous operation over the schedule horizon. Stern and Ceder (1983) improved this lower bound, based on the construction of a temporary timetable in which each trip's arrival time is extended to the time of the first trip that may feasibly follow it in the schedule.

The DF theory was extended by Ceder and Stern (1982) to include possible shifting in departure times within bounded tolerances. Basically, the shifting criteria are based on a defined tolerance time for maximum advance of the trip scheduled departure time (early departure), and maximum delay allowed (late departure). The interval of time over which $d(k,t)$ takes on its maximum value is then compared with the appropriate tolerance time elements for establishing conditions in which it is possible to reduce the fleet size by one via certain shifts.

Finally, all of the trips, including those that were shifted and the DH trips, are chained together for constructing the vehicle schedule (blocks). Two rules can be applied for creating the chains: first in – first out (FIFO), and a chain-extraction procedure described by Gertsbach and Gurevich (1977). The FIFO rule simply links the arrival time of a trip to the nearest departure time of another trip (at the same location), and continues to create a schedule until no connection can be made. The trips considered are deleted and the process continues. The chain-extraction procedure allows an arrival-departure connection for any pair within a given hollow (on each DF). The pairs considered are deleted and the procedure continues. Both methods end with the minimum derived number of vehicles (blocks).

3   Scope and Framework

It is the purpose of this work to use a method for better matching the passenger demand with a given timetable while attempting to minimize the fleet size (one of the main resources). This will result in a more reliable and comfortable service. This research follows the studies of Ceder (1984, 1987), where the average passenger load is counted at the max load point on an hourly basis, and the division of this load by the desired occupancy results in the frequency unless the minimum required frequency is not reached. The non-integer value of the frequency is then kept and based on the accumulative frequency curve (adding the frequency at each hour with respect to time), the departure times at the hourly max load point are determined with even headways. In the study of Ceder (1987), the average passenger loads are counted at the max load point of the route for each vehicle separately. These loads are accumulated with respect to time and based on the desired occupancy values the departure times at the route maximum load point are determined with uneven headways and even average loads only at this route point. In a new research (Ceder (2000)) the average passenger loads are counted at each vehicle's max load point as opposed to the route max load point. In order to derive the departure times with even load at the critical max load point of each vehicle an algorithm is developed in Ceder (2000). This algorithm is then applied graphically in Section 5. All the derived timetables are also based on a smoothing procedure between the time periods such that there is no need to round any number, and the desired occupancy is kept in these transition periods.

The study presented in this work is shown in a flowchart format in Figure 1. This flowchart has three columns: input, component, and output. The output of the components is also served as an input to a next component. Basically Figure 1 is the framework of the study with the following input: Network of transit routes; set of time periods; average loads on the transit vehicles at their max load points; average trip travel times; average trip layover times; average DH travel times; tolerances for the departure time shifting; and tolerances for the desired occupancies (load factors). The overall study process starts with the derivation of timetables with even average loads and smoothing consideration in the transition between time periods based on Ceder (2000). Then a new set of possible departure time shifting is determined. The asterisk in the first output indicates that this output is used as an input (in two input parts) indicated by asterisks. The next step is the construction of what is described in Section 2 (known as Deficit Functions) where in every departure (at a certain terminal or major stop) the DF is moving up by one and every arrival the DF is moving down by one. Then the DFs are going through both shifting and dead-heading trip insertion procedures, and the timetable is adjusted while complying with the tolerance constraints. The final step is the establishment of vehicle schedules (blocks).

4 Integration with Vehicle Scheduling

Once the timetables are constructed according to Ceder (2000) in order to meet best the fluctuated passenger demand, consideration should be given on how to execute them in an efficient manner. Each trip in the timetable becomes one element of a daily chain of trips to be carried out by a single vehicle. The problem is then to find the minimum number of chains (vehicle schedules, blocks) that contain all the trips in all timetables. Usually this minimum is fully required during peak hours and, therefore, represents the fleet size. The less is the fleet size, the higher is the saving in capital resources.

There is no doubt that interlinings (vehicles are allowed to switch from one route to another) can further reduce the fleet size. The DF theory in § 2.2 provides the procedures for determining the minimum fleet size with interlinings. In transit systems without interlinings the fleet size can be optimized by the shortturning strategies described by Ceder (1990, 1991).

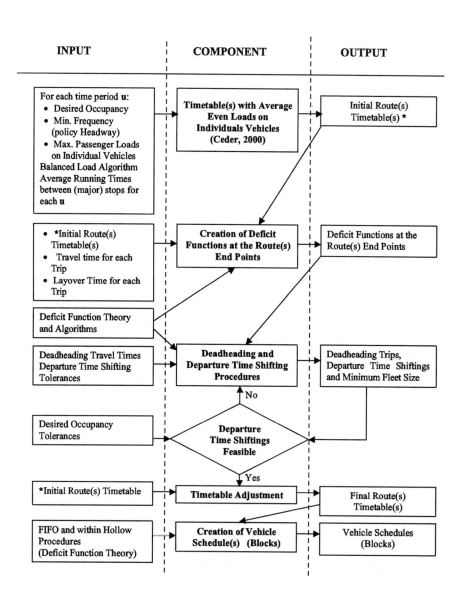

Figure 1. Framework of study.

This section presents a procedure to integrate the derived timetables and vehicle scheduling with interlinings using some fine tuning. Let:

Δ_{eu} a given positive tolerance (in minutes) for maximum shifting t_{1j}^* to the left (early departure), for each interval u, $j = 1, 2, \ldots, m$; $u = 1, 2, \ldots, v$

Δ_{lu} a given positive tolerance (in minutes) for maximum shifting t_{1j}^* to the right (late departure), for each interval u, $j = 1, 2, \ldots, m$; $u = 1, 2, \ldots, v$

$L_i(t)$ accumulative load (# of passengers) curve at stop i w.r.t. time t, where changes in the slope are made at the departure times of the transit vehicles at $i, i = 1, 2, \ldots, n$, and the last slope is extrapolated to the schedule (timetable) horizon.

$SL_i(t)$ the slope of $L_i(t)$ at $t, i = 1, 2, \ldots, n$

Δd_u a given positive tolerance (in passengers) of the desired occupancy d_u at time interval $u, u = 1, 2, \ldots, v$.

4.1 Tolerances of Departure Times

The DF procedures include possible shifting in departure (dep.) times within bounded tolerances. It is based on a defined tolerance time $[t_{1j}^* - \Delta_{eu}, t_{1j}^* + \Delta_{lu}]$ where $\Delta_{eu} > 0$ is the maximum advance (early dep.) of the trip scheduled departure time, t_{1j}^*, and $\Delta_{lu} > 0$ is the maximum delay allowed (late dep.). These tolerances are used wherever shifting in dep. times is allowed, and can be a necessary element in saving vehicles.

There are three ways to use Δ_{eu} and Δ_{lu} for saving one vehicle: (a) to shift t_{1j}^* to the left (early dep.) or to the right (late dep.) by value of time less than or equal Δ_{eu}, and Δ_{lu}, respectively, in order to reduce a given DF by one; (b) to shift two departures in opposite directions (left and right) within the bounded tolerances; and (c) to shift t_{1j}^* to left or right in order to allow for an insertion of a DH trip. Further explanation appears in the example in Section 5.

4.2 Tolerances of Desired Occupancies and Their Analysis

Once a shift has been made in a trip departure time it violates the balanced load criterion of this trip and the one to follow. That is, the attained desired occupancy at the trip's max load point is dependent on this trip's dep. time, and by changing it the trip's average max load will increase or decrease as well as for the next trip.

Let t_{1j}^* be a departure time determined at stop q based on even load at the max load point of each vehicle. That is, by adding d_u (the desired occupancy, or load factor) the minimum time that intersects one of the accumulative load curves is at q. Hence, $d_u = L_q(t_{1j}^*) - L_q(t_{1,j-1}^*)$. The shifts in departure times made in the DF procedure are defined as $t_{1j}^- = t_{1j}^* - \Delta_{eu}$ and $t_{1j}^+ = t_{1j}^* + \Delta_{eu}$ $\forall\, j = 1, 2, \ldots, m$, and relevant u for t_{1j}^- and t_{1j}^+.

In order to avoid excess average load, beyond $d_u + \Delta_{du}$, at each trip's critical point two criteria are established below for early and late departures.

Early Departure Criterion: Based on the above definitions:

$$L_i(t_{1j}^-) = L_i(t_{1j}^*) - \Delta_{eu}SL(t_{1j}^*), \quad \text{for } SL_i(t_{1j}^*) = SL_i(t_{1j}^-) \tag{2}$$
$$i = 1, 2, \ldots, n; \quad t_{1j}^- \text{ belongs to } u$$

Note 1: In case that the slope is charged within Δ_{eu} shift for any stop i, Equation (2) should consider two (or more) decreased portions, each is related to a different slope and its associated part of Δ_{eu}. The loads at t_{1j}^- can be expressed as:

$$L_i(t_{1j}^-) - L_i(t_{1,j-1}^*) < d_u, \quad i = 1, 2, \ldots, n; \qquad t_{1j}^- \text{ belongs to } u$$

These loads of the new t_{1j}^- departure across all stops are based on even loads at the max load point of each vehicle, in which only at q the desired occupancy d_u is attained for t_{1j}^- and in all other stops the load is less than d_u. Using Δ_{eu} shift to the left (early dep.) will further reduce these loads. However, the loads at each stop i, for the adjacent dep. to t_{1j}^- at $t_{1,j+1}^*$, will increase the loads by the Δ_{eu} shift. This increase is $\Delta_{eu}SL_i(t_{1j}^*)\ \forall\ i$ and relevant u, or it is the sum of portions according to Note 1.

The increased new loads at $t_{1,j+1}^*$ need to be checked against $d_u + \Delta_{du}$ across all stops. Certainly this check is applied for the maximum increase of load, and hence the early dep. criterion for accepting Δ_{eu} is:

$$\underset{i=1,2,\ldots,n}{\text{Max}} \left[L_i(t_{1,j+1}^*) - L_i(t_{1j}^-) \right] \leq d_u + \Delta d_u \tag{3}$$

where $t_{1,j+1}^*$ belongs to interval u, and $L_i(t_{1j}^-)$ is obtained by Equation (2) while considering Note 1.

Late Departure Criterion: Following similar arguments to the Δ_{eu} criterion in Equation (3), one can derive the Δ_{lu} criterion. That is,

$$L_i(t_{1j}^+) = L_i(t_{1j}^*) + \Delta_{lu}SL(t_{1j}^*), \quad \text{for } SL_i(t_{1j}^*) = SL_i(t_{1j}^+) \tag{4}$$
$$i = 1, 2, \ldots, n; \quad t_{1j}^* \text{ belongs to } u$$

Note 2: In case that the slope is changed within the Δ_{lu} shift for any i, Equation (4) should consider two (or more) increased portions; each is related to a different slope and its associated part of Δ_{lu}. The loads at t_{1j}^+ can be expressed as:

$$L_i(t_{1,j+1}^*) - L_i(t_{1j}^+) < d_u, \quad i = 1, 2, \ldots, n; \qquad t_{1,j+1}^* \text{ belongs to } u$$

In this case of late departure all the average loads on the $t^*_{1,j+1}$ dep. will be decreased. However, the average loads on the t^+_{1j} dep. will be increased in comparison with the average loads of the t^*_{1j} dep. across all stops. This increase is $\Delta_{lu}SL_i(t^*_{1j}) \; \forall \; i$ and relevant u, or it is by the sum of some portions according to Note 2.

The late departure criterion for accepting Δ_{eu} is, therefore:

$$\underset{i=1,2,\ldots,n}{\text{Max}} \left[L_i(t^+_{1j}) - L_i(t^*_{1,j-1}) \right] \leq d_u + \Delta d_u \qquad (5)$$

where t^+_{1j} belongs to interval u and is obtained by Equation (4) with the consideration of Note 2.

5   Example

The examples in this work are used as an explanatory device for the developed procedures. Prior to wrapping up the presented methodology further clarity can be obtained by a complete example of what is illustrated in a flow-chart form in Figure 1.

Table 1 contains the necessary information and data for a 3-hour example of a transit line from A to B and B to A. Point B can be perceived as the Central Business District (CBD) that attracts the majority of the demand between 6-9 am. There are 14 and 8 departures for A to B and B to A, respectively. The average observed max load on each trip, service and DH travel times, and desired occupancies are shown in Table 1. For all hours for both directions we have a minimum frequency of 2 vehicles per hour, an early dep. tolerance of 2 minutes, a late dep. tolerance of 3 minutes, and a desired occupancy tolerance of 8 passengers.

In order to construct the balanced load timetable, these Table 1 data are used for running the algorithm in Ceder (2000). Assuming that the max load is observed at the same stop for each direction, the algorithm in Ceder (2000) determines the new departure times shown in Figures 2 and 3. There are 14 new departures for direction A to B, in Figure 2, that are based on desired occupancies of 50 and 65 passengers. There are 3 new departures for direction B to A in Figure 3. For the A-B direction, in Figure 2, $d_u = 50$, $u = 1$ (6-7 am), and $d_u = 65$, $u = 2,3$ (7-9 am). The third dep. check for $L_p(t - T_{m1}) = 150$, where p is the single max load point (both for A-B, and B-A directions) and T_{m1} is the average travel time from A to m, results in a departure over 7:00 am. Therefore, d_u is changed to 65 and the third dep. is coordinated with $L_p(t - t_{m1}) = 165$ to be 7:07 am.

For the B-A direction, in Figure 3, at first $d_{u=1} = 50$ is set and results in one departure between 6-7 am whereas $F_m = 2$ in Table 1. That is, the second dep. at $L_p(t - T'_{m1}) = 100$ is beyond 7:00 am (7:04 am). Consequently, there is only one departure between 6-7 am which does not comply with $F_m = 2$, and hence the minimum frequency complementary component of the algorithm developed by Ceder (2000), is applied. Step (b) of this complementary

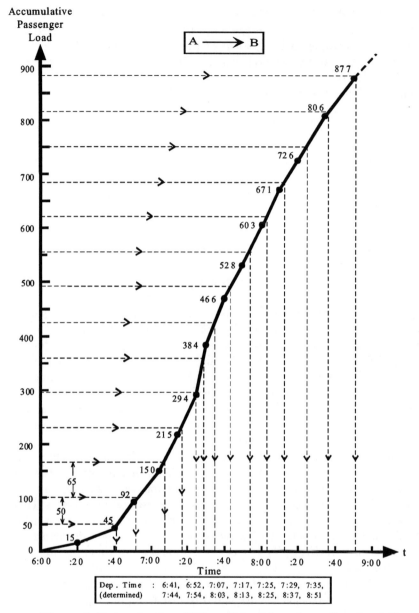

Figure 2. Determination of balanced load dep. times, direction A-B.

Time	Dep. time at the route dep. point A→B	B→A	Average observed max number of pass. on-board A→B	B→A	Travel time including layover time (min.) Service A→B	B→A	Deadheading A→B	B→A	Desired occupancy (pass.) A→B	B→A
6-7 am	6:20 6:40 6:50	6:30 6:45	15 30 47	22 38	60	50	40	35	50	50
7-8 am	7:05 7:15 7:25 7:30 7:40 7:50	7:10 7:25 7:45	58 65 79 90 82 62	52 43 59	75	60	45	40	65	50
8-9 am	8:00 8:10 8:20 8:35 8:50	8:25 8:40 8:55	75 68 55 80 71	23 51 28	70	60	45	40	65	50

Table 1. Given data for the example problem.

component provides the formula for the new desired occupancy:

$$d_{1m} = \frac{L_p(7:00)}{2+1} = \frac{91}{3}$$

That is, the first two departures are at load levels of 30 and 60, and the third at level of $60 + 50 = 110$ passengers. Another observation in Figure 3 is related to the policy headways. Since $F_m = 2$ is the only requirement, there is a large headway between 7:44 and 8:32 departures. However, if there is a policy (max) headway criterion then an adjustment can be made based on Ceder (2000).

Once the departure times are set at both route end points, the vehicle scheduling component can be integrated into the two-direction timetables. First, two DFs are constructed at A and B as it is shown in Figure 4. These DFs are based on the schedule of 21 trips (14 of A-B, 7 of B-A), presented with respect to their travel times in the upper part of Figure 4. Second, the DF theory leads to save one vehicle at $d(A, t)$ through a shifting of trip # 17 by one minute forward (late dep.), and inserting a DH trip from B to A (7:52 to 8:32 am). The total fleet required is then $8 + 4 = 12$ vehicles.

Since $\Delta_{lu} = 3$ minutes $\forall\, u$ in Table 1, the shifting of one minute is allowed of trip # 17 from A (dep. 8:26) to B (arrival 9:36). However, the feasibility

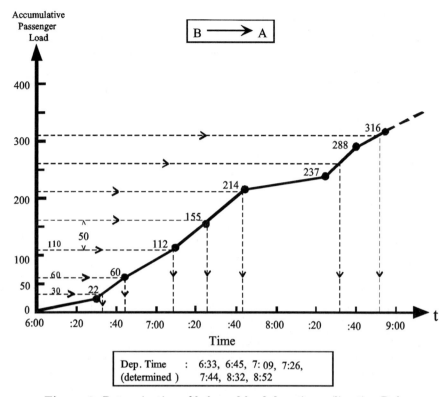

Figure 3. Determination of balanced load dep. times, direction B-A.

of this shift must be checked against the allowed tolerance for the desired occupancy change. That is, the check based on Equation (5):

$$L_A(8:26) - L_A(8:13) \leq 65 + 8$$

where $L_A(8{:}26) = 758$ and $L_A(8{:}13) = 687.5$ are derived from Figure 2 and Table 1. Thus $758 - 687.5 = 70.5 < 73$ complies with $\Delta d_3 = 8$. Another way to find this compliance is to look at the relevant slope of $L_A(8{:}25)$ in Figure 2 between $L_A(8{:}20) = 726$ and $L_A(8{:}35) = 806$. This $SL_A(8{:}25) = (806 - 726)/(15) = 5.3$ pass/minute will increase the average load on the 8:25 dep. by 5.3 since the shift is one minute. It means that the balanced max load of 65 will change to 70.3 (this is not exactly 70.5 like in Equation (5) due to the rounding of departure times to integer minutes). Having done the check for the desired occupancy tolerance criterion, the final efficient schedule can be set for both balancing the passenger loads at each trip's critical point, and for the minimum fleet size required. The timetables at the route's end points appear in the left part of Table 2, and the blocks in its right part. The blocks (vehicle schedules) are contracted from the timetable using the FIFO rule. The first block, e.g., starts with trip # 1 which linked with its first feasible connection at A, trip # 8 (7:23 links to 7:25), and the trip # 21 at B

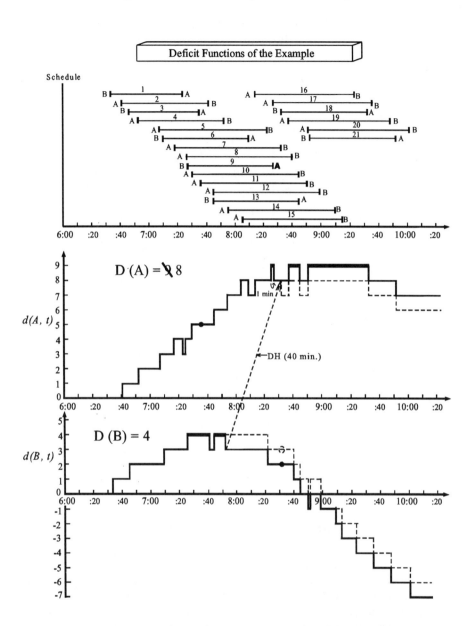

Figure 4. Deficit function analysis for the example problem.

(8:40 links to 8:52). This FIFO rule can be replaced by the chain-extraction procedure that allows an arrival-departure connection for any pair within a given hollow on the DF.

A → B			B → A			Block	Trips in block
Trip #	Departure time	Arrival time	Trip #	Departure time	Arrival time	number	(in sequence, via FIFO)
2	6:41	7:41	1	6:33	7:23	1	1-8-21
4	6:52	7:52	3	6:45	7:35	2	2-13-20
5	7:07	8:22	6	7:09	8:09	3	3-11
7	7:17	8:32	9	7:26	8:26	4	4-DH-19
8	7:25	8:40	13	7:44	8:44	5	5-18
10	7:29	8:44	DH*	7:52	8:32	6	6-16
11	7:35	8:50	18	8:32	9:32	7	7
12	7:44	8:59	21	8:52	9:52	8	9-17
14	7:54	9:09				9	10
15	8:03	9:13				10	12
16	8:13	9:23				11	14
17	8:26**	9:36				12	15
19	8:37	9:47					
20	8:51	10:01	(see Figure 4)				

* Inserted deadheading trip
** Trip 17 was shifted by one minute

Table 2. Timetable and vehicle schedule (blocks) of the example problem.

6   Discussion and Concluding Remarks

Different public transport agencies use different scheduling strategies based primarily on their own schedulers' experience, and secondarily on their scheduling software (if any). As a result, it is unlikely that two independent public transport agencies will use exactly the same scheduling procedures, at the detailed level. In addition, even at the same public transport agency, the schedulers may use different scheduling procedures for different groups of routes. Consequently, there is a need when developing computerized procedures to supply the schedulers with alternative schedule options along with interpretation and explanation of each alternative. One such alternative is developed in this work. Also, undoubtedly, it is desirable that one of the alternatives will coincide with the scheduler manual procedure. In this way, the scheduler will be in a position not only to expedite manual tasks but also to compare methods with others regarding the trade-off between passenger comfort and operating cost.

This work presents the integration of public transport timetables and vehicle scheduling for attaining the minimum fleet size. Average even loads on individual vehicles can be approached by relaxing the evenly spaced headways pattern (rearrangement of departure times). It is known that passenger demand varies even within one hour, reflecting the business, industrial, educational, cultural, social and recreational public transport needs of the community. This dynamic behavior can be detected through passenger load counts, and information provided by road supervisors. The adjustments of departure times, made by Ceder (1987, 2000), form the basis to improve the correspondence of vehicle departure times with the fluctuated passenger demand. These adjustments resulting in a balanced load timetables are based on a given vehicle desired occupancy at the maximum load point. This allows for introducing optional timetables with the consideration of even average loads. The construction of such timetables takes into account, in essence, the passenger perspective. A complementary measure to that is to consider the minimum number of public transport vehicles that are needed for the execution of the timetables.

This work provides a procedure to integrate these two components (timetabling and vehicle scheduling) based on the deficit function theory and given tolerances. The outcome is a set of efficient schedules from both the passenger and operator perspectives. The stepwise and graphical procedures to attain it allow some man-machine intervention and dialogue.

Bibliography

Ceder, A. (1984). Bus frequency determination using passenger count data. *Transportation Research A 18*, 439–453.

Ceder, A. (1987). Methods for creating bus timetables. *Transportation Research A 21*, 59–83.

Ceder, A. (1990). Optimal design of transit short-turn trips. *Transportation Research Record 1221*, 8–22.

Ceder, A. (1991). A procedure to adjust transit trip departure times through minimizing the maximum headways. *Computers & Operations Research 18*, 417–431.

Ceder, A. (2000). *Public transport timetables and vehicle scheduling with balanced passenger loads*. Working paper, Institute of Transport Studies, Faculty of Economics and Business, University of Sydney, Sydney.

Ceder, A. and H.I. Stern (1981). Deficit function bus scheduling with deadheading trip insertion for fleet size reduction. *Transportation Science 15*, 338–363.

Ceder, A. and H.I. Stern (1982). Graphical person-machine interactive approach for bus scheduling. *Transportation Research Record 857*, 69–72.

Ceder, A. and H.I. Stern (1984). Optimal transit timetables for a fixed vehicle fleet. In J. Volmuller and R. Hammerslag (Eds.), *Proceedings of the 10th International Symposium on Transportation and Traffic Theory*, UNU Science Press, Holland, 331–355.

Ceder, A. and O. Tal (1999). Timetable synchronization for buses. In N.H.M. Wilson (Ed.), *Computer-Aided Scheduling of Public Transport*, Springer, Berlin, 245–258.

Daduna, J.R., I. Branco, and J.M.P. Paixão (Eds.) (1995). *Computer-Aided Transit Scheduling, Lecture Notes in Economics and Mathematical Systems*, 430. Springer, Berlin.

Daduna, J.R. and A. Wren (Eds.) (1988). *Computer-Aided Transit Scheduling, Lecture Notes in Economics and Mathematical Systems*, 308. Springer, Berlin.

Desrochers, M. and J.-M. Rousseau (Eds.) (1992). *Computer-Aided Transit Scheduling, Lecture Notes in Economics and Mathematical Systems*, 386. Springer, Berlin.

Furth, P.G. and N.H.M. Wilson (1981). Setting frequencies on bus routes: Theory and practice. *Transportation Research Board 818*, 1–7.

Gertsbach, I. and Y. Gurevich (1977). Constructing an optimal fleet for transportation schedule. *Transportation Science 11*, 20–36.

Koutsopoulos, H.N., A. Odoni, and N.H.M. Wilson (1985). Determination of headways as function of time varying characteristics on a transit network. In J.M. Rousseau (Ed.), *Computer Scheduling of Public Transport 2*, North-Holland, Amsterdam, 391–414.

Rousseau, J.M. (Ed.) (1985). *Computer Scheduling of Public Transport 2*. North-Holland, Amsterdam.

Stern, H.I. and A. Ceder (1983). An improved lower bound to the minimum fleet size problem. *Transportation Science 17*, 471–477.

Wilson, N.H.M. (Ed.) (1999). *Computer-Aided Transit Scheduling, Lecture Notes in Economics and Mathematical Systems*, 471. Springer, Berlin.

Wren, A. (Ed.) (1981). *Computer Scheduling of Public Transport*. North-Holland, Amsterdam.

Rail Crew Scheduling and Rostering Optimization Algorithms

Andreas Ernst[1], Houyuan Jiang[1], Mohan Krishnamoorthy[1], Helen Nott[2], and David Sier[1]

[1] CSIRO Mathematical and Information Sciences,
Private Bags 10, Clayton South, MDC Clayton, VIC 3169, Australia
{Andreas.Ernst, Mohan.Krishnamoorthy, David.Sier}@cmis.csiro.au
E-mail for correspondence: Houyuan.Jiang@cmis.csiro.au
[2] PriceWaterhouseCoopers,
Helen.Nott@au.pwcglobal.com

Abstract. Train crew rostering involves the development of a duty timetable for each of the drivers of a rail transport organization. This duty timetable is spread over a period known as the roster planning horizon. We propose an optimization approach for an instance of a train crew rostering problem arising from a practical application. The problem involves developing continuous and cyclic work lines (or rosters) for train drivers spread over several depots in a rail network. The rail timetable includes over 1300 trips a week. The rosters for all the crew must conform to complex industrial regulations and work rules. The main objective in this model is to minimize the overall roster cost accrued from using the available crew while providing the requisite number of crew for each train trip in the schedule. The rosters must also satisfy quality standards for all drivers, by attempting to satisfy their personal preferences. We describe the problem and provide optimization formulations and solution approaches. We also present some computational results.

1 Introduction

Crew scheduling and rostering are two major tasks of crew management, particularly in railway systems. These problems are concerned with the development of duty schedules for crew, in order to cover a given *timetable* or *operational schedule*. *Crew scheduling* involves the short-term (typically one day or one week) tactical scheduling of crew, with the aim of developing a set of duties that will be performed by each crew to adequately cover the timetable. *Crew rostering* is a long-term, strategic problem that involves a larger planning horizon (typically of a few weeks) in which, the duties assigned to a particular crew are sequenced together to form a roster that conforms to corporate goals, work regulations and individual preferences. The goal might be to minimize roster cost, while work regulations might encompass safety requirements and enterprise bargaining agreements. For example, see Caprara et al. (1999) or Ernst et al. (1998) for more details.

Crew scheduling and rostering can arise at either the planning stage, in which the total number of crew is to be determined, or the operational stage, in which scheduling and rostering are done using a given number of available crew. Crew scheduling and rostering problems are difficult to solve due to their typically large size. Moreover, practical instances of crew scheduling and rostering applications involve many complex constraints arising from operational requirements, crew preferences and work regulations that need to be taken into account. In recent years, operations research has become a useful tool for tackling these difficult problems.

In this paper, we are interested in operational crew rostering in a freight train system, arising from a practical application. We develop optimization models in an attempt to solve the operational crew scheduling and rostering problems, with greater emphasis placed on solving the crew scheduling problem. We start with a known weekly *train timetable*, in which each train *journey* originates at a *starting location* on a given day, date and time, and terminates at an *ending location* on a specified day, date and time. A train journey may be further split into *segments*, (or legs or trips). In our application, these segments are between depots (or major cities) along the *route* (or journey). While the train continues its journey from its start location to its terminal location via several intermediate depots, drivers may get on and off a train journey at different depots. Therefore, the daily duty of a train driver may consist of several segments.

A *crew complement*, consisting of at least one but usually two staff members, covers each leg of the journey. In this paper we use the terms crew and crew complement interchangeably. The crew scheduling problem is to assign crew complement to each segment of each journey, or to ensure that each segment is covered by a crew complement. In crew rostering, the task is to construct a roster over the planning horizon for all crew such that each crew carries out a sequence of duties (crewing segments). A good roster should not only take into account many different regulations required by the employer, the government and the industry, but also reflect workers' preferences.

In the airline industry crew scheduling and rostering have received much attention. Airline crew scheduling and rostering have been studied in, e.g., Arabeyre et al. (1969); Barnhart et al. (1994); Day and Ryan (1997); Etschmaier and Mathaisel (1985); Gamache and Soumis (1998); Gamache et al. (1999); Gershkoff (1989); Hoffman and Padberg (1993); Mason et al. (1998) and Vance et al. (1997). Much of this research concentrates on set partitioning and/or set covering models for solving these problems. Train crew scheduling and train crew rostering have received much less attention until recently. Some examples can be found in Anbari (1987); Bianco et al. (1992); Caprara et al. (1997, 1998, 1999); Ernst et al. (1998); Fores et al. (1998); Larcher and Sinay (1982); Morgado and Martins (1992, 1993) and Tykulsker et al. (1985).

This paper is motivated by the work of Ernst et al. (1998) who presented a simulated annealing heuristic method for a train crew rostering application for National Rail, the Australian freight train system, as well as proposing a mathematical model for generating rosters. However, the model in Ernst et al. (1998) inadequately represents all aspects of the real application due to several assumptions. Later Ernst et al. (2000) proposed an integrated optimization model for solving the planning crew scheduling and rostering problem. While Ernst et al.'s model has certain similarities to the model presented in this paper, that model is used in the planning stage to determine crew numbers, whereas here we are concerned with the operational stage of covering a train timetable using existing crew.

We first describe the problem and recall some central concepts introduced in Ernst et al. (1998), before introducing some necessary notation, and presenting our optimization model for crew scheduling. We report our computational experience before making some conclusions.

2 Problem Description

We represent the rail network as a set of nodes and arcs. The nodes represent depots (or stations, or cities) and the arcs represent segments of train tracks that connect depots. In the Australian rail network, most of the depots are spaced about $5 - 8$ hours of travel time apart. Each depot has a number of crew complements, that consider it their *home depot*. Each crew complement involves two drivers, and the same two people are rostered together through the rostering period.

The majority of train crew duties involve driving trains according to a timetable. A critical concept introduced in Ernst et al. (1998) is that of a *roundtrip*. A roundtrip, which is similar to a pairing in the airline crew scheduling literature, consists of a sequence of segments beginning and ending at the home depot. A *roster* for a train crew involves several *tours of duty* or roundtrips. Roundtrips may either start and finish a tour of duty at its home base in a single shift without any long rest (such as in European networks), or involve crew taking rests, or *barracking* between two shifts in a roundtrip (as is most common in the sparse Australian network). A barracking roundtrip consists of two consecutive shifts between which a long rest (or overnight sleep) is required. The freight train timetable in the Australian rail network is repeated weekly, rather than daily as in the European network. Hence a weekly planning horizon for crew scheduling is a natural choice.

From this discussion it is clear that we cannot apply the methods for train crew scheduling and rostering methods developed for some European railway systems (see Caprara et al. (1997, 1999, 1998); Fores et al. (1998)), and for airline crew scheduling and rostering (see Barnhart et al. (1994); Day and Ryan (1997); Gamache and Soumis (1998); Gamache et al. (1999); Hoffman and Padberg (1993) and Mason et al. (1998)).

Assumptions and Input Data: We assume that the weekly timetable for all trains is known and remains unchanged during the planning horizon. We also assume that we have full knowledge of all the crew complements at each depot, and that the number of crew employed at each depot remains unaltered during the planning horizon.

Objective and Constraints: The main objective of rail crew rostering is to obtain a minimum cost sequence of roundtrips for each crew, and thus form a *line of work* for each crew. The lines of work form the *master roster* and this must conform to various constraints, such as:

- All train timetable segments must be covered using the available crew complements.
- A line of work should not violate major employment conditions and work regulations.
- The roster should satisfy several constraints that are designed to collectively improve the *quality of life* (QOL) for crew members.

Some of these are handled as constraints, while others are considered in the objective function.

In Ernst et al. (1998) each constraint is classified as either soft or hard. Hard constraints must be satisfied. While it is also desirable to satisfy the soft constraints, violation penalties are not as severe. The constraints of the model are:

   (i) Each shift cannot exceed a specified maximum duration.
  (ii) Any roundtrip with barracking must have a rest time within acceptable bounds.
 (iii) Each roundtrip must not have an excessive number of shifts.
  (iv) Each roundtrip must not have an excessive number of trips.
   (v) Crew complements must have a minimum rest time between two consecutive shifts in a roundtrip.
  (vi) Minimum time at home between two consecutive roundtrips must be respected.
 (vii) Each trip in the planning horizon should be covered.
(viii) Average weekly work content in a roster for each crew complement must be within acceptable bounds.

We relax the constraint (vii), whilst the rest remain as hard constraints. Consequently *undercovers* and *overcovers* may occur for some trips, although these are minimized by penalization in the objective function. The introduction of overcovers aims to use crew more efficiently. In reality, overcovers are achieved using *paxing* (or *dead heading*) in which the extra crew are treated as passengers. In practice crew take a bus to the next depot, as the freight trains only have room for a single crew. Paxing represents trips without performing real tasks for sake of bringing crew to the next depot to perform the next task.

An additional constraint that may be imposed is for the rosters to be cyclic. In this paper we consider both cyclic and non-cyclic rosters. The differences between cyclic and non-cyclic rosters are explained in Section 4.

Given the sparseness of the rail network in our problem, it is possible to precalculate and enumerate the full set of all feasible roundtrips as specified by constraints (i)-(v). In crew scheduling this is generally performed in a *crew pairing generation* phase. See Caprara et al. (1999). If we have a full set of feasible roundtrips to choose from, only constraints (vi), (vii) and (viii) need to be considered in the mathematical model. The degree of satisfaction of constraint (vi) or (viii) can be evaluated for a line of work (formed from several roundtrips), and the degree of satisfaction or violation of constraint (vii) can be evaluated for a roster (formed from all lines of work). In addition to satisfying the constraints, the roster must also deliver adequate QOL to all crew. This goodness of a line of work is deduced from a number of measures, such as:

(a) **Home base stay:** More and longer stays at the crew's home base are preferred.

(b) **Similar shifts:** Each crew member would prefer to work shifts that follow a similar pattern. In other words, each crew prefers to start consecutive shifts at roughly the same time of day.

(c) **Barracking lengths:** While barracking lengths are bounded by constraint (ii), there is a preferred barracking length which imposes much tighter bounds on the duration between shifts away from the home depot.

These are implicitly considered in the objective and constraints of the model. The objective function consists of a linear combination of real costs and penalty costs for constraint violations as detailed below:

(I) **Roundtrip cost**: This includes accommodation and allowances at the barracking depots included in a roundtrip, with different barracking costs for different depots. Excessive barracking is discouraged, particularly at the more expensive barracking locations. As all the segments as well as the barracking depots in a roundtrip are known at the time of generating the roundtrips, this (real) cost for each roundtrip can be pre-calculated.

(II) **Overcover penalty**: Whenever a trip is covered by more than one crew complement, an overcover penalty is applied. This can also be used to include the cost of paxing that may be required.

(III) **Undercover penalty**: Whenever a trip is not covered by any crew complement, an undercover penalty is applied. This is much more severe than overcover penalty.

(IV) **Workspread penalty**: As workload may vary for different shifts a penalty is applied to deviations from the average or expected workspread for a depot.

Clearly the penalty weights are different for each of the above requirements based on the relative importance of satisfying each of these.

3   Crew Scheduling Optimization Model

In this section, we present an optimization model for the crew scheduling problem. We introduce some mathematical notation and parameters before presenting our mathematical model.

Notation: Let $[0, L]$ define the planning time window, where L is the length of the planning horizon, specified in days. In our application, we consider a weekly planning horizon for crew scheduling, i.e., $L = 7$. Let D be the set of depots, indexed by $d = 1, \ldots, |D|$. At each depot d, it is assumed there are N_d crew complements available. Let T be the set of all trips generated from the train timetable, indexed by $t = 1, \ldots, |T|$. Let R be the set of all legal roundtrips, indexed by $r = 1, \ldots, |R|$. Each roundtrip is uniquely associated with a home depot. Therefore, the roundtrip set R can be partitioned into the union of the roundtrip sets R_d for all $d \in D$. Each roundtrip may cover more than one trip. Let T_r be the set of all trips covered by the roundtrip r. Conversely, each trip may be covered by any one of a few roundtrips. For each trip t, let Θ_t be the set of such roundtrips.

There are many roundtrips in R_d that cannot be performed by the same crew complement due to the minimum rest time required between successive roundtrips. Any two such roundtrips are said to overlap in time. A subset Ω_d of R_d is a *clique* if any two elements of Ω_d overlap. A clique is a *maximal clique* if it is no longer a clique when any other roundtrip from the associated depot is added to it. Let $\Psi_d = \{\Omega_d, i, i = 1, \ldots, K_d\}$ be the set of all maximal cliques at the depot d (K_d is integer). As the roundtrips repeat weekly, some roundtrips starting at the end of the current week and finishing into the next week may overlap with roundtrips starting at the beginning of the next week. Cliques resulting from such overlaps are called *transition-cliques*. Clearly transition cliques span two consecutive weeks. We include maximal transition-cliques in the set Ψ_d.

We now define some of the parameters that we use in our model. Let c_r be the cost for the roundtrip r; U_t (V_t) is the undercover (overcover) penalty cost for the trip t; Y_d the penalty cost for the depot d for the deviation of its workload to the average workload W across the network. The average workload at each depot must be within the specified minimum and maximum bounds of $\underline{W}$ and $\overline{W}$. The workload at each depot is the sum of the working content for all the selected roundtrips at this depot. The working content for the roundtrip r is denoted by W_r. Let P denote the minimum rest time between two consecutive roundtrips if they are performed by the same crew complement.

We can calculate c_r, the cost of a roundtrip $r \in R$ as the sum of the actual wage cost of barracking at all the barracking depots contained in the roundtrip. Additionally, this cost also includes the QOL penalty for barracking and the QOL penalty for time spent away from home.

Our major decision variable is the binary variable x_r for each roundtrip r with $x_r = 1$ if the roundtrip r is selected in crew scheduling. Otherwise it is zero. For each trip t, the binary variable ξ_t and the integer variable η_t represent undercoverage and overcoverage, respectively. $\xi_t = 1$ if the trip t is undercovered. Otherwise it is zero. Similarly, $\eta_t \geq 1$ if the trip t is overcovered. Otherwise it is zero. To define the workload deviations precisely, we define the over/under performance variables δ_d and λ_d for each depot d.

Mathematical Model: Given the definitions in the previous sections, we can define the following mathematical model for the Train Crew Scheduling problem (TCS).

$$\text{Minimize} \quad \sum_{r \in R} c_r x_r + \sum_{t \in T} (U_t \xi_t + V_t \eta_t) + \sum_{d \in D} Y_d (\delta_d + \lambda_d) \tag{1}$$

$$\text{subject to} \quad \sum_{r \in \Theta_t} x_r + \xi_t - \eta_t = 1, \qquad \forall\, t \in T \tag{2}$$

$$\sum_{r \in \Omega_d} x_r \leq N_d, \qquad \forall\, \Omega_d \in \Psi_d, d \in D \tag{3}$$

$$\sum_{r \in R_d} W_r x_r + \delta_d - \lambda_d = W N_d, \quad \forall\, d \in D \tag{4}$$

$$0 \leq \delta_d \leq (W - \underline{W}) N_d, \qquad \forall\, d \in D \tag{5}$$

$$0 \leq \lambda_d \leq (\overline{W} - W) N_d, \qquad \forall\, d \in D \tag{6}$$

$$x_r \in \{0, 1\} \qquad \forall\, r \in R \tag{7}$$

$$\xi_t,\ \eta_t \geq 0 \qquad \forall\, t \in T. \tag{8}$$

The objective function (1) is to minimize a linear combination of several costs represented by three terms:

- The actual cost of carrying out a roundtrip. This is independent of the crew complement that carries out this round trip.
- The overcover (or undercover) penalty cost of covering a trip more than once (or of not covering a trip at all). Undercoverage and overcoverage cannot occur at the same time due to the positive penalty in the objective and the constraint structure in (2).
- The weekly workload deviation penalty for each depot.

In TCS, constraint (2) specifies coverage for trips. For trip t, if η_t is positive then trip t is overcovered (hence paxing is required) and the undercover variable ξ_t must be zero. Similarly, if ξ_t is positive (trip t is not covered) then η_t must be zero. Ideally, both η_t and ξ_t should be zero since this implies perfect cover for trip t. We remark that overcoverage is more desirable than undercoverage.

Constraint (3) ensures that the number of roundtrips that are used from a solution clique will not be more than the number of available crew in a depot. All possible roundtrips are generated using the method proposed in Ernst et al. (1998). The generation of cliques is discussed shortly.

Constraints (4), (5) and (6) are concerned with the minimum, maximum and average workload as well as fairness of work distribution. Specifically, these constraints ensure that on average over a week at each depot, crews work within the acceptable limits $\underline{W}$ and $\overline{W}$. If δ_d is positive, depot d under-performs on average. Similarly, if λ_d is positive then depot d over performs. Both under and over performance are penalized in the objective. Workload constraints are typically called crew base constraints which refer to the number of work credits allowed for each depot (or crew base). Constraint (7) specifies that the roundtrip variables x_r are binary.

Our crew scheduling model is more sophisticated than traditional crew scheduling models such as those used in Caprara et al. (1999, 1997, 1998) and Fores et al. (1998). However, it will allow us to not only obtain feasible rosters in the crew rostering stage, but also to easily generate rosters at each depot. In traditional set partitioning or set covering models arising from airlines, railways and public transportation areas, problem sizes are very large with many millions variables. In our real application, the number of variables is less than a quarter of a million due to the sparseness of the underlying train network.

Remarks on the Mathematical Model: Our mathematical programming model (TCS) is a generalization of the traditional set partitioning or set covering models used in crew scheduling. These usually contain only the first term and possibly the second term of our objective, the coverage constraint (2) and the integrality constraint (7). Also, the relaxation variables ξ and η are sometimes not included. While the traditional model would give us a better solution in terms of low roundtrip cost (the first term), it would not give us an overall good solution at the crew rostering stage.

For computational complexity reasons crew scheduling problems are generally formulated so that the number of crews at each depot is not limited, and hence constraint (3) is not applied. In our application, the number of crews at each depot is pre-specified and included in constraint (3). This means that we can guarantee a feasible solution in the crew rostering for noncyclic rosters, and obtain cyclic rosters in most instances.

In our application the planning horizon is one week and the typical maximum length of a roundtrip is about two days. Thus a crew complement can perform more than one roundtrip in a week. The elastic set covering model of Graves et al. (1993) contains a constraint specifying that the maximum number of roundtrips selected at a depot should not exceed the maximum number of available crews at the depot. This is different from our constraint (3). Our model allows the number of roundtrips allocated to a depot to be larger than the number of available crew and eliminates any possibility of allocating too many overlapping roundtrips to a depot.

Clique Generation: The overlapping constraint (3) requires that the roundtrip clique set Ψ_d be defined for each depot d. Given the sparseness of the rail network, the number of roundtrip cliques at each depot is not

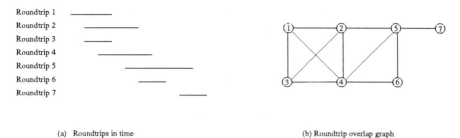

(a) Roundtrips in time   (b) Roundtrip overlap graph

Figure 1. Roundtrips in time and roundtrip overlap graph.

huge, hence the number of overlapping constraints is not too large. We define
an *overlap graph* $G = (N, A)$ to determine all roundtrip cliques at a depot.
Each node of the graph represents a roundtrip at the depot in the planning
horizon. An arc (i, j) connects two roundtrips if they overlap in time. A simple example is illustrated in Figure 1. Suppose there are seven roundtrips
in the planning horizon specified by starting and finishing time as shown in
Figure 1 (a) and suppose that the finishing time of a roundtrip is its actual
finish time plus the minimum rest time P. Then the roundtrip overlap graph
is shown in Figure 1 (b). Clearly the set of all maximal cliques in this graph is
$(1, 2, 3, 4), (2, 4, 5), (4, 5, 6), (5, 7)$. Furthermore, if we assume that roundtrips
5 and 7 finish in the following week, and that roundtrip 1 in the following week would overlap with roundtrips 5 and 7 in the current week, then
roundtrips 1, 5 and 7 should also be connected in the figure. Consequently,
$(1, 5, 7)$ becomes a maximal transition-clique and $(5, 7)$ is a clique but not a
maximal clique. See Figure 3 for further details on transition-cliques.

The clique generation algorithm (CGA) is as described in Figure 2. Clearly
the complexity of CGA for any depot d is $O(|\mathcal{R}_d| \log(|\mathcal{R}_d|))$, which implies
that generating all cliques at each depot is computationally inexpensive.

Step 0. Order all roundtrips at a depot in increasing order of their start times. Set
the roundtrip pool $\mathcal{R}_d$ to be the set of all roundtrips belonging to the depot d.
Step 1. Let $\mathcal{C}$ be a set, called the clique pool, and initialize as empty. Set the finish
time of $\mathcal{C}$ to be $+\infty$.
Step 2. If there are no roundtrips left in $\mathcal{R}_d$, terminate.
Step 3. Take the next roundtrip r with the earliest start time from the roundtrip
pool $\mathcal{R}_d$ and let $\mathcal{R}_d := \mathcal{R}_d \setminus \{r\}$. If the start time of the roundtrip r is less than
or equal to the finish time of the clique pool $\mathcal{C}$, go to step 5.
Step 4. The clique pool $\mathcal{C}$ makes a maximal clique. Remove any roundtrip from $\mathcal{C}$
if it finishes before or on the start time of the roundtrip r. Add the roundtrip
r into $\mathcal{C}$. Reset the finish time of the clique pool as the minimum finish time of
all roundtrips in the updated clique pool $\mathcal{C}$. Go to step 2.
Step 5. Add the roundtrip r to $\mathcal{C}$. Reset the finish time of the clique pool $\mathcal{C}$ as the
minimum finish time of all roundtrips in $\mathcal{C}$. Go to step 2.

Figure 2. Clique generation algorithm.

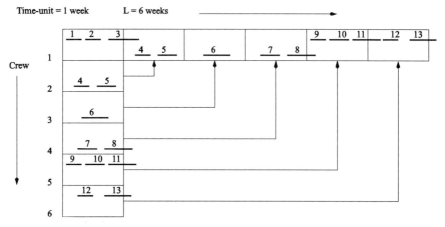

Figure 3. Overlaps of roundtrips in a cyclic roster.

4 Crew Rostering

From crew scheduling we have a set of best possible roundtrips at each depot. In crew rostering, we construct rosters for each depot. Rosters may be cyclic or non-cyclic. In a cyclic roster all crews have the same work pattern but may start at different weeks. In Figure 3 we see that roundtrips 3 and 4 do not overlap in time if the time horizon is only one week. Crew 1 can complete roundtrip 3 at the end of their weekly duty and crew 2 will have completed roundtrip 4 at the start of their weekly duty. However, if we consider an additional requirement that the rosters have to be cyclic, roundtrip 3 (performed by crew 1 at the end of their first week) will overlap with roundtrip 4 (performed by crew 1 at the start of their second week). To avoid such overlaps, we add one more arc connecting the roundtrips 3 and 4 in the roundtrip overlap graph. Similarly, arcs $(8, 9)$ and $(13, 1)$ are required in the roundtrip overlap graph.

We propose the algorithm in Figure 4 to generate non-cyclic rosters for each depot. Let $\mathcal{R}_d$ be all roundtrips selected from the depot d in the planning horizon. It is easy to establish that if the maximum size of all maximal cliques (including transition cliques, if the planning horizon is more than one week) in the set $\mathcal{R}_d$ is not greater than the number of crew complements at the depot, then the non-cyclic roster generation algorithm always generates a roster for each crew complement at the depot such that every roundtrip in $\mathcal{R}_d$ is performed. Furthermore, the complexity of the non-cyclic roster generation algorithm for the depot d is $O(|\mathcal{R}_d| \log(|\mathcal{R}_d|))$. That is, the non-cyclic roster generation is polynomially solvable.

This indicates that given a selection of roundtrips, the task of generating non-cyclic rosters for all crews at all depots is computationally inexpensive. However, it is more difficult to generate cyclic rosters. In fact, generating

While $\mathcal{R}_d \neq \emptyset$, **do**

Select a roundtrip $r \in \mathcal{R}_d$ with the earliest start time.

Let $\mathcal{R}_d := \mathcal{R}_d \setminus \{r\}$.

Assign the roundtrip r to a crew who can take it.

End

Figure 4. Non-cyclic roster generation algorithm.

a cyclic roster is equivalent to solving an asymmetric traveling salesman problem (ATSP) in a directed complete graph where each node represents a roundtrip and the cost of the arc (a, b) is equal to the minimum time gap in minutes between the starting times of roundtrips a and b, where b can be feasibly carried out after a by the same crew complement. The existence of a cyclic roster for N weeks is equivalent to the existence of a tour with the total cost less than or equal to $10080N$, where N is the number of crews at the depot and 10080 is the number of minutes in one week.

Cyclic Rosters: It is also possible to find a cyclic roster by solving an integer program. We introduce a binary variable z_{rc}, which is one if the roundtrip $r \in \mathcal{R}_d$ is assigned in week c. Otherwise it is zero. The integer program is

$$\text{Minimize} \quad f(z) \tag{9}$$

subject to

$$\sum_{c=1}^{N_d} z_{rc} = 1, \qquad\qquad \forall\, r \in \mathcal{R}_d \tag{10}$$

$$\sum_{r \in \mathcal{R}_d \cap \Omega_d} z_{rc} \leq 1, \qquad \forall\, \Omega_d \in \Psi_d, \quad c = 1, \ldots, N_d \tag{11}$$

$$\sum_{\substack{r_1, r_2 \in \mathcal{R}_d \cap \Omega_d \\ \text{Condition A}}} (z_{r_1 c} + z_{r_2 (c+1)}) \leq 1, \quad \forall\, \Omega_d \in \Psi_d, \quad c = 1, \ldots, N_d \tag{12}$$

$$z_{rc} \in \{0, 1\} \qquad\qquad \forall\, r \in \mathcal{R}_d, \quad c = 1, \ldots, N_d, \tag{13}$$

where Condition A in Constraint (12) is satisfied by a roundtrip r_1 that finishes in the following week and a roundtrip r_2 starting in the first two days of the current week. Constraint (10) says that each selected roundtrip in the crew scheduling stage must be covered exactly once in N_d weeks. Constraints (11) and (12) ensure that any crew at the depot d cannot take two roundtrips from a non-transition-clique and a transition-clique, respectively. Constraint (13) states that the variable z_{rc} is binary. In the model the objective function $f(z)$ is used to control the quality of the rosters. Clearly if $f(z) \equiv 0$, then any solution of the model gives a set of feasible rosters. One use of $f(z)$ is to balance the workload between weeks.

Step 0. Generate all maximal cliques (including transition-cliques)
of the set $\mathcal{R}_d$. Let $i := 1$.
Step 1. Assign roundtrips at week i $(i = 1, \ldots, N_d)$.
  Step 1a: If there is a maximal clique of size $N_d - i + 1$,
  then go to Step 1b. Otherwise go to Step 2.
  Step 1b: Choose the first occuring maximal clique of size $N_d - i + 1$
  from $\mathcal{R}_d$ that has not been checked in week i. Assign
  the roundtrip r from this clique that finishes first from this clique
  in week i. Remove r from the set $\mathcal{R}_d$. Go to Step 1a.
Step 2. Regenerate all maximal cliques of the set $\mathcal{R}_d$.
  Let $i := i + 1$. Go to Step 1.

Figure 5. Cyclic roster generation algorithm (CRGA).

It is known that the ATSP is a NP-hard problem (see Garey and Johnson (1979)). So we propose a polynomial heuristic that attempts to generate a cyclic roster at a depot. A very simple method is to just sort the roundtrips in chronological order, and then fill each crew's (or week's) roster in a greedy fashion using the remaining roundtrips not already scheduled previously.

If this heuristic fails, a slightly more complex method may be employed as described below. Let $\mathcal{R}_d$ be all the roundtrips selected from the depot d in one week. The heuristic algorithm in Figure 5 attempts to generate a cyclic roster for each depot.

If $\mathcal{R}_d$ is the set of selected roundtrips in one week at the depot d, and the maximum size of all maximal cliques (including transition cliques) in the set $\mathcal{R}_d$ is not greater than the number of crew complements at the depot, then the CRGA will generate a line of work (roster) over N_d weeks at the depot such that all roundtrips (except at most one) in $\mathcal{R}_d$ are assigned. Furthermore, the complexity of the CRGA for the depot d is $O(|\mathcal{R}_d| \log(|\mathcal{R}_d|))$.

In other words, at the beginning of each week i, the maximum size of maximal cliques in the set $\mathcal{R}_d$ is less than or equal to $N_d - i + 1$. This can be proved by induction. This clearly holds at week one by the assumption. We assume that this holds at week i, and prove that it will also hold for week $i + 1$. If there is no maximal clique with size $N_d - i + 1$ in $\mathcal{R}_d$ in week i, then the statement holds for week $i + 1$. Suppose there is a maximal clique with size $N_d - i + 1$ in $\mathcal{R}_d$ in week i. Then the first occurrence of a maximal clique of size $N_d - i + 1$ that has not been checked in week i, and the first finished roundtrip r from this clique to be assigned to week i, will be selected. Suppose that the finishing time of the roundtrip r is t. It follows that there are no more cliques of size of $N_d - i + 1$ before time t among unassigned roundtrips in $\mathcal{R}_d$. Also there is no clique with the size of $N_d - i + 1$ before the start of week $i + 1$ after assigning the last roundtrip in week i. Therefore, the statement holds for week $i + 1$. Note that the last assigned roundtrip in week i may finish in week i or week $i + 1$. Furthermore, if the last assigned roundtrip in week i finishes after the start of week $i + 1$, then in week $i + 1$ there is no maximal clique of size $N_d - (i + 1) + 1$ before the finishing time of the last assigned roundtrip in week i at the end of assigning roundtrips in week i.

Hence there is no maximal clique of size $N_d - N_d + 1 = 1$ in week N_d, and all remaining roundtrips in $\mathcal{R}_d$ can be assigned in week N_d. If the last assigned roundtrip finishes before the start of week $N_d + 1$, then we generate a work of line in N_d weeks to cover all selected roundtrips in the crew scheduling stage. Otherwise, the last assigned roundtrip in week N_d finishes after the start of week $N_d + 1$ but starts before the start of week $N_d + 1$.

It is easy to see that the complexity of the algorithm is $O(|\mathcal{R}_d| \log(|\mathcal{R}_d|))$ since each roundtrip is checked out and assigned exactly once and regenerating maximal cliques is $O(|\mathcal{R}_d| \log(|\mathcal{R}_d|))$ in each week.

There is no guarantee that the CRGA can generate a cyclic roster to cover all selected roundtrips from the optimization model. If the last roundtrip assigned by the CRGA finishes after the start of the week $N_d + 1$, one possible remedy to get a cyclic roster is to reassign the last assigned roundtrip at one of the weekends from week 1 to week $N_d - 1$. Once again, this approach may fail. However, in practice it is likely that CRGA will generate a cyclic roster as is discussed in the next section.

In order to be able to guarantee that a cyclic roster exists slightly stronger conditions are required. For example it is sufficient if there is a point in time where none of the roundtrips overlap. Alternatively a solution is guaranteed to exist if the maximum clique size is $N_d - 1$.

5   Computational Results

We have implemented the approaches to the crew scheduling and rostering problem discussed above on a computer using a 500MHz EV5.6 alpha processor. Version 6.5 of CPLEX (2000) has been employed to solve the MIP model, TCS. CPLEX is also used for solving the MIP model for the crew rostering problem at each depot. All possible roundtrips are generated using the method proposed in Ernst et al. (1998), and maximal roundtrip cliques are generated using the CGA method. Non-cyclic rosters at each depot can be generated in polynomial time, and a cyclic roster can be generated in polynomial time (allowing for the possibility that at most one roundtrip is uncovered).

We tested three real-world problems from Australian National Rail. These are denoted as A, B and C. The input data are specified in Table 1 and part of Table 2 in bold-face. The entries in Table 1 are the number of depots, the total number of crew complements, the total number of roundtrips generated using the method of Ernst et al. (1998), the total number of cliques generated by the CGA, the number of variables and the number of constraints in TCS, and the expected average weekly workload, in minutes, across the network.

The crew complement distributions for the three test problems are shown in the "Crew" columns in Table 2. Although there are 22 depots for problems B and C, there are crews at only the first 14 depots.

25

Prob	Depots	Crew	Trips	Roundtrips	Cliques	Var	Con	Workload
A	7	27	242	291	153	789	402	3273.00
B	22	278	1107	6065	466	8323	1595	1726.01
C	22	278	1309	15140	611	17801	1942	1741.56

Table 1. Input data for problems A, B and C.

In our numerical experiments, the penalty costs U_t, V_t and Y_d were set to 10000, 1000 and 0, respectively. For problems B and C, the minimum and maximum weekly workloads, $\underline{W}$ and $\overline{W}$ were set to 25 and 45 hours. These values were set to 25 and 65 hours for problem A as the expected average workload is 3273 minutes, which is more than 50 hours. In solving TCS we set the MIP stopping criteria in CPLEX as 0.4%.

The numerical results for problems A, B and C are shown in Tables 2 and 3. The roundtrip distributions and average weekly workload for the depots are shown in columns "R" and "W" in Table 2. We omit the statistics for those depots with no crews because no roundtrips are allocated to them.

	Problem A			Problem B			Problem C		
Depot	Crew	R	W	Crew	R	W	Crew	R	W
1	3	13	2648.0	28	42	1552.7	28	74	1656.7
2	5	22	3128.8	5	7	2648.8	5	7	2648.8
3	5	17	2935.2	20	50	1812.4	20	78	2436.7
4	0	0	0	19	41	1869.6	19	41	1929.5
5	5	18	3264.2	4	14	2534.5	4	14	2221.3
6	6	23	3245.8	39	58	1518.4	39	56	1515.8
7	3	9	2520.0	29	40	1851.2	29	39	1817.2
8				39	76	1500.1	39	88	1507.4
9				19	37	1571.8	19	98	2204.6
10				14	17	1517.6	14	35	1554.2
11				28	94	2275.5	28	99	2544.1
12				10	23	2697.0	10	24	2692.2
13				4	10	2440.5	4	10	2440.5
14				20	22	1501.3	20	32	1502.7

Table 2. Roundtrip and workload distributions for problems A, B and C.

We list some important performance indicators in Table 3, such as the total number of selected roundtrips, the total number of undercovers, the total

Problem	Roundtrips	Undercover	Overcover	CPU	Objective	Gap
A	102	23	2	2.83	264197.50	0.22%
B	531	40	96	974.22	667399.90	0.36%
C	695	44	51	8126.05	658914.74	0.36%

Table 3. Numerical outputs for problems A, B and C.

number of overcovers, the total CPU time in seconds, the TCS objective, and
the final relative gap between the best integer solution and the best lower
bound.

Cyclic rosters for all depots are generated using the optimization model
described in the previous section, so there is no guarantee that cyclic rosters
will be produced. However, the optimization model did generate a cyclic
roster for every depot in all three problems.

It may be possible to improve undercover, overcover and other performance indicators by experimenting with different values for parameters U_t,
V_t, Y_d, $\underline{W}$ and $\overline{W}$. The results in Table 3 show that some trips are not covered while some other trips are overcovered. We think that this is mainly
because the number of crews at each depot is fixed. In Ernst et al. (2000)
an optimization model has been proposed for solving a planning train crew
rostering problem where the number of crews at each depot is obtained from
the optimization model. The results in Ernst et al. (2000) show that the
optimization model for the planning train crew rostering problem not only
reduces undercoverage and overcoverage, but also reduces the total number
of crew in the train network. This implies that the crew distribution obtained
from the model in Ernst et al. (2000) is better than the pre-specified crew
distribution in this paper. However, in the operating stage, the organization
has to use its available crew at each depot and may employ part-time worker
to perform undercovered trips.

We also tested three randomly generated problems D, E and F, with
input data as given in Table 4. These three problems are generated using the
same real-world train network in problems B and C which has 22 real depots.
However, 4 of them are not allowed to be crew bases. All three problems have
18 depots but significantly different train timetables using perturbations of
the real train timetable. The number of crew complements assigned to each
depot is approximately equal to a certain average workload associated with
the depot. The results from problems D, E^1 and F are shown in Tables 5
and 6. These additional test problems show that the optimization model can
be solved in a reasonable time even for randomly generated data.

The heuristic for generating cyclic rosters was applied to all of the data
sets. We found that even the simple greedy heuristic suggested before CRGA

---

[1] The MIP stopping criterion in CPLEX is set to 0.7% for this test example.
When the stopping criterion was set to 0.4% CPLEX ran out of memory before
obtaining a solution.

Prob	Depots	Crew	Trips	Roundtrips	Cliques	Var	Con	Workload
D	18	314	1583	13496	764	16698	2365	2141.31
E	18	299	1501	10373	687	13411	2206	2168.60
F	18	307	1532	11071	671	14171	2221	2199.46

Table 4. Input data for problems D, E and F.

Depot	Problem D Crew	R	W	Problem E Crew	R	W	Problem F Crew	R	W
1	25	50	1536.5	22	53	2008	23	55	1852.61
2	7	11	2661.7	6	9	2532.8	4	6	2686.75
3	9	22	1612.4	8	21	1738.4	7	21	1968.57
4	38	106	2610.6	31	82	2461.1	34	81	2239.71
5	18	59	2337.6	17	57	2287.4	15	47	2269.53
6	45	142	2682.1	41	79	1738.3	45	136	2497.2
7	18	35	2619.6	24	48	2691.5	28	56	2684.0
8	39	73	1577.3	34	110	2648.7	37	77	1769
9	35	73	1526.8	32	87	2237.7	34	86	2109.4
10	6	12	2449.5	8	18	2695.5	9	18	2397.3
11	27	69	2034.3	25	64	1990.0	23	52	1806.0
12	32	69	2683.5	36	79	2611.0	37	77	2695.4
13	0	0	0	0	0	0	0	0	0
14	15	40	2510.7	15	25	1617.0	11	27	2360.1

Table 5. Roundtrip and workload distributions for problems D, E and F.

Problem	Roundtrips	Undercovers	Overcovers	CPU	Objective	Gap
D	761	2	29	347.82	278230.0	0.00%
E	732	4	59	5360.9	321480.0	0.66%
F	739	14	46	133.12	408750.0	0.00%

Table 6. Numerical outputs for problems D, E and F.

on page 64 already provides feasible solutions for all of the data sets A-F. The computational time ranged from 0.002 seconds for the smallest problem through to 0.030 seconds for problem D. Hence it appears that generating cyclic rosters is indeed very easy in practice even though it is theoretically possible to construct sets of roundtrips that are difficult or impossible to roster with the available crew. Since the resulting rosters are feasible, the roster constraints described in Section 2 are satisfied. The length of the cyclic rosters at each depot is equal to the number of its crew complements.

Finally, we address the issue of differences between our model and traditional crew scheduling models. If each depot had an unlimited number of crew complements, as in the traditional crew scheduling models, then roundtrip distribution should be quite different from Tables 2 and 5. As the feasible region in the traditional crew scheduling model is larger than that in our model, it would be much more economical in terms of the objective in the traditional model than that in our TCS. However, the model would not consider any overlapping constraints, which could result in many overlapping roundtrips at a single depot. For example, it would be possible for all selected roundtrips from the traditional crew scheduling model to overlap one another. Then in the crew rostering stage, the minimum number of required crew complements must not be less than the number of roundtrips allocated to this depot. This implies that each crew complement at most performs one roundtrip in a week, which is definitely not a good roster. This shows that our TCS model should give overall better rosters than the traditional crew scheduling and crew rostering approach. This is mainly due to the fact that our problem has a much longer planning horizon and that the network is extremely sparse.

6 Conclusions

We have proposed an optimization model for solving a class of operational train crew rostering problems. Our model can handle situations where traditional decomposed approaches to train crew rostering would not be appropriate. The main difference from the traditional model is that the number of crews at each depot is given a *priori* rather than optimized. We have also proposed exact and heuristic approaches for finding cyclic as well as non-cyclic rosters at each depot. The numerical results confirm that our optimization approach produces quality rosters in practice.

Bibliography

Anbari, F.T. (1987). Train and engine crew management system. *Computers and Railway Operations*, 267–284. Computational Mechanics Publications, Springer, Berlin.

Arabeyre, J.P., J. Fearnley, F.C. Steiger, and W. Teather (1969). The airline crew scheduling problem: A survey. *Transportation Science 3*, 140–163.

Barnhart, C., E.L. Johnson, R. Anbil, and L. Hatay (1994). A column generation technique for the long-haul crew assignment problem. In T.A. Ciriani and R.C. Leachman (Eds.), *Optimization in Industry 2*, Wiley, New York, 7–24.

Bianco, L., M. Bielli, A. Mingozzi, S. Ricciardelli, and M. Spadoni (1992). A heuristic procedure for the crew rostering problem. *European Journal of Operational Research 58*, 272–283.

Caprara, A., M. Fischetti, P.L. Guida, P. Toth, and D. Vigo (1999). Solution of large-scale railway crew planning problems: The Italian experience. In N.H.M. Wilson (Ed.), *Computer-Aided Transit Scheduling, Lecture Notes in Economics and Mathematical Systems*, 471, Springer, Berlin, 1–18.

Caprara, A., M. Fischetti, P. Toth, D. Vigo, and P.L. Guida (1997). Algorithms for railway crew management. *Mathematical Programming 79*, 125–141.

Caprara, A., P. Toth, D. Vigo, and M. Fischetti (1998). Modeling and solving the crew rostering problem. *Operations Research 46*, 820–830.

CPLEX (2000). *ILOG CPLEX 6.5 Reference Manual*. ILOG Inc., CPLEX Division.

Day, P.R. and D.M. Ryan (1997). Flight attendant rostering for the short-haul airline operations. *Operations Research 45*, 649–661.

Ernst, A.T., H. Jiang, M. Krishnamoorthy, H. Nott, and D. Sier (2000). *An integrated optimisation model for train crew management*. Technical report, CSIRO Mathematical and Information Science, Australia. Submitted to Annals of Operations Research.

Ernst, A.T., M. Krishnamoorthy, and D. Dowling (1998). Train crew rostering using simulated annealing. In *Proceedings of International Conference on Optimization Techniques and Applications*, 859–866.

Etschmaier, M.M. and D.F.X. Mathaisel (1985). Airline scheduling: An overview. *Transportation Science 19*, 127–138.

Fores, S., L. Proll, and A. Wren (1998). A column generation approach to bus driver scheduling. In M.H.G. Bell (Ed.), *Transportation Networks: Recent Methodological Advances*, Elsevier, Amsterdam, 195–208.

Gamache, M. and F. Soumis (1998). A method for optimally solving the rostering problem. In G. Yu (Ed.), *OR in the Airline Industry*, Kluwer, Boston, 124–157.

Gamache, M., F. Soumis, G. Marquis, and J. Desrosiers (1999). A column generation approach for large scale aircrew rostering problems. *Operations Research 47*, 247–263.

Garey, M.R. and D.S. Johnson (1979). *Computers and Intractability*. W.H. Freeman, New York.

Gershkoff, I. (1989). Optimizing flight crew schedules. *Interfaces 19*(4), 29–43.

Graves, G., R. McBride, I. Gershkoff, D. Anderson, and D. Mahidhara (1993). Flight crew scheduling. *Management Science 39*, 736–745.

Hoffman, K.L. and M.W. Padberg (1993). Solving airline crew scheduling problems by branch-and-cut. *Management Science 39*, 657–682.

Larcher, R. and M. Sinay (1982). Scheduling of railway crew for random arrivals. In *Research for Tomorrow's Transport Requirement: Proceedings of the World Conference on Transport Research*, British Columbia, Canada, 1156–1161.

Mason, A.J., D.M. Ryan, and D.M. Panton (1998). Integrated simulation, heuristic and optimisation approaches to staff scheduling. *Operations Research 46*, 161–175.

Morgado, E.M. and J.P. Martins (1992). Scheduling and managing crew in the Portuguese railways. *Expert Systems with Applications 5*, 301–321.

Morgado, E.M. and J.P. Martins (1993). An AI-based approach to crew scheduling. In *Proceedings of the 9th Conference on Artificial Intellegence for Applications*, IEEE Computer Society Press, 71–77.

Tykulsker, R.J., K.K. O'Neil, A. Ceder, and Y. Sheffi (1985). A commuter railway crew assignment/work rules model. In J.-M. Rousseau (Ed.), *Computer Scheduling of Public Transport 2*, Elsevier, Amsterdam, 233–246.

Vance, P.H., C. Barnhart, E.L. Johnson, and G.L. Nemhauser (1997). Airline crew scheduling: A new formulation and decomposition algorithm. *Operations Research 45*, 88–200.

Applying an Integrated Approach to Vehicle and Crew Scheduling in Practice

Richard Freling, Dennis Huisman, and Albert P.M. Wagelmans

Econometric Institute, Erasmus University Rotterdam,
PO BOX 1738, NL-3000 DR Rotterdam, The Netherlands
{freling,huisman,wagelmans}@few.eur.nl

Abstract. This paper deals with a practical application of an integrated approach to vehicle and crew scheduling, that we have developed previously. Computational results have shown that our approach can be applied to problems of practical size. However, application of the approach to the actual problems that one encounters in practice, is not always straightforward. This is mainly due to the existence of particular constraints that can be regarded as "house rules" of the public transport company under consideration.

In this paper we apply our approach to problems of individual bus lines of the RET, the Rotterdam public transport company, where particular constraints should be satisfied. Furthermore, we investigate the impact of allowing drivers to change vehicle during a break. Currently, the rule at the RET is that such changeovers are only allowed in split duties; they are never allowed in other types of duties. We show that it is already possible to save crews if for the non-split duties restricted changeovers are allowed.

1 Introduction

Commercially successful computer packages use a sequential approach for vehicle and crew scheduling. Sometimes integration is dealt with at the user level (e.g., see Darby-Dowman et al. (1988)) or crew considerations are taken into account in the vehicle scheduling phase (e.g., in HASTUS, see Rousseau and Blais (1985)). In the operations research literature, only a few publications address a simultaneous approach to vehicle and crew scheduling (see Freling et al. (1995a, 1999); Haase and Friberg (1999); Haase et al. (1999) and Gaffi and Nonato (1999)). For an extensive overview, we refer to Freling et al. (2000). None of the publications mentioned there makes a comparison between simultaneous and sequential scheduling. Furthermore, to the best of our knowledge, no integrated model has been applied to a practical situation before.

In Freling et al. (2000) some models and algorithms for integration of vehicle and crew scheduling were discussed and applied to problems of practical size. The test problems were based on real-life problems from the RET, the public transport company in the city of Rotterdam, the Netherlands. In these test problems, we omitted certain complicating constraints that are specific to the RET. In this paper, we will apply our approach to the actual RET problems. Although the RET also provides service on tram and metro lines, only bus lines will be considered. For bus lines we can expect a larger benefit of the integration, because the relative difference between crew and vehicle costs is higher than in tram and metro scheduling. Moreover, tram and metro scheduling problems are more restricted because of constraints with respect to the capacity of the rail track and the capacity at the endpoints of the lines.

We will compare our integrated approach with a sequential approach. Furthermore, we will investigate the impact of allowing drivers to change vehicle during a break. Currently, the rule at the RET is that such *changeovers* are only allowed in split duties; they are never allowed in other types of duties. *Split duties* have a long break between the two parts of the duty such that the driver can go home during this break. We show that it is already possible to save crews if for the non-split duties, resticted changeovers are allowed.

The paper is organized as follows. Section 2 describes the situation which holds for almost all bus lines at the RET and especially for the lines we consider in this paper. In Section 3, we discuss a sequential approach that takes all relevant RET constraints into account. The integrated approach that we use is discussed in Section 4. The paper is concluded with a computational study (Section 5). The main objective of this study is threefold:

1. We show that our integrated approach can be applied to a practical situation.
2. We make a comparison between the results of the sequential and the integrated approach.
3. We look at the effect of allowing (restricted) changeovers.

2   Problem Description at the RET

In Figure 1 we show the relation between four operational planning problems at the RET. Decisions about which routes or *lines* to operate and how frequently, are determined by local and regional authorities and are given to the RET. Also known are the travel times between various points on the route. Based on the lines and frequencies, timetables are determined resulting in *trips* with corresponding start and end locations and times. The second planning process is vehicle scheduling, which consists of assigning vehicles to trips, resulting in a schedule for each vehicle or *vehicle blocks*. On each vehicle block a sequence of tasks can be defined, where each task needs to be assigned to a working period for one crew (a *crew duty* or *duty*) in the crew scheduling process. The feasibility of a duty is dependent on a set of collective agreements and labour rules, that refer to sufficient rest time, etc. Crew scheduling is short term crew planning (one day) for assigning crews to vehicles, while the crew rostering process is long term crew planning (e.g. half a year) for constructing rosters from the crew duties.

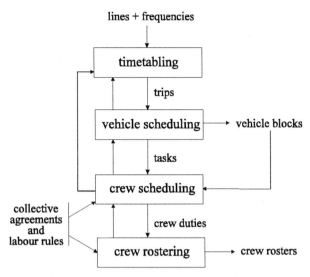

Figure 1. Four related planning problems.

In § 2.1, we discuss the definitions and restrictions at the RET and in § 2.2, we focus on one particular restriction, which will play a major role in the rest of this paper.

2.1 Definitions and Restrictions

At the RET the whole planning process is solved line-by-line, so there is no inter-lining and the size of the problems is relatively small (up to 259 trips). Almost all lines have two locations (lets call them A and B) and all the trips are from A to B or from B to A. In this paper we only consider lines with this property.

For every line there is one given depot and all buses must start and end in that depot. A *deadhead* is the driving of a vehicle without passengers. There are three kinds of deadheads: from the depot to the start of a trip, from the end of a trip to the depot and between two trips (from one endpoint of the line to the other one). The workday of one driver is called a *(crew) duty*. There are many different types of duties, for example early duties, late duties and split duties. Every duty consists of two pieces with in between a break. A *piece* consists of consecutive *tasks* (trips, layovers and deadheads) performed by one driver on the same bus. A *layover* is the time between two consecutive trips that a driver with bus waits at a start or end location of the line. For the layover there are several restrictions:

- After every trip and deadhead there is a layover of at least two minutes.
- There is a minimal layover in a *round-trip*, where a round-trip consists of two consecutive trips from location A to B and back to A; this is the so-called *round-trip-condition*, discussed in more detail in § 2.2.

- The total layover time in a duty is at least 10% of the length of the duty.
- If the longest piece in a split duty is more than 4 hours, then there is at least once a layover of at least 10 minutes.

All duties start or end at the depot or at so-called relief locations. A *relief location* is the start or the end location of the line. In some cases both the start and the end location is a relief location and in other cases only one of them is a relief location. Duties starting at the depot always start with a *sign-on* time and the other duties start with a *relief* time. Duties ending at the depot end with a *sign-off* time and the other duties end after a layover. These times are fixed and known in advance. The breaks are only allowed at relief locations. A break is only necessary in duties that have a length greater than a certain minimum.

Furthermore, there are restrictions for all 8 types of duties with respect to the earliest and latest starting time, the latest ending time, the minimum length of the break, the maximum spread and the maximum working time. On Sundays there are no split duties. For more details we refer to Huisman (1999).

Currently changeovers are only allowed in split duties and under no circumstances in other duties. In this paper we compare this situation with the variant where a changeover is also allowed during a break of a non-split duty, provided that there is enough time to change vehicle. In that case the first piece ends with a layover and the second piece starts with the relief time. Of course, the length of the last layover of the first piece is then at least the relief time, because the driver who takes over the first bus needs this amount of time. As a consequence there is continuous attendance of the vehicle, i.e., there is always a driver present when the bus is outside the depot.

2.2 Round-trip-condition

In this subsection, we discuss the most important and complicated constraint of the RET: the round-trip-condition. As mentioned before, the round-trip-condition is a restriction on the total layover in a round-trip (two consecutive trips such that a driver drives from location A via location B back to location A). The round-trip-condition requires that the total layover at the locations A and B in one round-trip is at least 10 minutes if the round-trip is 60 minutes or longer and 15% of the travel time if the round-trip is shorter than 60 minutes with a minimum of 5 minutes. It is allowed to violate this condition at most once in every crew duty, but only if the total layover is at least 5 minutes.

In advance, we can construct combinations of three trips that violate the round-trip-condition. We show this in Figure 2. We use these combinations in the sequential approach as well as in the integrated approach.

In a sequential approach it is not possible to incorporate the possibility of violating the round-trip-condition at most once in every crew duty. The reason is that we have to deal with the round-trip-condition in the Vehicle

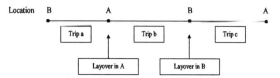

Figure 2. Violation of round-trip-condition.

Scheduling Problem (VSP) and at that stage we do not know anything yet about crew duties. Hence, if we violate the round-trip-condition more than once in the VSP, we may later on end up with an infeasible solution to the Crew Scheduling Problem (CSP), because some crew duty has more than one violation of the round-trip-condition.

As we will see in Section 4, in our integrated approach, we can exploit the possibility that the round-trip-condition may be violated once in every crew duty. This is actually the main reason why we can save vehicles by using an integrated approach instead of a sequential one.

3 Sequential Approach

In this section, we give a sequential approach for solving the VSP and the CSP. First we discuss in § 3.1 the VSP with round-trip-condition and after that we discuss in § 3.2 the CSP. For both problems we give a mathematical model and a solution method.

3.1 Vehicle Scheduling

Let $N = \{1, 2, ..., n\}$ be the set of trips, numbered according to increasing starting time, and let $E = \{(i, j) \mid i < j, \ i, j \text{ compatible}, \ i, j \in N\}$ be the set of arcs corresponding to deadheads and layovers. The nodes s and t both represent the depot. We define the vehicle scheduling network $G = (V, A)$, which is an acyclic directed network with nodes $V = N \cup \{s, t\}$, and arcs $A = E \cup (s \times N) \cup (N \times t)$. A path from s to t in the network represents a feasible schedule for one vehicle, and a complete feasible vehicle schedule is a set of disjoint paths from s to t such that each node in N is covered. Let R be the set of combinations of trips that violate the round-trip-condition. Every combination consists of three trips, that we will refer to as a, b, and c like in Figure 2.

Let c_{ij} be the vehicle cost of arc $(i, j) \in A$, which is usually some function of travel and idle time. Furthermore, fixed costs for using a vehicle can be added to the cost of arcs (s, i) or (j, t) for all $i, j \in N$. For the remainder of this paper, we assume that the primary objective is to minimize the number of vehicles. This means that the fixed costs are high enough to guarantee that this minimum number will be achieved.

Using decision variables y_{ij}, with $y_{ij} = 1$ if a vehicle covers trip j immediately after trip i, $y_{ij} = 0$ otherwise, the VSP with round-trip-condition can be formulated as follows:

$$\text{Minimize} \quad \sum_{(i,j)\in A} c_{ij}\, y_{ij} \tag{1}$$

$$\text{subject to} \quad \sum_{j:(i,j)\in A} y_{ij} = 1 \quad \forall i \in N, \tag{2}$$

$$\sum_{i:(i,j)\in A} y_{ij} = 1 \quad \forall j \in N, \tag{3}$$

$$y_{ab} + y_{bc} \leq 1 \quad \forall (a,b,c) \in R, \tag{4}$$

$$y_{ij} \in \{0,1\} \quad \forall (i,j) \in A. \tag{5}$$

Constraints (2) and (3) assure that each trip is assigned to exactly one predecessor and one successor, i.e., these constraints guarantee that the network is partitioned into a set of disjoint paths from s to t. Constraint (4) assures that the round-trip-condition is met, because the trips a, b, and c cannot be assigned to the same vehicle.

When a vehicle has an idle time between two consecutive trips which is long enough to let it return to the depot, it does so. In that case the arc between the trips is called a *long arc*; the other arcs are called *short arcs*. In general, a bottleneck for solving VSP when using a network such as G may be the large size of those networks due to a large number of arcs in E. For the problems that we consider in this paper, however, the number of arcs does not cause serious complications.

We solve this model with subgradient optimization and Lagrangian relaxation, where we relax constraint (4). Then the remaining subproblem is a quasi-assignment problem, which is solved in every iteration of the subgradient optimization with an auction algorithm (see Freling et al. (1995b)). In this way we get a lower bound on the optimal solution. Furthermore, we also compute a feasible solution in every iteration by changing the solution of the subproblem such that the round-trip-condition is not violated anymore. This requires at least one more vehicle. We terminate the subgradient optimization if the gap between the lower bound and the value of the best feasible solution indicates that we have found an (almost) optimal solution.

3.2 Crew Scheduling

We solve the CSP in two steps, first we generate all feasible duties and then we select the optimal duties by solving a set covering model.

Generation of Duties: After the computation of the optimal vehicle schedule we generate duties in three steps.

1. **Definition of all relief points.** A relief point is the point on the vehicle block, where a driver can have his/her break or can be relieved by another driver. Of course, the depot is also a relief point. For the other relief points

the following two properties hold: First the location of the relief point is a relief location, and second the relief point is after a layover whose length is greater than or equal to the relief time.

2. **Generation of all feasible pieces.** A piece is the work between two relief points on the same vehicle block. Here we take into account a maximum length of the piece. We use these pieces also as *trippers*, i.e., one-piece duties.

3. **Generation of all feasible duties.** Two pieces with a break in between are combined to construct a duty. Here we take care of restrictions on the length of the duty, the length and position of the break, the maximum spread and so on. In some variants the two pieces must belong to the same vehicle, but in some other variants this restriction is not present.

Mathematical Formulation CSP: Let d_l be the cost of duty $l \in L$, where L is the set of all feasible duties, and define $L(i) \in L$ as the set of duties covering task $i \in I$, where I is the set of all tasks. A task can be a trip or a deadhead. Consider a binary decision variable x_l indicating whether duty l is selected in the solution or not. In the set covering formulation of the CSP, the objective is to select a minimum cost set of feasible duties such that each task is included in one of these duties. This is the following 0-1 linear program:

$$\text{Minimize} \quad \sum_{l \in L} d_l \, x_l \tag{6}$$

$$\text{subject to} \quad \sum_{l \in L(i)} x_l \geq 1 \quad \forall \, i \in I \tag{7}$$

$$x_l \in \{0, 1\} \quad \forall \, l \in L \tag{8}$$

where constraints (7) assure that each task will be covered by at least one duty. The solution of the set covering problem can be translated into a solution of the CSP by simply deleting double trips. The most important advantage of using a set covering formulation instead of a set partitioning formulation is the fast computation time.

Solution Method CSP: The algorithm for solving the CSP is shown in Figure 3. For further details we refer the interested reader to Freling et al. (2000) and to Freling (1997).

4   Integrated Approach

In this section we discuss the integrated approach which we applied to the problems of the RET. First we describe in § 4.1 and 4.2 a model for the integrated Vehicle and Crew Scheduling Problem (VCSP) and a general solution method, respectively. The big difference between an integrated and a sequential approach is that it is not possible to generate all feasible duties in

Step 0: Initialization
Select a set of pieces such that each task can be covered by at least one piece.
The initial set of columns consists of these pieces.

Step 1: Computation of lower bound
Solve a Lagrangian dual problem with the current set of columns.

Step 2: Selecting of additional columns
Select columns from the previously generated duties with negative reduced cost.
If no such columns exist (or another termination criterion is satisfied), go to
Step 3; otherwise, return to Step 1.

Step 3: Construction of feasible solution
Use all the columns selected in Step 0 and Step 2 to construct a feasible solution.

Figure 3. Solution method for the CSP.

advance, because the number of feasible duties is too large. The problem is
to generate the pieces, such that we can combine two pieces to a duty and
in our case we also have to implement the round-trip-condition in the generation of pieces. We explain this in § 4.3 for the case where violation of the
round-trip-condition is not allowed and in § 4.4 where violation is allowed
at most once. Finally, we discuss in § 4.5 how we implement the restrictions
with respect to relieving.

4.1   Mathematical Model VCSP

The mathematical formulation we propose for the VCSP is a combination
of the quasi-assignment formulation for vehicle scheduling based on network
$G = (V, A)$ defined in § 3.1, and a set partitioning formulation for crew
scheduling. The quasi-assignment part assures the feasibility of vehicle schedules, while the set partitioning part assures that each trip is assigned to a duty
and each deadhead task is assigned to a duty if its corresponding deadhead is
part of the vehicle schedule. Before providing the mathematical formulation,
we need to recall and introduce some notation. As before, N denotes the set
of trips, K denotes the set of duties, and $A^s \subset A$ and $A^l \subset A$ denote the sets
of short and long arcs, respectively. Furthermore, $K(i)$ is the set of duties
covering trip $i \in N$, $K(i,j)$ denotes the set of duties covering deadhead tasks
corresponding to deadhead $(i,j) \in A^s$ and $K(i,t)$ and $K(s,j)$ denote the set
of duties covering the deadhead task from the end location of trip i, to the
depot and from the depot to the start location of trip j, respectively. Decision
variables y_{ij} and x_k are defined as before, i.e., y_{ij} indicates whether a vehicle
covers trip j directly after trip i or not, while x_k indicates whether duty k is
selected in the solution or not. The VCSP can be formulated as follows:

$$\text{Minimize} \quad \sum_{(i,j)\in A} c_{ij}\, y_{ij} + \sum_{k\in K} d_k\, x_k \qquad (9)$$

subject to

$$\sum_{j:(i,j)\in A} y_{ij} = 1 \qquad\qquad \forall i \in N \qquad\qquad (10)$$

$$\sum_{i:(i,j)\in A} y_{ij} = 1 \qquad\qquad \forall j \in N \qquad\qquad (11)$$

$$\sum_{k\in K(i)} x_k = 1 \qquad\qquad \forall i \in N \qquad\qquad (12)$$

$$\sum_{k\in K(i,j)} x_k - y_{ij} = 0 \qquad\qquad \forall (i,j) \in A^s \qquad\qquad (13)$$

$$\sum_{k\in K(i,t)} x_k - y_{it} - \sum_{j:(i,j)\in A^l} y_{ij} = 0 \quad \forall i \in N \qquad\qquad (14)$$

$$\sum_{k\in K(s,j)} x_k - y_{sj} - \sum_{i:(i,j)\in A^l} y_{ij} = 0 \quad \forall j \in N \qquad\qquad (15)$$

$$x_k, y_{ij} \in \{0,1\} \qquad\qquad \forall k \in K, \forall (i,j) \in A \qquad\qquad (16)$$

As before, the objective coefficients c_{ij} and d_k denote the vehicle cost of arc $(i,j) \in A$, and the crew cost of duty $k \in K$, respectively. The objective is to minimize the sum of total vehicle and crew costs. The first two sets of constraints (10) and (11) are equivalent to the quasi-assignment part of the formulation for the VSP discussed in § 3.1. Constraints (12) assure that each trip i will be covered by one duty in the set $K(i)$. Furthermore, constraints (13), (14) and (15) guarantee the link between deadhead tasks and deadheads in the solution, where deadheads corresponding to short and long arcs in A are considered separately. In particular, constraints (13) guarantee that each deadhead from i to j is covered by a duty in the set $K(i,j)$. The other two constraint sets (14) and (15) ensure that the deadheads from the end location of trip i to t and from s to the start location of trip j, possibly corresponding to long arc $(i,j) \in A$, are both covered by one duty. Note that the structure of these last three sets of constraints is such that each constraint corresponds to the selection of a duty from a large set of duties, if the corresponding y variable has value 1. The model contains $|A| + |K|$ variables and $5|N| + |A^s|$ constraints, which may already be quite large for instances with a small number of trips.

The difference between the variants with respect to changeovers is in the definition of K. If changeovers are not allowed, there are less possible duties than if changeovers are allowed.

4.2   General Solution Method VCSP

In Figure 4 we give an outline of our algorithm to compute a lower bound on the optimal solution and to get feasible solutions. We compute the lower bound in Step 1 by first replacing the equality signs in constraints (12) – (15)

Step 0: <u>Initialization</u>

Solve VSP and CSP (using the algorithm of Figure 3) and take as initial set of columns the duties in the CSP-solution.

Step 1: <u>Computation of lower bound</u>

Solve a Lagrangian dual problem with the current set of columns.

Step 2: <u>Generation of columns</u>

Generate columns with negative reduced cost. Compute an estimate of a lower bound for the overall problem. If the gap between this estimate and the lower bound found in Step 1 is small enough (or another termination criterion is satisfied), go to Step 3; otherwise, return to Step 1.

Step 3: <u>Construction of feasible solutions</u>

Based on feasible vehicle solutions from Step 1, construct corresponding feasible crew solutions using the algorithm of Figure 3.

Figure 4. Solution method for the VCSP.

by greater-than-or-equal signs, and then applying Lagrangian relaxation with respect to these constraints. To compute the best Lagrangian lower bound we apply subgradient optimization. In every iteration of the subgradient optimization we get a feasible vehicle schedule and we use these schedules to construct feasible (vehicle and crew) solutions at the end. For further details about the Lagrangian relaxation we refer again to Freling et al. (2000).

The main difference to the CSP is that we cannot generate the total set of duties before we start with the optimization algorithm, because we do not have a vehicle solution in advance. So we have to generate the duties during the process, where we have to take into account that the round-trip-condition can be violated at most once in every duty. For this reason we have to construct pieces with no violation and pieces with one violation. How this is done, is described in § 4.3 and 4.4. After that we just combine pairs of pieces into duties.

In every iteration i we compute an estimate LBT_i of the lower bound for the overall problem. Let LBS_i denote the value of the Lagrangian lower bound in iteration i, then the estimate is computed as $LBT_i = LBS_i + \sum_{k \in K_i} \overline{d}_k$ where K_i is the set of duties added in iteration i and $\overline{d}_k$ is the reduced cost of duty k.

LBT_i is a lower bound for the overall problem if all duties with negative reduced costs are added to the master problem. Of course, we can stop if LBT_i is equal to LBS_i, but in practice we stop earlier, namely if the relative difference is small or if there is no significant improvement in LBS_i during a number of iterations.

At the end we compute feasible crew schedules by solving instances of the CSP that are defined by the (feasible) vehicle schedules from the last iterations. It is also possible to do another subgradient optimization after the convergence and compute crew schedules based on a few vehicle schedules from the last iterations of this optimization.

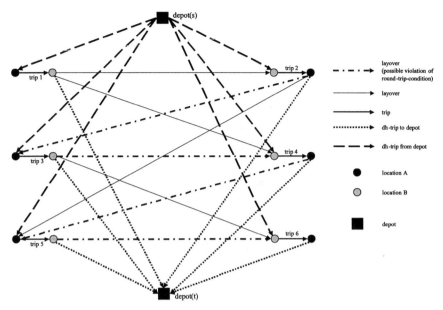

Figure 5. Piece network.

4.3   Generation of Pieces Without Violation of the Round-trip-condition

Recall that if we want to generate duties with at most one violation of the round-trip-condition, we have to generate pieces with no violation and pieces with at most one violation. In this subsection we look at pieces where the round-trip-condition is not violated and in the next subsection we look at pieces where the round-trip-condition can be violated at most once.

Network Structure: We generate pieces using a network that is an extension of the quasi-assignment network G for vehicle scheduling (see § 3.1). Let a *start point* (*end point*) be defined as the point corresponding to the start (end) of a vehicle trip. We define the acyclic network $G^p = (N^p, A^p)$, where nodes correspond to the start and end points of each trip, and a source s and a sink t representing the depot. Arcs in A^p correspond to trips, deadheads and layovers. Figure 5 illustrates network G^p with six trips, six deadheads from and six to the depot and eight layovers. The trips (2,3,4) and (4,5,6) are combinations that violate the round-trip-condition.

The cost associated with an arc $(i, j) \in A^p$ is equal to the value of the Lagrangian multiplier corresponding to this arc. Every arc corresponds to one of the relaxed restrictions in the model VCSP in § 4.1. Each path $P(u, v)$ between two nodes u and v on network G^p corresponds to a feasible piece of work if its duration is between the minimum and maximum allowed duration of a piece of work. The duration of a piece of work starting at s and/or ending at t is determined by incorporating travel, sign-on, sign-off and layover times.

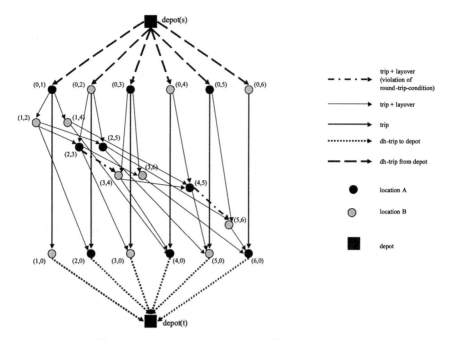

Figure 6. Better representation of piece network.

We do not need to generate all feasible pieces, but only pieces that correspond to a shortest path between every two pair of nodes. In Freling et al. (2000) we prove that if no duties with negative reduced cost can be constructed from these pieces, then the relaxation has been solved to optimality. If we would not have the problem of the round-trip-condition, finding all shortest paths is very easy (see Freling et al. (2000)), but in our case solving this restricted shortest path problem is not very straightforward. We have considered two ways of solving this problem, namely by an exact procedure or by a fast heuristic.

Exact Shortest Path Algorithm: Because we want to generate pieces where the round-trip-condition is never violated, we define a new acyclic network $G^{p2} = (N^{p2}, A^{p2})$. The set of nodes consists of two nodes for every trip (one which we can reach from the depot and one where we can go to the depot) and we have a node for every deadhead in A^p. The source and the sink correspond again to the depot and the set of arcs is now defined by deadheads from and to the depot and a combination of a trip and a layover or only a trip. For the same example as in Figure 5 we have the network G^{p2} in Figure 6.

We can generate the pieces now by just deleting the arcs corresponding to violations of the round-trip-condition (in the example the arcs from (2,3) to (3,4) and from (4,5) to (5,6)) and solve the shortest path problem between every node u of type (0,..) and every possible end node v. For all feasible paths from u to v, three additional paths are considered, namely path $s, u, ..., v$, path $u, ..., v, t$ and path $s, u, ..., v, t$.

Heuristic for Generating Pieces: Because we have to generate the pieces many times, we have developed a heuristic for generating the pieces. We do this by solving two "shortest" path problems in the network G^p between every pair of nodes. We explain this heuristic via the example in Figure 5. For solving the first "shortest" path problem we delete the arcs from 2 to 3 and from 4 to 5 (the parts of the violation of the round-trip-condition at location A) and compute after that the shortest path. For the other "shortest" path we delete the arcs from 3 to 4 and from 5 to 6 (parts corresponding to location B). Finally, we choose the best of the two "shortest" paths. In this way, we always get a piece that never violates the round-trip-condition, although it is possible that we do not get the piece with the lowest reduced cost. In the example, the piece with the lowest reduced cost can be a piece consisting of trips 3, 4, and 5, which can never be the solution of our heuristic. However, from computational experiments we have concluded that this heuristic works very well for the problems at the RET.

4.4 Generation of Pieces with at Most One Violation of the Round-trip-condition

In this case, the pieces can have at most one violation of the round-trip-condition. Of course, we can generate here also pieces with no violation, e.g., if there is no violation possible or if it is not attractive to violate the round-trip-condition. We can easily adapt the exact algorithm and the heuristic for this case. Previously, we computed the shortest path in an acyclic network from u to v by just computing the shortest path from u to every node u' connected with u and so on. In this case, we have to remember two paths to u', namely the shortest path with at most one violation of the round-trip-condition and the shortest path without violation of the round-trip-condition. We do this for all nodes between u and v such that we can add a violation at the moment we want.

4.5 Relief Restrictions

Recall that the RET has several restrictions with respect to relieving

1. a duty that does not start at the depot, starts with a relief time and at a relief location,
2. a duty that does not end at the depot, ends after a layover and at a relief location.

These restrictions also hold for the start/end of every piece in the variants where changeovers are not always allowed. The difficulty of these restrictions is that a duty cannot end at every relief location, because the time between the departure and the arrival must be at least the relief time. We solve this problem in the following way: First we construct a set of all possible layovers where a duty cannot end. We can do this in advance and when constructing the pieces we assure that a piece never ends at such a layover. Of course, we also introduce this restriction now for the pieces before the break, although that is not required. This is not a problem, because if a duty ends with a layover before the break, there is another duty with the same properties that ends with a trip.

5   Computational Results on RET Data

This section deals with a computational study of the integration of bus and driver scheduling. We have used data of two individual bus lines from the RET (lines 35 and 38). These lines are used to compare the different approaches for bus and driver scheduling, i.e., we investigate the effectiveness of integration as compared to the traditional sequential approach and we compare the different variants of allowing changeovers. The following points are relevant for all the different approaches proposed in this section:

1. The objective is to minimize the number of drivers and buses in the schedule, therefore, we use fixed costs. At the RET, the fixed cost for a driver is about equal to the fixed cost for two buses. As an indication, we take as fixed cost for the bus 600 cost units and for the driver 1,200 cost units. For a fair comparison between the sequential and integrated approach, we also need variable vehicle costs per minute for every deadhead, because if we do not add variable vehicle costs, the VSP will only minimize the number of buses during peak hours and the number of buses during off-peak hours may be much too large. Because every bus needs a driver, in the off-peak hours the number of drivers is far from optimal. Therefore, the total number of drivers is much higher than if we add variable vehicle costs. We take as variable vehicle cost one cost unit per minute.

2. The different variants, as described in § 2.1, are called a and b, where a is changeovers are only allowed if there is enough time to relief one driver by another and b is the current RET rule that changeovers are allowed for split duties and not for other types of duties.

3. The column generation is terminated once the difference between the real and the current lower bound is less than 0.1% or if no significant improvement in the current lower bound is obtained for a certain number ($max\_tailing\_off$) of iterations; this is called $tailing\text{-}off$ $criterion$. We have considered an improvement significant if it was at least 5%.

4. For the integrated approach we generate 10 feasible solutions, where 5 follow from the last iterations of the column generation and the other 5 follow from the last iterations of the last subgradient optimization (see § 4.2).

5. The strategies for generating columns in the pricing problem are obtained after extensive testing and tuning (see Freling (1997)).

	Line 35	Line 38
number of trips	131	259
number of round-trip-condition restrictions	6	39
number of relief restrictions	48	93
types of duties	8	8
number of relief locations	1	1
deadheads between two endlocations allowed	yes	no
max_tailing_off	50	15

Table 1. Important properties of lines 35 and 38.

		Line 35		Line 38		Line 38, conv	
		var. a	var. b	var. a	var. b	var. a	var. b
se- quen- tial	lower: buses	10	10	14	14	14	14
	lower: drivers	16	17	21	23	21	23
	lower: total costs	26634	27834	35653	38053	35653	38053
	number of buses	10	10	14	14	14	14
	number of drivers	16	17	23	24	23	24
	total costs	26638	27838	38073	39273	38073	39273
inte- gra- ted	lower: total costs	24945	24968	35553	36040	33152	33273
	iterations	24	20	19	19	105	113
	cpu (sec.)	279	232	2433	2502	16920	17640
	number of buses	10	10	13	13	13	13
	number of drivers	15	15	22	23	21	22
	total costs	25495	25573	36438	37755	35040	36311
	gap (%)	2.16	2.37	-	-	5.39	8.37

Table 2. Results of lines 35 and 38, and line 38 after convergence (conv).

6. All tests are executed on a Pentium II 350 PC with 64Mb of computer memory.

In Table 1 we summarize the size of the problem and the important properties for lines 35 and 38. We have only considered the problems for weekdays.

The lines 35 and 38 are representative for the RET, where line 35 is of middle size and line 38 is of large size (e.g., line 38 is the largest line of the RET if we look at the number of trips). In Table 2, we show the results of lines 35 and 38. For these results we have used the heuristic for generation the pieces described in § 4.3.

The table consists of four parts, two for the sequential and two for the integrated approach. In the first part we summarize the results for the lower bound phase of the VSP (number of buses), the CSP (number of drivers) and of the total vehicle and crew costs. In the second part we show the characteristics of the feasible solution found by the sequential approach. We do not give the computation times for the sequential approach, because they can be neglected. For the integrated approach we give for line 35 the lower bound after convergence and for line 38 the lower bound of the last iteration. The lower bound phase for line 38 terminated, because the tailing-off criterion is satisfied. We also give the number of column generating iterations and the total computation time for the lower bound phase. In the last part we give the number of buses, drivers and the total costs for the best of the ten feasible solutions. For line 35 we also give the relative gap between the lower and the upper bound. We cannot do this for line 38, because there we do not have a lower bound for the total problem but only for a subproblem, since the algorithm terminated before convergence. This is the reason why we have also computed a "real" lower bound, obtained when the tailing-off criterion was not used. These results are given in the right columns of Table 2. That is, we can even compute for the largest RET problem a "real" lower bound within 5 hours. Furthermore, in both variants the solutions are better than the solution that was obtained when the tailing-off criterion was used.

For all the variants the integrated approach gave better results than the sequential approach and the gap for the integrated approach is always lower than 10%. For example in Table 2, we saved two drivers and one bus (about 8% reduction in costs) for the variants a and b and we still had a gap of 5.39% and 8.37%, respectively. Because of the possibility to violate the round-trip-condition once in every crew duty, we can even save buses in the integrated approach. (Without this possibility only drivers could have been saved.) Another important result is that we find solutions with less drivers if more flexibility for changeovers is allowed. We have to be careful with interpreting these results, however, because the difference in the lower bounds is negligible. Hence, it is possible that our approach simply works better in these variants, while the optimal solutions are actually the same.

In Huisman (1999) one can find more results with some other variants and cost structures. For example, we took different costs for split duties and we added an extra restriction on the maximum length of the break. We also looked at a problem where no changeovers are allowed at all and where changeover are always allowed. For the first situation we applied a different model (see also Freling et al. (2000)).

Exact Computation of the Lower Bound: When interpreting the above results, one should keep in mind that we used a heuristic for generating the pieces, i.e., we did not necessarily take the pieces with the lowest reduced cost. So it is possible that we found that there were no duties left with negative reduced cost with the pieces we generated, whereas a combination of two pieces with negative reduced cost did exist. In that case, we did not

	Heuristic	**Exact (normal)**	**Exact (equality)**
lower: total costs	24968	24846	25052
iterations	20	40	66
cpu (sec.)	232	2198	4138

Table 3. Comparison of the heuristic and exact method to compute the lower bound.

find a real lower bound for the total problem, but only an approximation of this lower bound. Therefore, we have compared this approximation with the lower bound obtained by using the exact shortest path algorithm, described in § 4.3, to generate the pieces. This was done for several lines and variants. A typical example of our findings is given in Table 3, which summarizes the results for line 35, variant b.

The column "exact (normal)" refers to the situation that the method and all parameters are the same as for the heuristic. This means that the exact lower bound is lower than the lower bound computed with the heuristic, although not much. Moreover, we can obtain a better exact lower bound if we do not compute the Lagrangian relaxation with "greater-than-or-equal-to" signs in constraints (12) – (15), but with equality signs instead. The results of this approach are given in the column "exact (equality)." We see that the lower bound of the heuristic is actually a real lower bound.

6 Conclusions

In this paper we applied an integrated approach to vehicle and crew scheduling at the RET. We handled complicating constraints, which have not been considered in the literature before. We showed the results for two individual bus lines, including the RET bus line with the largest number of trips. For these lines the integrated problem could be solved in a reasonable amount of time, where the gap between the lower bound and the best feasible solution was less than 10% in all cases. The main conclusion is that we can save vehicles and/or crews by integrating the vehicle and crew scheduling problem, which may lead to a big decrease in costs. Especially, in the case that (almost) no changeovers are allowed, integration is very attractive.

Another important result is that sometimes it is indeed possible to reduce the total costs by allowing changeovers more often.

Acknowledgements

From 1996 until 1999, the second author was also affiliated with the RET, the public transport company in Rotterdam, the Netherlands. The authors are grateful to the RET company for providing the data and for supporting the research. They would also like to thank two anonymous referees for their comments on an earlier draft of this paper.

Bibliography

Darby-Dowman, K., J.K. Jachnik, R.L. Lewis, and G. Mitra (1988). Integrated decision support systems for urban transport scheduling: Discussion of implementation and experience. In J.R. Daduna and A. Wren (Eds.), *Computer-Aided Transit Scheduling, Lecture Notes in Economics and Mathematical Systems*, 308, Springer, Berlin, 226–239.

Freling, R. (1997). *Models and Techniques for Integrating Vehicle and Crew Scheduling*. Thesis Publishers, Amsterdam.

Freling, R., C.G.E Boender, and J.M.P. Paixão (1995a). *An integrated approach to vehicle and crew scheduling*. Technical Report EI9503/A, Econometric Institute, Erasmus University, Rotterdam.

Freling, R., D. Huisman, and A.P.M. Wagelmans (2000). *Models and algorithms for integration of vehicle and crew scheduling*. Technical Report EI2000-10/A, Econometric Institute, Erasmus University, Rotterdam.

Freling, R., J.M.P. Paixão, and A.P.M. Wagelmans (1995b). *Models and algorithms for vehicle scheduling*. Technical Report to appear in Transportation Science.

Freling, R., A.P.M. Wagelmans, and J.M.P. Paixão (1999). An overview of models and techniques for integrating vehicle and crew scheduling. In N.H.M. Wilson (Ed.), *Computer-Aided Transit Scheduling, Lecture Notes in Economics and Mathematical Systems*, 471, Springer, Berlin, 441–460.

Gaffi, A. and M. Nonato (1999). An integrated approach to extra-urban crew and vehicle scheduling. In N.H.M. Wilson (Ed.), *Computer-Aided Transit Scheduling, Lecture Notes in Economics and Mathematical Systems*, 471, Springer, Berlin, 103–128.

Haase, K., G. Desaulniers, and J. Desrosiers (1999). *Simultaneous vehicle and crew scheduling in urban mass transit systems*. Technical Report GERAD G-98-58, École des Hautes Études Commerciales, Montréal.

Haase, K. and C. Friberg (1999). An exact branch and cut algorithm for the vehicle and crew scheduling problem. In N.H.M. Wilson (Ed.), *Computer-Aided Transit Scheduling, Lecture Notes in Economics and Mathematical Systems*, 471, Springer, Berlin, 63–80.

Huisman, D. (1999). Sequentiële en geïntegreerde benaderingen voor voertuig- en personeelsplanning bij de RET (in dutch). Master's thesis, Faculty of Economics, Erasmus University, Rotterdam.

Rousseau, J.-M. and J.Y. Blais (1985). HASTUS: An interactive system for buses and crew scheduling. In J.-M. Rousseau (Ed.), *Computer Scheduling of Public Transport 2*, North-Holland, Amsterdam, 473–491.

A Network Flow Approach to Crew Scheduling Based on an Analogy to an Aircraft/Train Maintenance Routing Problem

Taïeb Mellouli

Decision Support & OR Laboratory, University of Paderborn,
Warburger Str. 100, D-33098 Paderborn, Germany
mellouli@uni-paderborn.de

Abstract. Airlines' and railways' expensive resources, especially crews and aircraft or trains are to be optimally scheduled to cover flights or trips of timetables. Aircraft and trains require regular servicing. They are to be routed as to regularly pass through one of the few maintenance bases, e.g., every three to four operation days for inspection. Apart from complicating workrules, crews are to be scheduled so as to "pass through" their home bases weekly for a two-day rest. This analogy is utilized in order to recognize opportunities for integrating classical planning processes for crew scheduling, and to transfer solution methodologies. A mixed-integer flow model based on a state-expanded aggregated time-space network is developed. This mathematical model, used to solve large-scale maintenance routing problems for German Rail's intercity trains, is extended to the airline crew scheduling problem where *maintenance states* are replaced by *crew states*. The resulting network flow approach to an integrated crew scheduling process involving multiple crew domiciles and various crew requests is tested with problems from a European airline. A decision support system and computational results are presented.

1 Introduction

In today's highly competitive markets, computer-aided systems for production planning, scheduling, and control have become a critical success factor for airlines and railways. Costs to operate a timetable of flights/trips can be substantially reduced by finding the most efficient use of company's expensive resources, especially vehicles (aircraft/trains) and crews. Here, the satisfaction of crew personnel is to be regarded as well. For increasing size of timetables, real-life resource scheduling problems are becoming more complex. This is a source of new challenges for mathematical programming.

Providers of public transportation systems are faced with a very complex planning and control process consisting of several phases: product planning, production planning/scheduling and resource allocation, as well as operations control. Using forecasts of passenger demand, the transportation network is designed as a set of lines with capacities and frequencies of service. Flight/trip duration on these lines is determined. Based on this the products, each as a subset of flights/trips with same departure time on one/several/all days of the week, are planned for winter or summer term. The courses and schedules of flights/trips of the resulting timetable may differ during weekends and vacations, or for shorter planning horizons of a few weeks (e.g., demand-driven tourists airlines). Usually departure times of flights/trips are fixed at the end of the product planning phase. For some airlines, a flight scheduling step is combined with subsequent production planning steps (cf. Suhl (1995)). Also a fleet assignment step is needed whenever some flights/trips can be operated with aircraft/trains of different types or passenger capacities.

A main task of production planning is to schedule and assign resources to flights/trips in order to operate a given timetable. This is classically performed in two steps for each resource type—for vehicles as well as for crews. For trains/aircraft, a set of logical *rotations* covering all trips/flights of the timetable is generated. Each rotation consists of a sequence of flights/trips that can be performed with the same aircraft/train. Parts of these rotations are then assigned to physical aircraft/trains taking into account various maintenance requirements. For crews, a set of *pairings* covering all flights of the timetable is generated, where each pairing consists of a work-and-daily-rest schedule for crews of up to five days length, starting and ending at the same home base. These pairings (which in some sense correspond to rotation parts) are then assigned to individual crew members taking into account scheduled (office or simulator) activities and requested flights or off-duty days.

In the next two sections this two-step procedure, respectively for scheduling aircraft/trains and crews, is discussed together with terminology, underlying problems, and existing solution approaches. In Section 4 we propose an integration of the considered planning steps for both resources. This is based on a recognized analogy between the integrated planning process for vehicles and that for crews. In spite of decisive differences of planning these two resources, this analogy is utilized in order to transfer solution methodologies.

A flow model based on a state-expanded aggregated time-space network is proposed for the train/aircraft maintenance routing problem in Section 5, then transferred and extended to the airline crew scheduling problem in Section 6. The ideas are tested with real-life maintenance routing problems of German Rail and crew scheduling problems of a major airline. Computational results, a prototype, and model extensions are presented in Sections 7 and 8.

Our network flow approach to crew scheduling involves multiple domiciles, proceedings of crews, and various crew requests and, thus, seems to be a genuine alternative to commonly used set partitioning models (cf. § 3.2). Our current investigations are generally related to long-distance service with applications to railways and airlines. The problem of routing German Rail's trains requiring regular maintenance, being tightly related to an aircraft maintenance routing problem, is studied in Mellouli (1998). For crew scheduling, we focus on the airline domain in this paper with a rather European problem setting. In the predecessor conference of CASPT 2000, research at our institute was initiated which deals with the design and realization of operations control support systems for railways (Suhl and Mellouli (1999)).

2 Building Rotations and Maintenance Requirements

The task of building rotations for aircraft and trains is now defined as a special case of vehicle scheduling problems different from those of bus transit. The drawbacks of the classical sequential proceeding for scheduling and assigning aircraft/trains are illustrated and an integrated planning process is discussed.

2.1 Rotations – Vehicle Scheduling Problems

A timetable T is given as a finite set of trips/flights i ($\in T$) with known start station $orig(i)$ and end station $dest(i)$ as well as departure time $start\text{-}time(i)$ and duration $duration(i)$. The set of used terminal stations is denoted by TS. A bus, train or aircraft (vehicle) arriving at $dest(i)$ after having served trip/flight i ($arrival\text{-}time(i) := start\text{-}time(i) + duration(i)$) needs a ground or turn time $turn\text{-}time(dest(i))$, in order to serve another trip/flight. Thus, this vehicle is available at $end\text{-}time(i) := arrival\text{-}time(i) + turn\text{-}time(dest(i))$.

A *rotation* (or a route) is a sequence of trips which can be performed by the same vehicle within the given planning horizon. Every two consecutive trips i and j of a rotation ($i, j \in T$) must be *compatible*, either without dead-heading if $dest(i) = orig(j)$ and $end\text{-}time(i) \leq start\text{-}time(j)$ or with a dead-head trip from $dest(i)$ to $orig(j)$ during $dh\text{-}time(dest(i), orig(j))$ if $end\text{-}time(i) + dh\text{-}time(dest(i), orig(j)) \leq start\text{-}time(j)$.

The *Vehicle Scheduling Problem (VSP)* is the task of building a collection of rotations (*vehicle schedule*), such that each trip of T is covered exactly once by a rotation. The requirements and optimization criteria may differ from a problem setting to another. For the *single-depot* and *multiple-depot VSP* for bus transit, each one-day rotation must start and end at the same depot (cf., e.g., Löbel (1998) for an overview of solution approaches). Here, dead-head trips including those from/to depots are needed in order to serve all trips of the timetable by a given/minimum number of buses. The minimization of dead-head effort is an integral part of planning bus rotations.

The task of building rotations for aircraft or trains differs in many respects: Depots are not required and an aircraft or a (long-distance) train may start and end its multiple-day route at any pair of stations. For railway and airline applications, timetables are designed in a balanced way in terms of equality of the numbers of ending and starting trips at stations over several days. Thus, dead-head trips for aircraft and trains are not used or only needed in exceptions—their insertion is usually controlled by planners. Therefore, the basic VSP version for airline and railway applications requires neither depots nor dead-heading which we call the *simple VSP*. The main goal here is to use a minimum/given number of vehicles (fleet size).

The simple VSP with *periodicity* arises whenever a periodic timetable is given for a certain period, e.g., one week, and the vehicle schedule should work repeatedly for all periods of the planning horizon. In this case, a rotation corresponds to a cyclic sequence of one-period routes.

2.2 Maintenance Requirements for Aircraft and Trains

Since safety of passengers is immediately at risk when using not regularly checked vehicles, legal rules for maintenance are imposed for each transportation type. For German Rail, each intercity train requires a 4-hour inspection at a maintenance base every three to four operation days. The difficulty of scheduling trains with this maintenance requirement is tightly related to the transportation network structure. Only 10 of the 60 terminal stations are close to one of the company's maintenance bases.

For airlines, the guidelines of the Federal Aviation Administration (FAA) stipulate that a so-called A-check of aircraft at a maintenance base is required every 65 flight hours or about once a week. Heavier checks are spaced at relatively large intervals: B-checks are required every 300 to 600 flight hours as well as C- and D-checks about once every one to four years (cf. Feo and Bard (1989)). To comply with FAA guidelines or with national authorities' guidelines, each airline company adopts a certain maintenance policy. It is common practice in airline industry to require regular routine maintenance for aircraft (equivalent to A-checks), e.g., every at most 35 to 40 hours of flying or every 3, 4, or 3-4 operation days. Maintenance policies for routine and heavy checks may differ from an airline company to another.

Many airlines, especially European tourists airlines operate world-wide without a clear hub-and-spoke structure. Flights are scheduled with returns not necessarily to the origin airport. It is then a difficult task to bring all aircraft on a frequent regular basis to one of the few airline's maintenance bases (mb) for inspection without expensive dead-head flights. One has to combine all of the non-mb-to-non-mb flights or flight pairs with mb-to-non-mb and non-mb-to-mb ones very carefully. The problem has the same structure as the above mentioned maintenance problem for trains, and becomes, though, an easier special case when checks are performed only at night.

2.3 Logical Rotations First – Assignment Second

Usually each airline has a department for technical maintenance control (TMC) keeping under surveillance both the maintenance requirements of physical aircraft and the capacity of technical stations (personnel, space, times). TMC often prepares a maintenance plan that fixes in advance when and where each of the physical aircraft has to be serviced. The various checks are scheduled before assigning flights to physical aircraft (tail assignment).

A common airlines' practice in scheduling physical aircraft is based on a sequential procedure which we call "logical rotations first – assignment second." The basic version of this method consists of two steps:

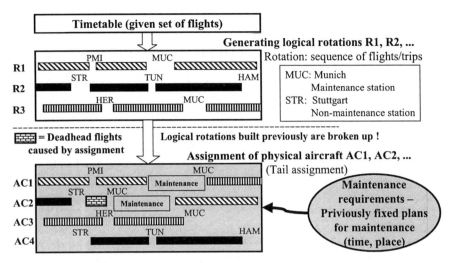

Figure 1. Logical rotations first – assignment second.

1. Generate *logical rotations* covering all flights, where rotations are thought to be assigned to logical aircraft without special requirements
2. Assign parts of these rotations to physical aircraft where requirements and previously fixed plans of maintenance are taken into account

The drawbacks of this sequential procedure are illustrated by a simple example in Figure 1. Rotations built in a first step are broken up in order to link rotation parts together and integrate maintenance checks scheduled previously on aircraft. Sometimes, dead-head flights are inserted in order to proceed aircraft to maintenance bases, or "overflow aircraft" are needed to assign remaining flights (cf. AC4 in Figure 1). If possible, these flights are integrated into the gaps of some aircraft rotations, often causing planning and dead-head costs. Owing to the high combinatorial degree of the problem, finding even a feasible solution necessitates considerable planning effort. The planners often have no time to perform what-if-analyses by constructing alternative plans in order to maximize utilization of aircraft, enhance rotation quality, or handle unscheduled aircraft-downtimes due to technical reasons.

The procedure for building logical rotations is a computationally easy task and corresponds to solving the simple VSP. This can be performed as follows: Firstly sort all flights chronologically with respect to departure times. Then, maintain a list of partly used logical aircraft (initially empty), that are represented by the corresponding partly constructed rotations $R_1, R_2, ..., R_k$, each as a list of flights to be served by an aircraft. If the next processed flight fits into some R_j $(j = 1, ..., k)$, then choose one of them (to insert the new flight) according to a certain dispatching strategy, e.g., choose the aircraft whose last flight arrives earliest (FIFO: first-in first-out) or latest (LIFO: last-in first-out). Otherwise, a new rotation R_{k+1} is created containing the processed flight as first element.

2.4   Integrating Maintenance into the Rotation Building Process

Rotation planners are used to define previously assigned maintenance checks by TMC as *dummy flights* starting and ending at the same station with corresponding duration as well as start and end times. Integrating these dummy flights into the timetable, a "greedy" procedure such as that described at the end of § 2.3 is applied. In the resulting schedule for logical aircraft, the dummy flights for the maintenance checks are integrated into the constructed rotations. However, maintenance checks of different physical aircraft may occur within one logical rotation while other rotations contain no or only fewer checks than required for a physical aircraft. Suppose that an A-check is required every three days for each aircraft. Three checks on consecutive nights may be scheduled for aircraft AC_1, AC_2, and AC_3, respectively. A constructed logical rotation, say R_1, may include two or three of the corresponding dummy flights while R_2 and R_3 include none or only one of them. Thus all rotations R_1, R_2, and R_3 are broken up into rotation parts of up to one day length in order to be able to assign the checks to the right aircraft, respectively.

To avoid the drawbacks of the sequential approach (cf. Figure 1), we rather propose an integrated process for scheduling and assigning physical aircraft which is based on a main scheduling and routing step:

- Build (logical) rotations that are feasible with respect to maintenance requirements of frequent regular type, e.g., routine or A-checks
- Each rotation involves regular checks of an aircraft *which are spaced evenly* according to a given maintenance rule, e.g., every 3-4 days
- The remaining irregular or heavier checks—scheduled in advance—are integrated as dummy flights into the constructed rotations

The resulting rotations are of good quality with respect to the second assignment step: They fit into physical aircraft if no heavy check is scheduled, otherwise only longer parts of them have to be coupled in order to place scheduled irregular checks on the right physical aircraft (cf. Section 4).

The main scheduling and routing step corresponds to solving an extension of the *simple VSP* with *maintenance routing*, a computationally much more difficult task than the simple VSP: The rotations covering all trips of the timetable are constructed such that the corresponding (logical) aircraft are *routed* as to regularly pass through one of the few company's maintenance bases for inspection every given operation time interval, in terms of numbers of flight hours or operation days. A network flow approach capable of solving large-scale maintenance routing problems is given in Section 5.

The proposed integrated planning process requires a tight collaboration between TMC and rotation builder. Time and place of regular routine checks should not be fixed in advance on aircraft. Rather an exact specification of the rules related to regular maintenance is required: duration of operation interval and of inspection as well as the available maintenance bases and their capacities in terms of maximum number of processed aircraft.

3 Crew Scheduling

In this section an overview of the task of scheduling airline crews is given. The commonly used sequential approach which we call "crew pairings first – chaining and assignment second" is illustrated. The classical set partitioning approach and the importance of developing alternative network flow approaches are discussed. This section gives the necessary background in order to discuss the analogy of crew scheduling and vehicle scheduling with maintenance routing as well as the proposed integration of crew scheduling steps in the next section. Based on this, a network flow approach to crew scheduling is developed in the remaining methodology part of this paper.

3.1 Airline Crew Scheduling and Sequential Approach

Given an airline's timetable and aircraft rotations, the task of crew scheduling is to assign work schedules to individual crew members during a given planning horizon such that each flight of the timetable is served once and the total costs are minimized. Expenses are incurred by overnight stays of crews at hotels and duration outside their home bases, by their proceedings as passengers to/from work bases, as well as by pay-and-credit (flight hours over or under duty flight hours guaranteed by monthly base pay).

A multitude of workrules is to be taken into account, first to build legal (flight) duties—a duty is a sequence of flights without daily crew rest (up to 4 or 6 landings, maximum duty time including check-in/-out times 10 or up to 14 hours with an excess)—and then to build pairings. A pairing is a sequence of duties over up to 5 calendar days starting and ending at the same home base (possibly with crew proceedings). The minimum daily rest time between two consecutive duties of a pairing amounts to 10 up to 16 hours depending on the duration of the preceding duty, particularly in case of excess times.

A logical work schedule of a crew member is something like a "crew rotation" for the planning period (called bid line or crew roster). It consists of a sequence of pairings with weekly rests (2 calendar days or daily rest time plus 36 hours) in-between. All pairings of a bid line should start and end at the same home base of a crew member. Building pairings and bid lines is also further complicated by some rules. In our airline application, e.g., it is not allowed, within every 7-day time span, to have more than 4 duty time excesses, more than 8 excess hours, or more than 3 night flights.

The airline crew scheduling problem is considered to be one of the hardest practical problems owing to its high combinatorial complexity. Even existing computerized approaches solve the problem within two steps:

1. Build a collection of pairings for crews, such that each flight is covered by exactly one pairing (Crew Pairing Problem: CPP).
2. Construct crew work schedules: Chain pairings into bid lines for a given planning period and assign them to individual crew members.

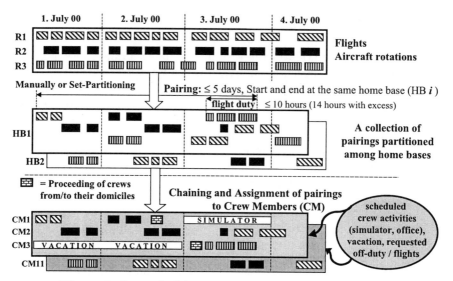

Figure 2. Crew scheduling and classical two-phased approach.

Figure 2 illustrates the sequential approach to crew scheduling. Using the given flights and the aircraft rotations, a collection of crew pairings is built manually or by using a set partitioning model to solve the CPP (cf. § 3.2). Daily duties are preferred to be on the same aircraft rotation avoiding change-overs of crews, which may cause crew rescheduling in case of delays.

While the second step is influenced by the seniority principle in North America, some European airlines try to take account of previously scheduled crew activities (office, simulator) as well as crew requests (requested off-duty or vacation days, requested flights to sunny destinations with several overnight stays). In this case the second step seems to be more complex since the constructed crew rosters should not only comply with all workrules but also fit into the "fingerprints" of individual crew members, in terms of scheduled activities as well as individual off-duty and flight requests.

In spite of the many facets of the problem, some drawbacks of the classical two-step procedure can be addressed here in a short form. Crew pairings built in a first step may turn out to be inconvenient when trying to chain them into rosters and especially when assigning them to individual crew members. Figure 2 shows an extreme example: Because of the previously scheduled activities and crew requests, some pairings must be broken up causing additional crew proceedings. Further, bottlenecks may occur, e.g., whenever two pairings remain without sufficient time for weekly rest in-between. Such situations may accumulate in the presence of several crew domiciles with restricted numbers of crews—available at certain days. In this case, swapping pairings or parts of pairings from a home base to another causes considerable costs in terms of crew proceedings and overnight stays outside domicile.

3.2   Approaches Based on the Classical Set Partitioning Model

Most research works on the CPP are based on the classical set partitioning model (cf. Garfinkel and Nemhauser (1969)). An optimal partition of the flight set (timetable) into pairings is computed as a subset of all possible pairings—a selection or a selected collection of pairings—covering each flight exactly once. The integer linear program used in literature to solve the set partitioning problem has a simple structure:

- To each possible pairing, a $0-1$ variable (column) is associated, indicating whether the pairing is selected in the computed solution
- To each flight, an equality constraint (row) is associated: In the matrix a coefficient 1 is set if the flight (of the row) is included in the pairing (of the column) and 0, otherwise. The right hand side of the equality is set to 1 in order to ensure that each flight is covered by exactly one pairing.
- The total costs incurred by the selected pairings are minimized

While the number of feasible duties does not exceed 4-5 times the number of (long-distance) flights or only 2-3 times in case of restricting cities of overnight stays, the number of feasible chains of flight duties (of a certain quality) over up to 5 calendar days increases drastically, e.g., $20,000 - 30,000$ for less than 200 flights. This number increases to 1/3 and 1/2 million pairings, if

1. Crews have several home bases: The quality as well as hotel and proceedings costs of a pairing depend on the home base of the crews
2. Proceedings of crews are allowed at the extremities (and in some cases within pairings) in order to proceed crews from/to their home bases

These practical requirements harden the crew scheduling problems and cause some complications to the set partitioning approach. Some pairings which seem to be of bad quality may turn out to decrease the total costs in the computed selection of pairings. Further, pairings are in some sense "domicile-dependent:" The same sequence of duties can be served by crews of different home bases, inducing different operating costs in terms of proceedings and overnight costs. Thus for both requirements, the total number of possible pairings (columns of the model) becomes considerably larger.

With several crew domiciles, a straightforward necessary requirement is that the computed selection of pairings should be evenly partitioned among all bases according to the number of crew members stationed, respectively, especially in terms of flight hours. To achieve such a partition, additional *base constraints* can be inserted into the set partitioning model setting a minimum and maximum total flight hours to each of the home bases, respectively.

In order to avoid computing all possible pairings, the above set partitioning model is often solved using *column generation*: Starting with a subset of all possible pairings, the set partitioning model is solved. Using dual prices of an optimal solution of this partial problem, new pairings *guaranteeing* solution improvement are built and added into the partial model. The task of building new pairings carried out by the column generator is modeled as a shortest path problem in an expanded graph in Lavoie et al. (1988) and as a time-constrained shortest path problem in, e.g., Desrosiers et al. (1995).

A different branch-and-cut approach is developed by Hoffman and Padberg (1993). They handle set partitioning problems with base constraints as well, and conclude that these problems are more difficult than those without base constraints. Based on this branch-and-cut approach, Borndörfer (1998) succeeded in solving large-scale set partitioning problems by adding special preprocessing and separation techniques based on graph theoretic results. The set partitioning/covering problems can also be relaxed to network, matching, graph covering as well as assignment, shortest route or minimum spanning tree problems (cf. El-Darzi and Mitra (1992, 1995)). These relaxations are then solved at nodes of tree search (or branch-and-bound) algorithms.

In our opinion, research is still required to solve the crew scheduling problem not only for increasing size of timetables. Most solution approaches deal with the one-domicile case and only solve the CPP without sufficient consideration—especially in the multiple domicile case—of time-dependent crew availability at bases and of crew requests (cf. § 3.1). An even partition of pairings and flight hours among home bases is only a first step. An improved approach to crew scheduling should also take into account the satisfaction of crews and their time-dependent availability at home bases. The ultimate goal is to optimally solve the entire crew scheduling problem including the second step, see Figure 2. Our contribution on this matter is to integrate crew scheduling steps and to provide an alternative network flow approach to the proposed integrated process (cf. Sections 4 and 6).

3.3 Advantages of an Alternative Network Flow Approach

Bodin and Golden (1981) note (p. 104): "The crew scheduling problem is generally broken down into two parts—generating pairings and constructing bid lines. ... . Since both problems have complicated workrules for constraints, the construction of an effective network flow model seems unlikely. With multiple domiciles, the problems become even more intractable."

The aim of our work is to develop such an effective network flow model for crew scheduling problems, notably with multiple domiciles (home bases), taking into account crew requests as well as time- and day-dependent crew availability at different domiciles. The idea of our network flow model is to implicitly represent all possible partitions of the flight set into chained crew pairings (bid lines) as feasible flows within a special network. Thus this model tries to combine the CPP with a part of the second chaining and assignment step of crew scheduling (cf. § 3.1) by constructing bid lines (not only pairings) whose number is equal to the number of crew members for each home base, respectively. Note that the number of possible bid lines for a planning horizon of, say, two weeks is much higher than that of pairings. Further, availability of crew members is taken into account in the developed model by satisfying crew requests and scheduled activities cumulatively for each home base.

The effectiveness of our flow model is that it can be directly solved by standard mathematical optimization software even for larger timetable size. Thus, our network flow design primarily is an investigation on *alternative models for crew scheduling problems*. This stands in contrast with graph/network constructions or theoretic results mentioned in § 3.2 to obtain relaxations, separations, or efficient column generators—all of them are investigations on *the solution of the classical set partitioning model* by various techniques.

4 Analogy and Integrated Planning Process

In this section, a recognized analogy of crew scheduling to scheduling/routing vehicles with regular maintenance requirements is illustrated and utilized in order to integrate planning processes, especially for crew scheduling.

The analogy: While vehicles are routed as to pass through a maintenance base every 3-4 days for inspection, crews have to "pass through" their home bases every at most five days in order to accomplish a weekly 2-day rest. Thus a rotation or a route of a vehicle integrating regular maintenance requirements is in some sense an analogue of what we call a *pairing chain*—a sequence of pairings spaced by weekly rest periods which can be performed by the same crew member over a given planning horizon, e.g., two weeks for a concrete case study with a European airline. Here an inspection of a vehicle at a maintenance base corresponds to a weekly rest of a crew member at his/her home base. This analogy can be extended by viewing previously scheduled crew activities and requests as analogies of scheduled irregular checks for vehicles. Figure 3 illustrates the whole analogy (compare Figures 1 and 2).

Integrated crew scheduling process: The crew scheduling task that will be handled by a flow model analogous to that for the VSP with maintenance routing has the following characteristics (cf. Figure 3):

1. A collection of *pairing chains* is constructed such that each flight of the timetable is covered exactly once by a pairing chain. Each pairing chain represents a full (logical) bid line/crew schedule/roster for crews without regard to its complete match to a concrete individual crew member.
2. Each pairing chain in the solution is assigned to a home base and contains feasible pairings spaced by weekly rests at that same home base.
3. The number of pairing chains constructed for a home base is equal to the number of crew members stationed at that home base.
4. The total flight hours are partitioned evenly among home bases, and, within each home base, among different sections of the planning horizon.
5. The *availability* of crews at the beginning of the planning period is taken into account—at which station they are and which "state" they have within a pairing or within off-duty days of a weekly rest.
6. For each home base hb, the scheduled activities (office, simulator) and requests (vacancy/off-duty days, special flights) of all crew members stationed at hb are integrated into the pairing chains generated for hb crews.

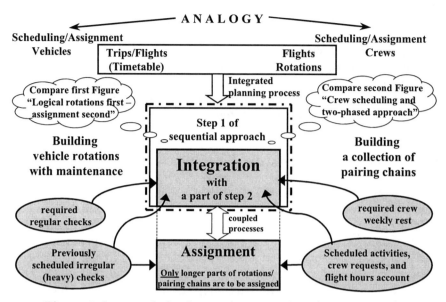

Figure 3. Integrated planning processes – analogy (vehicles, crews).

These characteristics show that the considered crew scheduling problem setting combines the CPP (step 1) with tasks actually accomplished in step 2 of the sequential approach, cf. § 3.1 and Figure 2, notably:

- linking pairings into pairing chains having the structure of bid lines for crew members over the planning horizon (cf. points 1 and 2)
- ensuring that constructed pairings fit in a larger extent to the available crew members stationed at different home bases (cf. points 3 and 4)
- taking into account considerations related to crew availability at the beginning of and within the planning period (cf. points 5 and 6)

This defines a new integrated crew scheduling task (cf. integration in Figure 3): A *crew pairing chain problem* (CPCP)—a strict extension of the CPP—*with multiple domiciles and crew availability considerations*. The remaining step is an easier assignment task of pairing chains to individual crew members. Its main feature is that only longer parts of pairing chains (often more than one pairing) are assigned to individual crews. Finally, we propose a certain coupling of the integrated crew scheduling task with the remaining assignment step: When bottlenecks occur within the assignment step, parts of the timetable can be reconsidered by solving the CPCP again.

5   A Flow Model for Maintenance Routing Based on a State-expanded Aggregated Time-space Network

An in-depth presentation of our network flow approach to aircraft/train maintenance routing is provided in this section, as it will be transferred and extended in the next section in order to obtain a network flow approach to crew scheduling. The formal mathematical model is given in Mellouli (1998).

Our solution methodology is based on integrating maintenance requirements within a flow model by extending an appropriate time-space network for building rotations (cf. § 5.2) in order to handle *maintenance states* of vehicles, introduced in § 5.1. A feasible flow represents a vehicle schedule, as a set of rotations, including evenly distributed sessions for regular maintenance (cf. § 2.1 and 2.3) by means of regularly serving *maintenance dummy trips*, being integrated in the resulting state-expanded flow network (cf. § 5.4).

Flows representing "equivalent" vehicle schedules are *aggregated* in order to reduce the network and model size. Therefore, each flow within the developed *state-expanded aggregated time-space network* represents a *class of aggregated* vehicle schedules (cf. § 5.3). Out of an optimal flow computed by a mathematical optimizer, rotations fulfilling maintenance of a concrete vehicle schedule are extracted by an efficient *flow decomposition* algorithm.

5.1 Maintenance States

Let us first consider a simple case of maintenance rule "servicing can only take place at night and a fixed number μ of operation days between two consecutive inspections is required." We define μ *maintenance states* $ms = 1, 2, ..., \mu$ that vehicles may take at points in time. A vehicle taking a maintenance state ms at a certain point in time is within its ms-th day of operation after being serviced. A vehicle just finishing a maintenance session at night receives the maintenance state $ms = 1$ in the following day. Every start of a new calendar day, the maintenance states ms of vehicles are increased by one, from ms to $ms + 1$ (for $ms = 1, ..., \mu - 1$). A vehicle in maintenance state $ms = \mu$ has completed its required duty duration and must be routed to a maintenance station at the end of that day. Finally, servicing a vehicle switches its maintenance state from $ms = \mu$ to $ms = 1$ (reset). Figure 4 shows the cases of a 3-day and 4-day maintenance rule.

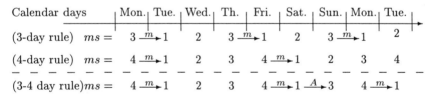

Figure 4. 3-day, 4-day, and 3-4 day maintenance rule, servicing only overnight.

For aircraft maintenance routing, a common rule for regular A-checks is "every three to four (operation) days" (cf. § 2.2). If servicing is allowed only at night, either 3-day or 4-day intervals of operation between consecutive checks may occur. To represent this, the 4-day rule is modified (cf. Figure 4): In addition to the normal transition of maintenance states from

ms to $ms + 1$, an alternative transition from $ms = 1$ to $ms = 3$ is allowed which we call *A-transition* (the increase of maintenance states is *accelerated* from +1 in the normal case to +2). In this way, a shortened 3-day interval of operation becomes possible (see Figure 4). The occurrences of the 3-day operation intervals can be favored, minimized, and controlled in the model by bonuses/penalty costs/bounds on A-transitions.

When servicing can take place also during the day, the adequate counterparts of the rules in Figure 4 are of the form "every approximately μ days" (cf. Figure 5) or "every approximately 3-4 days" (cf. Figure 6).

Observing for the first rule type that there is a little difference in terms of operation times whether a vehicle is serviced at a certain night, the evening before, or the morning after, we partition the planning period in so-called *period sections* $ps = 0, 1, 2, ...$ of one day length each starting at say 14:00 (middle of 4:00 and 24:00). Now a reset of maintenance states by servicing normally takes place *within* the same period section, i.e., before beginning a new period section of operation. For this, a maintenance state 0 is added, getting maintenance states $ms = 0, 1, ..., \mu$. A reset of maintenance states normally within a period section is now done from $ms = \mu$ to $ms = 0$ (maintenance state of vehicle just before and just after servicing, respectively). This leads to the strategy shown in Figure 5 for the cases $\mu = 3$ and $\mu = 4$.

Calendar days		Sun.	Mon.	Tue.	Wed.	Th.	Fri.	Sat.	Sun.
		14:00			*14:00*			*14:00*	
Period sections	$ps =$	0	1	2	3	4	5	6	
(approx. 3 days) $ms =$		2	$3 \xrightarrow{m} 0$	1	2	$3 \xrightarrow{m} 0$	1	2	
(approx. 4 days) $ms =$		3	$4 \xrightarrow{m} 0$	1	2	3	$4 \xrightarrow{m} 0$	1	

Figure 5. "Approximate 3 days/4 days" rule servicing also during the day.

To model the "approximate 3-4 day" rule (requiring two checks every week), the "approximate 3-day" rule is modified to an "approximate 3.5 day" rule by enlarging the length of each period section by 4 hours (a week has 6 period sections, each of 28 hours length). The result is depicted in Figure 6.

Period sections	$ps =$	0	1	2	3	4	5
1.week	$ms =$	2	$3 \xrightarrow{m} 0$	1	2	$3 \xrightarrow{m} 0$	1
2.week	$ms =$	2	$3 \xrightarrow{m} 0$	1	2	$3 \xrightarrow{m} 0$	1

Figure 6. "Approximate 3-4 days" rule, servicing overnight and during the day.

5.2 Aggregated Time-space Networks: A Basic Flow Formulation

A timetable T is a given set of trips (or flights). For each station $k \in TS$, let E^k be the list of all arriving trips of T at station k and S^k be the list of all starting trips of T from station k. Thus, each trip $i \in T$ occurs twice, namely, as an end event in $E^{dest(i)}$ and as a start event in $S^{orig(i)}$.

Preprocessing: For each station $k \in TS$, the trips in the lists E^k and S^k are now chronologically sorted with respect to their *end times* (cf. § 2.1) and their *start times*, respectively. Let us partition E^k and S^k into *end blocks* and *start blocks*, respectively, as in Figure 7, such that end times of E_l^k trips are less than or equal to start times of S_l^k trips which in turn are strictly less than end times of E_{l+1}^k trips. This partition is unique, so let ν_k be the resulting number of end blocks (= number of start blocks) at station $k \in TS$. Each end block $E_l^k (l = 1, ..., \nu_k)$ precedes the corresponding start block. (Only E_1^k and/or $S_{\nu_k}^k$ may be empty for some stations k.)

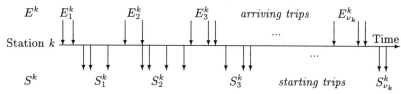

Figure 7. Preprocessing: building end blocks and corresponding start blocks.

For each station $k \in TS$ and $l = 1, ..., \nu_k$, each ending trip of E_l^k is compatible with each of the starting trips of S_l^k. Representing trips by arcs and vehicles by flow units, this compatibility is established through the *connection node* n_l^k, see Figure 8. Observe that each ending trip of E_l^k is also compatible with each starting trip of $S_{l'}^k$ for each $l' > l$. This is established rather implicitly by traversing the connection nodes $n_l^k, n_{l+1}^k, ..., n_{l'}^k$ in-between along horizontal *waiting arcs* following the time axis.

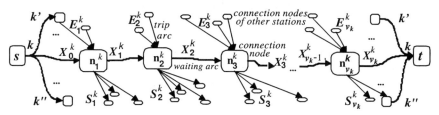

Figure 8. Connection line $CL(k)$ of an aggregated time-space network.

For each station $k \in TS$, let us call the line constituted by the connection nodes together with the waiting arcs a *connection line* $CL(k)$. Thus, a connection line $CL(k)$ regulates all possible trip connections at station k; *direct* connections (for aggregated ending trips and aggregated connection trips) through one connection node, and *indirect* connections through several connection nodes linked by waiting arcs. Here, vehicles, in the same station, waiting at the same time for indirect connections between two connection nodes are aggregated. An *aggregated time-space network* is generally a collection of connection lines which are properly linked with each other.

Network construction: The nodes are source s, sink t, and the connection nodes n_l^k for each $k \in TS$ and $l = 1, ..., \nu_k$. For each $i \in T$, a trip arc emanates from one of the connection nodes of $orig(i)$ and terminates at one of the connection nodes of $dest(i)$. Since each trip is to be served by exactly one vehicle, the flow value on trip arcs is set to 1 for the simple VSP.

E_l^k trip arcs are aggregated to terminate at a *single* node n_l^k while S_l^k trip arcs are aggregated to emanate from n_l^k. Further, for each $k \in TS$ and $l = 0, 1, ..., \nu_k$, a *waiting arc* from n_l^k to n_{l+1}^k of connection line $CL(k)$ is introduced ($n_0^k := s$ and $n_{\nu_k+1}^k := t$). The flow value $X_l^k \geq 0$ on this arc represents the number of vehicles standing (and waiting for indirect connections) at station k at points in time between the l-th and $l + 1$-st start/end blocks. Especially, X_0^k (respectively, $X_{\nu_k}^k$) represents the number of vehicles standing at the start (end) of the period at station k.

The basic flow model for VSPs on the resulting aggregated time-space network is formulated as follows:

$$Mfsz\text{--}sVSP = \text{Minimize} \sum_{k \in TS} X_0^k \quad \left(= \sum_{k \in TS} X_{\nu_k}^k \right)$$

subject to

$$X_{l-1}^k + \left| E_l^k \right| = \left| S_l^k \right| + X_l^k \quad \forall \, k \in TS \quad \forall \, l = 1, \dots, \nu_k \quad\quad (1)$$

$$X_l^k \geq 0 \quad\quad\quad\quad\quad\quad\quad\quad\quad \forall \, k \in TS \quad \forall \, l = 0, 1, \dots, \nu_k \quad\quad (2)$$

The objective function minimizes the *fleet size* = total number of vehicles initially taken from the *pool* s to the stations that then terminate at t. The flow through the network represents active vehicles (on trip arcs) or standing vehicles (on waiting arcs). Constraints (1) model the flow conservation at each connection node n_l^k. For the simple VSP, X_l^k (for $l = 0, 1, ..., \nu_k$) represent local minima of a stair-wise function of the number of standing vehicles at station k with respect to the time axis (X_0^k is a local minimum unless E_1^k is empty). Figure 9 shows this for the case $X_0^k = 3$.

Denoting by δ_l^k the *difference of cumulative starting and cumulative ending trips* at station k until (including trips of) the l-th end and start blocks ($\delta_0^k = 0$), it follows from summing up constraints (1) that $X_0^k = \delta_l^k + X_l^k$ for $l = 0, 1, ..., \nu_k$ (cf. Figure 9). Note that δ_l^k takes its maximum whenever X_l^k takes its global minimum, being equal to 0 for X_0^k being minimized. Thus, $X_0^k = \max_{l=0,1,...,\nu_k} \delta_l^k := \Delta^k$, being the *maximum difference of cumulative starting and cumulative ending trips* at station k. Thus the *minimum fleet size* for the simple VSP is given by:

$$Mfsz\text{--}sVSP = \text{Minimize} \sum_{k \in TS} X_0^k = \sum_{k \in TS} \Delta^k$$

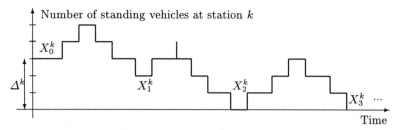

Figure 9. Stair-wise function of numbers of standing vehicles with local minima.

The *optimal flow* for the simple VSP on aggregated time-space networks can then be computed by: $X_l^k = \Delta^k - \delta_l^k \quad \forall\, k \in TS \quad \forall\, l = 0, 1, \ldots, \nu_k.$

5.3 Flow as a Class of Aggregated Schedules – Flow Decomposition

Given a certain flow on the constructed network, the liberties one can take when building a concrete vehicle schedule become clear. Suppose that a computed optimal flow delivers flow values $X_0^k = 3, X_1^k = 2$, and $X_2^k = 0$ at a station $k \in TS$ for ending and starting trips as in Figures 7–9. That is, three vehicles stand at station k at the beginning and two other vehicles enter station k through trips of E_1^k before trips of S_1^k start. At this moment, three arbitrary vehicles from the standing five vehicles may be chosen to serve the trips of S_1^k. The remaining two vehicles together with the two vehicles entering station k through trips of E_2^k are then used to serve the trips of S_2^k.

Two extreme strategies can be used: FIFO—the first vehicle entering the station is used for the next starting trip from that station—and LIFO—the last vehicle entering the station is used. FIFO or LIFO can also be applied to the X_0^k (new) vehicles initially standing at a station k: They can be used for the first starting X_0^k trips (here, for the three trips of S_1^k) or only whenever needed (here, for the last trip of S_1^k and for the last two trips of S_2^k).

Thus, there are several ways of matching trips within a route, beginning new routes, or linking routes into rotations, ranging from a FIFO to a LIFO strategy for both cases while satisfying a given (optimal) flow. This applies to each of the terminal stations of given trips. Thus, each single flow implicitly represents a large *class of aggregated vehicle schedules* of same fleet size. This constitutes a main property of flow models on aggregated time-space networks which we utilize to obtain a two-step solution methodology:

1. Compute an optimal flow representing a class of aggregated schedules
2. Decompose this flow into rotations of a concrete optimal vehicle schedule

These two steps apply to the case of VSPs without and with maintenance routing and later to crew scheduling. The first step is performed by an efficient computation for the simple VSP (see § 5.2) and by solving a mixed-integer mathematical model for the case of difficult extensions such as maintenance routing as well as for the case of crew scheduling.

The second step is always carried out by an efficient flow decomposition algorithm. This algorithm is presented in Mellouli (1998) for the case of vehicle scheduling with maintenance routing and is similar for crew scheduling. The principle idea is to generalize the proceeding for the above example. After some preprocessing, each connection line of the network is processed. A connection line will generally include some ending and starting activities (scheduled by the computed solution of the mathematical model of step 1). These activities are linked according to a chosen strategy, e.g., FIFO.

5.4   Extension of the Flow Model to Maintenance Routing

According to a given maintenance strategy (cf. § 5.1), here as in Figure 6 for the case study of German Rail's intercity trains, we now integrate maintenance states and their handling within the flow model of § 5.2 on aggregated time-space networks. Servicing is possible at bases near to some stations $TS_M \subset TS$. A vehicle entering a maintenance station $k \in TS_M$ at a point in time t can be serviced and returned to station k at *maintenance-finish-time(k,t)*. The main ideas of the model extension are summarized in the following three points (cf. Figure 10):

(a) View vehicles with different maintenance states as different commodities, which flow separately within the network *as long as* vehicles remain within the same period section and without interim maintenance. For this, nodes and arcs are duplicated as in Figure 10 (a), resulting in connection lines $CL(k, ps, ms)$ at different maintenance state levels *(ms-levels)* 0, 1, 2, 3, separately for each station $k \in TS$ and period section $ps = 0, 1, 2, ...$ (The basic model had only one connection line $CL(k)$ for each station k, cf. Figure 8.)

(b) Whenever a new period section begins, increment maintenance states of vehicles by diverting arcs. For this, waiting arcs between nodes of successive period sections are diverted to the next higher ms-level for $ms = 0, 1, 2$, cf. Figure 10 (b). This concerns each change of period sections (or *ps-change*). Analogously trip arcs are diverted for *ps-change trips*, trips over a ps-change. When periodicity is required a period change is a special case of ps-change.

(c) Whenever a vehicle enters a maintenance station $k \in TS_M$ with maintenance state 3 after serving a trip, its state can be reset to 0 after servicing at the maintenance base close to k. For each trip $i \in T$ ending at a maintenance station, $dest(i) \in TS_M$, we associate *a maintenance dummy trip* m_i with $start\text{-}time(m_i) := end\text{-}time(i)$ and $end\text{-}time(m_i) := maintenance\text{-}finish\text{-}time(dest(i), end\text{-}time(i))$, cf. Figure 10 (c). In order to reset the maintenance state of a vehicle (from $ms = 3$ to $ms = 0$) after accomplishing a trip i, the arc associated with m_i emanates from a connection node at ms-level 3 and terminates at a connection node at ms-level 0.

For point (a), note that the arc where a vehicle 'flows' determines the maintenance state of that vehicle either when serving a trip (for a trip arc) or when standing at a station (for a waiting arc). Separate flow balance constraints for nodes at different ms-levels 0, 1, 2, 3 ensure a correct transmission of the maintenance state information of vehicles. A flow value on a trip arc,

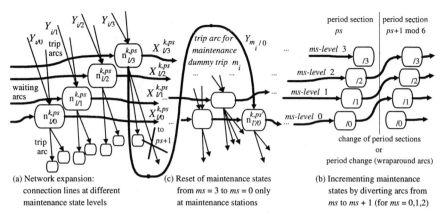

Figure 10. Flow model on a state-expanded aggregated time-space network.

$Y_{i/ms}$, equal to 1 for a state ms means that the vehicle serving trip i ends at station $k = dest(i)$ in its ms-th period section after the period section of latter servicing. Declaring $Y_{i/ms}$ as 0/1-variables, cover constraints in the model will ensure that the sum of these variables for a fixed trip i is equal to 1, because each trip must be served by exactly one vehicle. Unlike the case of the basic model of § 5.2, these constraints involving 0/1-variables (called special ordered sets), are not of flow balance or flow bound type and primarily make the resulting mixed-integer model computationally difficult.

For each trip $i \in T$ and maintenance state $ms = 0, 1, 2, 3$, we associate a so-called *p-activity* denoted by i/ms as the *possible activity* of serving trip i by a vehicle having maintenance state ms at *end-time(i)*. A 0/1-variable $Y_{i/ms}$ is associated to each p-activity i/ms. Because of cover constraints for each trip i, $Y_{i/ms}$ is set to 1 for a unique ms in the model solution, and in this case, the p-activity i/ms defines a (real) *activity* of a vehicle (within the computed vehicle schedule), which serves trip i and ends with maintenance state ms. Similarly, a maintenance dummy trip m_i (defined in point (c)) associated to each trip $i \in T$ with $dest(i) \in TS_M$ is viewed as a p-activity $m_i/0$ (ending at ms-level 0). If the associated 0/1-variable $Y_{m_i/0}$ is equal to 1 in the model solution, a vehicle is directly driven to the maintenance base close to $dest(i)$ after having served trip i at ms-level 3 (activity $i/3$).

This dependency in serving p-activities $i/3$ and $m_i/0$, for a fixed trip i, is exploited in the model to avoid splitting the connection node at ms-level 3, at which the trip arc associated to $i/3$ terminates and from which the maintenance dummy trip arc associated to $m_i/0$ emanates (cf. Figure 10). For this, we add *logical constraints* which ensure that the flow $Y_{m_i/0}$ on m_i arc *can only* take value 1 *if* the flow $Y_{i/3}$ takes value 1. This is achieved by constraints of the form $Y_{i/3} \geq Y_{m_i/0}$ for each trip i with $dest(i) \in TS_M$. At ms-level 0, a node is split whenever *end-time(m_i)* occurs within a start block.

The mathematical model is an extension of the basic formulation of § 5.2. The sum of flows X_0^k over all stations k minimized in the objective function is now taken over all possible maintenance states ms as well (sum of $X_{l/ms}^{k,ps}$ for $ps = 0$ and $l = 0$), further, the number of period change trips and the sum of flows on period change maintenance dummy trips are added. Altogether the flow value at period start ($=$ fleet size) is minimized here. Penalty costs are set on $Y_{m_i/0}$ with unfavorable servicing times. Besides cover and logical constraints, flow conservation at each connection node is stated.

The following steps build these network nodes together with sets of incoming and outgoing p-activity arcs:

1. For each connection line $CL(k, ps, ms)$, define the sets of all ending and of all starting p-activities j/ms (j is a trip or maintenance dummy trip).
2. Perform the preprocessing steps in § 5.2 in order to partition these sets into end and start blocks (cf. Figure 7) of p-activities. A connection node is associated to each end and corresponding start block (cf. Figure 8). Arcs for p-activities in the end block (start block) are incoming (outgoing).

Linking connection lines: Connection nodes are linked by waiting arcs (cf. Figure 8). Waiting arcs at the extremities (first and last waiting arcs) of each connection line regulate how the different connection lines are linked with each other. For the simple VSP in Figure 8, the first waiting arcs of connection lines simply emanate from a super source and the last waiting arcs terminate at a sink. For the simple VSP with periodicity, the last and first waiting arcs of each connection line collapse, building *wraparound arcs*. For maintenance routing, cf. point (b) and Figure 10 (b) above, the last waiting arc of each connection line $CL(k, ps, ms)$ with $ms = 0, 1, 2$ is diverted to the next higher ms-level for a ps-change, i.e., it collapses with the first waiting arc of $CL(k, ps+1 \bmod 6, ms+1)$, respectively ("mod 6" is inserted to ensure periodicity of schedules). A collapse of two arcs is a simple constraint in the model stating that the flow values on these two arcs (X-variables) are equal. The first waiting arc of connection line $CL(k, ps, 0)$ and the last waiting arc of connection line $CL(k, ps, 3)$ are omitted. This structure justifies that all X-variables can be declared as continuous variables (cf. Mellouli (1998)).

6   Network Flow Approach to Crew Scheduling

In § 6.1 the problem characteristics motivated by a concrete airline application are given. Using the analogy shown in Section 4, the maintenance routing model of Section 5 is transferred to the airline crew scheduling problem in § 6.2, and in § 6.3, base and crew considerations are incorporated into the resulting network flow framework. Discussing aspects related to crew scheduling with multiple domiciles, specific workrules, and airline's preferences, our currently realized, slightly modified network flow model for an integrated crew scheduling process is presented in § 6.4.

6.1 Problem Characteristics and Case Study

Our work on crew scheduling is guided by a real application of a European airline having about ten home bases with different sizes, where crew personnel (cockpit and cabin attendants) of various categories are stationed. So we investigate the crew scheduling problem with multiple crew domiciles. In the case study, each cockpit-crew member works only on one specified aircraft type. We first handle the case of scheduling cockpit-crews, pilots or co-pilots (first officer), and note in the conclusion (Section 8) on extensions of our approach to schedule cabin-crews who can work on different aircraft types and to dynamically obtain "crew teams" in a multiple-domicile setting.

At the partner airline crew schedules are currently built manually on a semi-monthly basis following the classical sequential approach illustrated in § 3.1. Crew requests and activities scheduled previously are considered in a second assignment step. The network flow model developed in this section handles their crew scheduling problem in an integrated form (CPCP with multiple domiciles and crew availability considerations) as proposed in Section 4. We experiment with semi-monthly as well as monthly planning periods and handle workrules mentioned in § 3.1 corresponding to the airline's rules. The main objective is to minimize the total operational costs, notably, costs of crew proceedings as passengers and of overnight stays at hotels together with expenses paid to crews for times spent outside home base.

6.2 Network Design by Analogy to Maintenance Routing

Let us require for the moment that each crew member serves only one, possibly compound, flight in each duty period. (Consecutive flights of an aircraft rotation can be linked into compound flights in a preprocessing step, in order to disallow overnight stays at certain cities specified by the airline.) For each flight f, set $start\text{-}time(f) := start\text{-}time(f) - check\text{-}in\text{-}time(orig(f))$ and $end\text{-}time(f) := arrival\text{-}time(f) + check\text{-}out\text{-}time(dest(f)) + daily\text{-}rest\text{-}after(f)$. Check-in and check-out times are needed by crews before and after a flight duty, e.g., 60 and 15-25 min, respectively. In analogy to end-times of trips incorporating turn-times for vehicles, cf. § 2.1, the daily rest time is added to the end-time of a duty, which corresponds to the point in time where crews are ready to serve a next duty. The workrule stipulating that daily rest time is dependent of the preceding duty duration can be handled in this way.

Applying a procedure for the simple VSP to this set of flights with the defined end-times, one generates a minimal number of rotations, each consisting of a duty chain properly spaced by daily rests. Now, how to avoid obtaining duty sub-chains of more than 5 days length (infeasible pairings) and how to enforce weekly rests at crew domicile between the pairings?

Now the full analogy illustrated in Section 4 is utilized in order to apply the network flow model of Section 5 to crews. The analogue of "regular servicing every 3-4 days at maintenance bases" for vehicles is "regular weekly rest every at most 5 calendar days at domicile" for crews. Thus a *vehicle rotation* incorporating 4-hour checks according to this requirement corresponds to a *pairing chain* for crews with 2-day weekly rests between pairings.

We define *crew states* $cs := 1,2,3,4,5$ instead of maintenance states, cf. § 5.1. A crew member having state cs at a certain point in time is in its cs-th day in operation after the latter weekly rest, i.e., cs = number of days within the current pairing. A transition from cs to $cs + 1$ is done every night (day change) for $cs := 1,2,3,4$. For the moment, connection lines $CL(b, cd, ms)$ are needed (like $CL(k, ps, ms)$ used for maintenance routing in § 5.4), for each (overnight) base b where crews can have a daily rest, each calendar day cd of the planning period, and each crew state $cs := 1,2,3,4,5$.

As pairings may have 1 up to 5 days length, a reset of crew states (terminating a pairing) can be done from each $cs = 1,2,3,4,5$ to $cs = 1$. A reset of crew states ensures weekly rest by means of serving an *off-duty dummy trip or flight* (analogue of a maintenance dummy trip, cf. § 5.4) starting and ending at the same home base. In order to get weekly rests of two consecutive calendar days (preferred type), off-duty arcs aggregating dummy flight arcs start at each day cd at say 12:00 from connection lines of home bases with $cs = 2,3,4,5$ and terminate at 0:00 of day $cd + 2$. As end-time of flights already incorporates a daily rest, 12:00 is used instead of 0:00 at day cd.

At this moment, at least one complication occurs. Unlike aircraft/trains that can be serviced in different maintenance bases, crews do not spend their weekly rest at *any* of the home bases, but only at their corresponding one. To ensure this, the flow network is duplicated for each hb such that crews of different home bases are not merged. Thus the complete network comprises different network layers, each layer includes the connection lines $CL(hb, b, cd, cs)$ for one hb. In this way flows on connection lines aggregate only those crew members having the same home base and being at a certain point in time with the same crew state in the same airport. This defines a multi-commodity state-expanded aggregated time-space flow network.

In order to link the connection lines within the network, the last waiting arc of $CL(hb, b, cd, cs)$, for $cs = 0,1,2,3,4$, is diverted to the next higher crew state level (cs-level) and collapses with the first waiting arc of $CL(hb, b, cd + 1, cs + 1)$, as shown in Figure 10 for the maintenance routing problem. For the case $hb \neq b$, hotel costs are assigned to the flow (X-type) variable on these overnight waiting arcs. The first waiting arc of connection line $CL(hb, b, cd, 1)$ and the last waiting arc of connection line $CL(hb, b, cd, 5)$ are omitted.

To each flight (duty) f, p-activities $f/hb, cs$ for each hb and cs are associated, indicating the possibility of serving flight f by a crew member stationed at home base hb with crew state cs, i.e., starting the duty at his/her cs-th day within a pairing. For each of these p-activities an arc is associated as in § 5.4 with 0/1-variable as flow value. Owing to the added daily rest time to end-times of duties, $day(start\text{-}time(f)) \neq day(end\text{-}time(f))$ often holds and the associated arc is diverted to the next higher cs-level. In this case, hotel costs are associated to the flow on this arc for the case $dest(f) \neq hb$.

Cover constraints as in § 5.4 set equal to 1 the sum of flows on arcs for all p-activities $f/hb, cs$ for each fixed flight f, over all crew states cs and home bases hb. Only one of these p-activities for f becomes a real activity by "exclusively receiving" the flow value 1 in the model solution.

To extend our model to duties of several flights, all possible duties are built—their number is only up to 3-4 times that of flights—then considered for p-activities instead of flights. The sum of flows in a cover constraint for flight f now includes p-activities $fd/hb, cs$ not only of all crew states cs and home bases hb, but also over all duties fd containing the flight f.

Returning to the off-duty arcs of two days (cd and $cd + 1$), introduced above, they also correspond to p-activities $OFF(cd)/hb$, indicating the possibility that one or several crew members of home base hb begin a weekly rest at day cd. Their number is equal to the integer flow value associated to the $OFF(cd)/hb$-arc, emanating from a node in $CL(hb, hb, cd, cs)$, $cs = 2, 3, 4, 5$, and terminating at the first node of $CL(hb, hb, cd + 2, 1)$—both are connection lines with $hb = b$. In this way, pairings during up to 4 days can be constructed. To get pairings of 5-day length, flight arcs with $cs = 5$ for p-activity $f/hb, 5$ are simply delayed according to weekly rest duration and terminate in a connection line with $b = hb$. Here costs of proceeding the crew member from $dest(f)$ to hb are associated ($=0$, if $dest(f) = hb$).

In this way proceedings after a 5-day pairing are modeled. To model crew proceedings at the start of pairings, p-activities $f/hb, 1$ at cs-level 1 are assigned costs of proceeding crews from hb to $orig(f)$ ($=0$, if $hb = orig(f)$). We also integrate so-called IP arcs (indirect proceedings) as p-activities, aggregating potential proceedings after duties at different cs-levels, in order to proceed crews from any overnight base to the corresponding home base. IP arcs model the opportunity to proceed crews within pairings in order to obtain overnight stays at home bases. These IP arcs can be used together with off-duty arcs to terminate pairings of up to 4 days length with proceedings as well. Since IP arcs are used after end-time of duties, the modeled proceedings are linked to the end of those duties as *direct* proceedings in a post-processing phase, unless workrules are violated. In the latter rare case proceedings remain *indirect*, i.e., after an overnight stay.

Finally, a hidden problem remains: If a crew member accomplishes its weekly rest and does not directly start a pairing, its crew state will be unnecessarily increased within the network. To avoid this, we introduce a second type of off-duty arcs emanating from the first node of connection line $CL(hb, hb, cd, 1)$ and terminating at the first node of $CL(hb, hb, cd + 1, 1)$. The integration of these *extra-off arcs* associated with an integer flow value does not only solve the problem but enhance the model as well: Favoring the usage of extra-off arcs by associating bonuses, the optimizer tries to include as much as possible joined off-duty days of more than 2 days (off-duty followed by extra-offs), thus decreasing the number of *layover days* within pairings possibly outside domiciles, which are not preferred by crews.

6.3   Flow Model with Crew Considerations and Base Constraints

Recall that each connection line $CL(hb, b, cd, cs)$ contains a certain number of connection nodes which are linked by waiting arcs (as in Figure 8). In order to construct the network, the nodes and waiting arcs of connection lines are built by performing the preprocessing steps indicated in § 5.2 as follows: The lists $E(hb, b, cd, cs)$ and $S(hb, b, cd, cs)$ of all ending/starting p-activities of connection line $CL(hb, b, cd, cs)$ are constructed, then chronologically sorted with respect to end-times and start-times, respectively, and finally partitioned into end blocks and corresponding start blocks as shown in Figure 7. Each connection line looks like $CL(k)$ depicted in Figure 8, from which the nodes and the corresponding inflow=outflow constraints are deduced.

After solving the flow model by a mathematical optimizer, the flow on each network layer for hb can be decomposed in several ways into paths (cf. § 5.3), each corresponding to a pairing chain for a hb crew member. Including crew considerations (and base constraints) into our network flow model for crew scheduling aims at enforcing some conditions to the flow solution such that a decomposition exists from which crew schedules can be constructed with minimum additional costs (by the remaining assignment step of Figure 3). To obtain this, the following refinements of the model are performed.

Availability of crews at period start: For each crew member being of domicile hb, ready to serve new flights of the current period at *ready-time*, in its cs-th day within a pairing at *ready-time*, and finishing from serving flights of the previous period at airport b (possibly $\neq hb$), we associate a so-called *crew arc* with flow value fixed to 1 entering the connection line $CL(hb, b, day(ready\text{-}time), cs)$ at *ready-time*. Thus, especially, pairings starting in the previous period can be completed with flight duties of the current period.

The number of pairing chains is equal to the number of crew members for each home base: This main requirement of the integrated crew scheduling process proposed in Section 4 (cf. Figure 3) is guaranteed by generating a flow of value 1 to each crew member as done above. In this way the value of the flow circulating within the network layer for each home base hb is equal to the number of crew members on- or off-duty with domicile hb.

Scheduled crew activities and crew requests constitute another source for new p-activities and arcs in the model network. Scheduled office or simulator activities of hb crews are handled like flight duties, which are to be served only by hb crews. Thus for each scheduled office and simulator activity sa, p-activities sa/cs for each crew state cs are generated only for one hb. The sum of flows on the corresponding arcs is set to 1 in order to serve sa.

Crew requests: For each request of consecutive off-duty or vacation days from day cd to day cd' ($> cd + 1$) for a hb crew, a p-activity arc emanating from the first node of $CL(hb, hb, cd + 2, 1)$ and terminating at the first node of $CL(hb, hb, cd' + 1, 1)$ is added—a preceding weekly rest of two days is guaranteed by the model. If the requested off-duty days have to be served, the flow value on corresponding arc is set to 1, otherwise a user-defined bonus is associated to the flow value. Finally, a request of flight f for a hb crew is modeled either by setting a bonus on flow values of duty p-activities $fd/hb, cs$ (for fd containing f) or by deleting all $fd/hb', cs$ for $hb' \neq hb$.

Distributing flight hours evenly among home bases: The sum of flow values on duty p-activities related to a hb, weighted by the flight minutes of duties, is restricted to be $\geq$ a given minimum and $\leq$ a given maximum of flight hours for hb. The minima and maxima for these *base constraints* depend on the number of crews at each home base, their current account of flight minutes, and their scheduled activities or vacation days.

6.4   The Realized Network Flow Model for Crew Scheduling

The network flow approach currently realized is a refined and simplified version of the framework given in the latter two subsections. It takes care of both airline's specific workrules and preferences as well as computational aspects for problems with several domiciles and crew considerations.

We presented a network flow approach to crew scheduling which deals with typical workrules of airline's crews—the majority of rules in our application—namely, those for constructing feasible duties and those stipulating that the daily rest time within pairings is dependent from the preceding duty and its excess hours and that pairing lengths do not exceed five calendar days.

In order to construct a selection of pairing chains (not only of pairings) simultaneously by a flow, the number of days within pairings is recorded by introducing five crew states. A state reset ensures weekly rests between pairings. If one is only concerned about the recognition or generation of feasible pairings, one can use a set of a few states whose state transition graph is cyclic (as known in automata theory) and break the search after constructing a 5-days pairing. This cyclic transition graph may, however, generate long duty chains (infeasible pairings) in a solution within a flow framework.

For problems with one domicile, states can be precisely differentiated for different days of pairings. For our application with several domiciles, not only the number of states but also the number of home bases is relevant for the model size, which still increases only linearly in terms of the number of flight duties. For this type of problems, we decided to experiment with gradually more complex models in order to get a feeling of the computational bottlenecks of typical airline's problems in our application. The type of "complex rules" for pairings/pairing chains which are not integrated deals rather with exceptions and restricts the number and hours of duty excesses as well as the number and succession of night flights within 7-day blocks (cf. § 3.1).

The first model integrates network layers for several domiciles hb with base constraints for flight hours' distribution. Instead of crew states, off-duty constraints are added, enforcing for each hb that the sum of flows on off-duty arcs of connection lines with $b = hb$ starting within *each* 7-days span is $\geq$ the number of crew members stationed at hb. For 350 flights and 6 domiciles, problems were solved in less than one minute. Surprisingly, only a few duty chains were longer than five days in the solution. The observed common property (*) is: They are the main cause for violating complex rules and all included overnight stays are outside domicile. By adding some problem-dependent constraints, nearly optimal solutions could be produced.

The second step was to use the five crew states to systematically avoid getting infeasible duty chains. Here, the user can disallow duty excesses and night flights for certain crew states (or days within pairings), getting a good quality class of pairings. Such restrictions practically do not affect optimality.

However, the partner airline was satisfied with the produced solutions, since they evenly distribute flight hours and duties among home bases, and, by extra constraints, over the period as well. (This property is not guaranteed when only solving the CPP.) Our partners rather expressed an additional wish: Crews prefer not to change the hotel each night within a pairing. We integrated this requirement together with the complex rules without exploding the number of states by still utilizing flow aggregation.

Let us call an *atomic pairing* a feasible pairing without overnight stays at home base. Owing to property (*), we are convinced that care is to be taken only for atomic pairings of 1 to 5 days length to be feasible. These are automatically linked into pairings and pairing chains *properly* owing to the powerful off-duty constraints together with extended/refined base constraints. The latter distribute flight minutes, and thus duties, not only among different domiciles hb, but also among different sections of the planning period.

In order to avoid the generation of all atomic pairings, we utilized flow aggregation to link, over connection lines with $b \neq hb$, previously generated pairing prefixes (of up to 3 days length) starting at hb and pairing suffixes (of up to 2 days length) ending at hb. The generation procedure takes account of the "same hotel" wish (for pairing prefixes) and of complex rules as well.

The analogue of duty p-activity becomes a pairing prefix or a pairing suffix related to a home base hb. The arc associated to a prefix emanates from, and that of a suffix terminates at, a connection line with $hb = b$. Suffix arcs can only emanate from a connection line with $hb \neq b$, being possible *connections* to prefix arcs terminating there. To avoid the sum of flows of cover constraints to get longer, that harden the mathematical models computationally, we introduced crew states within prefixes and suffixes as follows.

Pairing prefixes/suffixes beginning with the same flight can be represented in a *tree form*, as shown in the middle of Figure 11 (see p. 119). This recognized *structure sharing* of pairings within the *flight/duty trees* is utilized in order to replace a p-activity arc by a path of smaller p-activity arcs where flights (or duties) are considered instead of pairing prefixes/suffixes. This constitutes a method for *aggregating pairing flows*, which is enhanced by special techniques (being not handled here). We get a similar structure of connection levels with different crew states as in the model of § 6.2. Crew states are introduced whenever they are needed. That is, it may happen that a duty with excess hours or night flight does not appear in each of the possible crew state levels.

7   Computational Results and Model Refinements

The developed models for maintenance routing and crew scheduling (cf. § 5.4
and § 6.4) have been implemented and tested with real-life data. For German
Rail's test problems, 1,496 trips are to be served repeatedly every week by a
minimum number of intercity trains requiring servicing every 3-4 days. For
a European airline, we tested with semi-monthly/monthly periods: 350/650
flights are to be served by a team of pilots stationed at 6 home bases.

All results in Table 1 are obtained on an AMD Athlon at 1.2 GHz with 256
MB RAM. Model name, numbers of trips/flights, terminal stations/overnight
bases, and maintenance/home bases are indicated together with numbers of
columns/rows/nonzeros of the model, as well as computing times (min:sec)
for finding optimal LP, then best IP solution, and IP total time. The optimizer
MOPS (Suhl (1994)) is used to solve the generated mixed-IP models.

Model	$|T|$	ts/b	m/hb	Cols	Rows	NonZs	LP t.	bestIP	IP t.t.
MR0	656	38	8	5,573	3,405	17,317	0:25	0:00	no gap
MR1	1,496	57	8	11,947	7,018	36,602	4:31	0:34	3:43
CS1	300	10	6	6,662	3,321	22,032	0:09	1:44	1:45
CS2	342	10	6	8,198	3,900	27,632	0:11	1:00	1:01
CS1-2	642	10	6	14,905	7,309	51,060	1:05	23:29	23:31

Table 1. Computational results – MR: Maintenance Routing, CS: Crew Scheduling.

Although the timetable size of CS test problems is relatively smaller than
that of MR problems, the complexity of CS problems becomes apparent:
They need more solution time in the IP Branch-and-Bound phase, not only
depending upon $|T|$, but also upon the distribution of crew requests. Still a
nice feature of our aggregated flow models is the observed small gap between
LP and IP optimal solutions. For the semi-monthly periods required by the
airline, the CS problems 1 and 2 are solved within a few minutes. A test with
the combined monthly period indicates the robustness of our model and an
increase in computing time (first good IP solution found within 10 min).

In order to recognize situations when the MR problems are not feasible
in trivial ways, our model handles the maintenance rule "3-4 days" as a *soft
constraint*, allowing the operation time interval to fall below 3 days (by using
an A-transition, cf. § 5.1) or to exceed 4 days by an analogous B-transition
from maintenance state $ms = 1$ to $ms = 1$ with higher penalty costs. With the
latter B-transition even rotations without checks ($ms = 1$ over the period)
could be generated and the model becomes feasible with any timetable and
set of maintenance bases. Violations of the rules can be localized by the model
solution and analyzed by planners as shown in Mellouli (1998).

Goal programming can be used for this kind of problem to allow under- and over-covering of flights at certain penalty costs, as applied in Darby-Dowman and Mitra (1985) to the set partitioning model. We avoided this technique as the computationally difficult cover constraints become longer. We rather integrated an *extra flow*—with associated fix costs—circulating within our crew scheduling network in addition to the "limited flow" induced by the given crews. This extra flow representing "overflow crews" would cover "bottleneck flights," which though did not occur in our test cases.

We used goal programming in another context, namely in distributing flight times among home bases and different sections of the planning period. We allow that amounts of flight minutes fall below/exceed the specified minima/maxima at appropriately chosen penalty costs. This technique turned out to be useful in recognizing good quality crew schedules, saving proceedings and overnight costs with only minimal shift of flight hours.

8   Conclusion – Prototype and Model Extensions

A novel network flow approach to crew scheduling based on a state-expanded aggregated time-space network with application to airlines is developed utilizing a recognized analogy to maintenance routing problems. Our method differs from existing approaches based on the classical set partitioning model and handles an integrated planning process, where all flights are covered by a collection of pairing chains partitioned for crews stationed at different home bases. Time-dependent crew availability/requests are taken into account.

The developed model is integrated into a decision support system for crew scheduling (cf. Figure 11). A graphically interactive user interface allows for preprocessing flights in aircraft rotations, flight connections, as well as generated duties and pairing prefixes/suffixes. The latter are represented as flight/duty trees (structure sharing) allowing a clear view to planners. After solving the model, pairing chains, partitioned among home bases, are extracted from the solution and represented as specified by the airline.

Our model can be extended for scheduling cabin-crews who can work on different aircraft-types. Here, flights f of different aircraft-sizes requires different crew-sizes csz_f. In our flow models, the right hand sides of cover constraints for flights f are to be set to csz_f instead of 1, respectively.

A leading principle of our airline partners is to *"keep crew teams together as much as possible."* Both cockpit and cabin-crews prefer not to change the team within a pairing for one, let alone, for several times. In order to integrate this aim, we proposed a *hierarchical* proceeding, which is approved by our partners. The idea is to firstly schedule pilots, then to try to prefer pairings, atomic pairings, or duty chains used by pilots as (parts of) pairings for co-pilots and then for CDC crews (chef-de-cabin). The commonly used pairing parts by Cockpit/CDC are then given as input or favored when scheduling cabin attendants. By hierarchical crew scheduling, free capacities of CDC crews can be used when scheduling cabin attendants.

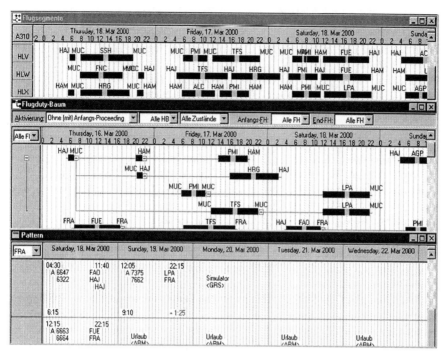

Figure 11. A decision support system for crew scheduling.

In Carl and Gesing (2000) a coupling of the flight planning process with scheduling and rescheduling of resources is described. Integrating planning tasks for aircraft and crews, as proposed in Section 4 and supported by the models in Section 5 and Section 6, is very useful to enhance this coupling. Our crew scheduling model involves scheduled activities and requested off-duty days/vacations for crews. For short term crew rescheduling, absence of crews due to illness can be integrated as scheduled off-duty days, and our model can be applied to parts of the flight set obtaining a modified schedule. Furthermore, our refined model (cf. Section 7) recognizes "bottleneck flights" which may be considered for flight rescheduling or cancellation. Using the analogy of Section 4, this applies to unscheduled aircraft-downtimes due to technical reasons as well.

Finally, *middle* and *long term planning* tasks related to crew scheduling can be approached. Questions such as "where to optimally station new crew members" or "how to find the optimal airline's crew size with a best crew partition among home bases" can be answered by solving our flow models: The number of (new) crew members needed at a home base hb is equal to the value of (extra) flow (cf. Section 7) circulating within hb network layer. Applying our model to different sets of candidate home bases and timetables delivers valuable what-if-analyses for long term home base planning tasks. By analogy this applies to maintenance base planning as well.

Bibliography

Bodin, L. and B. Golden (1981). Classification in vehicle routing and scheduling. *Networks 11*, 97–108.

Borndörfer, R. (1998). *Aspects of Set Packing, Partitioning, and Covering.* Shaker, Aachen. PhD thesis.

Carl, G. and T. Gesing (2000). Flugplanung als Instrument des Informationsmanagements zur Ressourcenplanung und -steuerung einer Linienfluggesellschaft. In J.R. Daduna and S. Voß (Eds.), *Informationsmanagement im Verkehr*, Physica, Heidelberg, 167–198.

Darby-Dowman, K. and G. Mitra (1985). An extension of set partitioning with application to crew scheduling. *European Journal of Operational Research 72*, 312–322.

Desrosiers, J., Y. Dumas, M.M. Solomon, and F. Soumis (1995). Time constrained routing and scheduling. In M.O. Ball, T.L. Magnanti, C.L. Monma, and G.L. Nemhauser (Eds.), *Network Routing, Handbooks in Operations Research and Management Science*, 8, Elsevier, Amsterdam, 35–139.

El-Darzi, E. and G. Mitra (1992). Solution of set-covering and set-partitioning problems using assignment relaxations. *Journal of the Operational Research Society 43*, 483–493.

El-Darzi, E. and G. Mitra (1995). Graph theoretic relaxations of set covering and set partitioning problems. *European Journal of Operational Research 87*, 109–121.

Feo, T.A. and J.F. Bard (1989). Flight scheduling and maintenance base planning. *Management Science 35*, 1415–1432.

Garfinkel, R.S. and G.L. Nemhauser (1969). Set partitioning problem: Set covering with equality constraints. *Operations Research 17*, 848–856.

Hoffman, K.L. and M.W. Padberg (1993). Solving airline crew scheduling problems by branch-and-cut. *Management Science 39*, 657–682.

Lavoie, S., M. Minoux, and E. Odier (1988). A new approach to crew pairing problems by colomn generation with application to air transportation. *European Journal of Operational Research 35*, 45–58.

Löbel, A. (1998). *Optimal Vehicle Scheduling in Public Transit.* Shaker, Aachen. PhD thesis.

Mellouli, T. (1998). Periodic maintenance routing of German rail's intercity trains by a flow model based on a state-expanded aggregated time-space network. In *6th Meeting of the EURO WG on Transportation, Gothenburg, Sweden, September 9-11, 1998.* To appear in Transportation Research.

Suhl, L. (1995). *Computer-Aided Scheduling: An Airline Perspective.* Deutscher Universitäts-Verlag, Wiesbaden.

Suhl, L. and T. Mellouli (1999). Requirements for, and design of, an operations control system for railways. In N.H.M. Wilson (Ed.), *Computer-Aided Transit Scheduling, Lecture Notes in Economics and Mathematical Systems*, 471, Springer, Berlin, 371–390.

Suhl, U. (1994). MOPS: A mathematical optimization system. *European Journal of Operational Research 72*, 312–322.

Tabu Search for Driver Scheduling

Yindong Shen and Raymond S.K. Kwan

School of Computing, University of Leeds, Leeds, LS2 9JT, UK
{yindong,rsk}@comp.leeds.ac.uk

Abstract. This paper presents a Tabu Search heuristic for driver scheduling problems, which are known to be NP-hard. Multi-neighbourhoods and an appropriate memory scheme, which are essential elements of Tabu Search, have been designed and tailored for the driver scheduling problem. Alternative designs have been tested and compared with best known solutions drawn from real-life data sets. The algorithm is very fast, has achieved results comparable to those based on mathematical programming approaches, and has many potentials for future developments.

1 Introduction

The bus and train driver scheduling problem presented in this paper is the process of partitioning a predetermined bus or train vehicle work into a set of driver duties, i.e., a driver schedule. The number of possible combinations for partitioning the vehicle work is usually astronomical. There is a set of restrictions on efficient provision of driver schedules. For example, daily working time for a driver has to be limited to a certain number of hours. These restrictions are called *labour agreement rules*, which vary a great deal between different operators. The most common rules for bus drivers are listed in Smith and Wren (1988). These rules governing the legality of a driver duty affect profoundly the complexity of duty compilation. The criteria are usually that the schedule must be legal according to the rules and should have the minimum number of duties and lowest total hours of work.

Using computers to solve the driver scheduling problem started in the 1960's. Since then a great deal of progress has been made and many approaches have been developed and reported in this series of conferences (see, e.g., Wilson (1999)). The earliest approaches are heuristics-based, which were heavily reliant on domain knowledge, such that the algorithms were complex, not readily portable to different transport operators, and frequently failed to escape from local optima.

By the 1980's it was generally recognised that heuristics alone were not suitable for general use (see Wren and Rousseau (1995)). The purely heuristic approach was, therefore, abandoned in favour of mathematical programming approaches aided by heuristics. The principle mathematical programming approach, e.g., TRACS II (see Kwan et al. (2000); Fores et al. (1999); Kwan et al. (1999)), first generates all the valid potential driver duties according to labour agreement rules, then if necessary reduces the very large potential duty set using heuristics, and lastly uses a set covering formulation to select a cheapest subset of duties that covers each piece of vehicle work. Although the application of column generation techniques (see Fores et al. (1999); Rousseau and Desrosiers (1995)) enables mathematical programming solvers to handle larger problem instances, mathematically-based systems still cannot be seen as black boxes that solve a large problem in one go, and problem subdivision inevitably leads to loss of optimality in the recombined solution. Integer solution techniques such as Branch-and-Bound are limited by the amount of search space that could be explored within reasonable time. For this reason Integer Linear Programming (ILP) solution processes, e.g., may sometimes be terminated before any integer solution could be found. In recent years, meta-heuristics (see Aarts and Lenstra (1997); Reeves (1993)) have been widely and successfully used for seeking practical near-optimal solutions to NP-hard problems. The main advantages of meta-heuristics are that they are usually very efficient in searching through very large solution spaces, always returning a solution, and each class of meta-heuristics has a methodical and strategic structure that is problem domain independent.

This paper describes research into the tabu search meta-heuristic approach (see Glover and Laguna (1997); Aarts and Lenstra (1997)) for driver scheduling. We first define the driver scheduling problem and its solution. Next, the HACS (Heuristics for Automatic Crew Scheduling) approach is overviewed. We then introduce the tabu search technique tailored for driver scheduling problems, in which multi-neighbourhoods and an appropriate memory scheme are designed. Some experiments in exploiting the neighbourhoods designed are described, and comparison of results is discussed.

2 Driver Scheduling

In bus and train vehicle schedules, vehicle work is usually presented by a set of *blocks*. A *block* presents a sequence of journeys to be operated by one vehicle during one day. For driver scheduling purposes, a sequence of *relief opportunities* (RO for short), in which drivers can be relieved, is usually identified in a block.

A *schedule* is a solution to the driver scheduling problem, which consists of a set of legal *driver duties* that together cover all the vehicle work. Allowing intermediate solutions to contain some infeasibility, we define a *schedule* simply as a set of driver duties that together cover all the vehicle work. A *duty* is the work to be performed by a driver during one day from signing on until signing off at depot. Each duty consists of a sequence of driver activities, i.e., a *sign-on activity*, a set of *spells*, and a *sign-off activity*. A *spell* is continuous vehicle work to be operated by a driver without break. Each spell includes at least two ROs. The first and the last ROs of a spell are called *active relief opportunities* (AROs for short), which are actually used

Figure 1. Illustration of a driver duty.

to relieve drivers. According to this definition of a duty, a duty may be illegal because either there is not enough time to make a connection between two consecutive spells, or the duty has violated some labour agreement rules. Figure 1 illustrates a duty, where an arrow denotes a link between two driver activities. *Sign-on time* at *depot* and *sign-off time* at *depot* are treated similar to AROs when linkages are considered.

From Figure 1, a duty can be defined by a sequence of *links*. Each *link* consists of an ordered pair of AROs (the first is called an *arrival-ARO* while the second is called a *departure-ARO*).

Given a duty D_i, a *cost function* $f(D_i)$ returns the value of wage cost, the calculation of which is company specific. A *penalty function* $g(D_i)$ is defined as a value to reflect how well D_i conforms with the labour agreement rules. In particular, if D_i is a valid duty then $g(D_i) = 0$.

It is generally not easy to assess precisely how much individual spells and links are contributing to the cost and penalty of a duty. For example, a duty might attract a penalty if there is not a suitable meal break between the third and fifth hours and all duties under eight hours might cost the same. Hence, all the links in the same duty will be assigned the cost and penalty of the duty.

The objectives of HACS are:

1. to remove any infeasibility,
2. to minimise the total number of duties,
3. to minimise the total wage cost.

The first objective is represented by minimising $g(S) = \sum_{i=1}^{n} g(D_i)$ until $g(S) = 0$, where $g(S)$ is the total penalty of solution schedule S and n is the number of duties in S. The second and the third objectives are sometimes in conflict. A trade-off is, therefore, needed dependent on the requirement of operators. In UK the second objective usually has priority over the third one. These two objectives can be merged and represented as minimising $f(S) = \sum_{i=1}^{n} f(D_i) + n * 5000$, where $f(D_i) << 5000$.

3   Overview of the HACS Approach

The general scheme of the HACS approach is as follows:

Step 1: *Construct an initial schedule*
Step 2: *Minimise total penalty using tabu search technique*
Step 3: *Minimise total cost using tabu search technique*
    If the total cost is reduced and the currently best schedule is still infeasible, go back to step 2; otherwise go to step 4.

Step 4: *Reduce penalty aided by using extra duties*

If the currently best schedule is feasible, then terminate; otherwise, adding extra duties is used to reduce total penalty and then go back to step 2.

This scheme attempts to improve the solution as much as possible by means of steps 2 and 3. Adding duties is only a last resort.

3.1 Construction of an Initial Schedule

An initial solution is the starting point of local search, and its construction is problem-oriented. It is generally recognised that a good initial solution is important, and, therefore, most approaches would try to build a good feasible solution as a starting point. However, the construction process would be complex and very sensitive to problem variations. Getting a very good initial solution is also time-consuming.

In contrast, the HACS approach does not aim to start from a good initial schedule. A simple and quick method is used to construct an initial schedule that covers all the vehicle work using a deliberately low target number of duties. The process consists of the following steps driven by some labour rule parameters:

1. Estimate a tight target number of duties N by the following formula:
$$N = V/(S - A - B - C)$$
   where V denotes the total vehicle work hours. S denotes the maximum length of spreadover stipulated in the labour agreement rules. A and B denote the signing on and signing off time allowances, respectively. C denotes the minimum meal break time.

   This formula guarantees that the target number is lower than that achievable. The reason of getting such a low target number can be found below from the method of forming spells.

2. Partition blocks into a set of spells as follows: The number of spells is assumed to be twice the target number of duties since only two-spell duties are considered. The average spell length can be deduced by dividing the total vehicle work by the number of spells. According to the average spell length, the blocks are partitioned into a set of spells without overlapping work. To ensure that all the spells are valid according to labour agreement rules and together cover all the vehicle work, the number of spells required is allowed to be increased when necessary.

   From this process of forming spells, a larger target number causes more spells formed with a shorter average length. HACS, therefore, uses a very low target number, leaving room for the refining process to improve the solution by adding duties incrementally.

3. Sort the spells by starting time.

4. Couple pairs of spells to form duties: Couple the spells in a simple manner to form driver duties such that the first spell has its starting time earlier than the starting time of the succeeding spell, and an arbitrary depot

(the first depot in the data file) is assigned to each duty for the driver to sign on and sign off. Some duties may be time infeasible or have violated some labour agreement rules.

5. Form the last one-spell duty, if the number of spells is odd.

3.2   Refinement of the Current Schedule

The initial schedule constructed is very likely to contain infeasible duties. The refining process will improve it iteratively. HACS focuses on the refining process, which contains three components: *minimising infeasibility, minimising cost*, and *adding duties*. The former two components are driven by the tabu search technique tailored for driver scheduling, which plays the most important role in the HACS approach and is presented in the next section. After the minimising processes using tabu search, the current best schedule may still be infeasible because tabu search cannot guarantee an optimal solution, and the target number may be lower than that achievable. Therefore, the *adding-duties* process is devised to reduce infeasibility until a feasible schedule is obtained:

- One duty is added at a time when tabu search cannot lead to any improvement.
- Some work from the infeasible duties is taken out to form the extra duty, such that the overall total infeasibility is reduced.

4   Tabu Search Technique for Driver Scheduling

Tabu search is a class of meta-heuristics proposed by Glover (1989, 1990) for solving hard combinatorial optimisation problems. The basic idea is to avoid being trapped at local optima by allowing the acceptance of non-improved solutions. Glover and Laguna (1993) indicate that a fundamental element underlying tabu search is the use of flexible memory, which creates and exploits structures for taking advantage of history. Strategic use of memory can make dramatic differences in the ability to solve a problem (see Glover and Laguna (1997)), but tabu search is a general search scheme that must be tailored to the problem at hand. This section presents the main components of the tabu search tailored for driver scheduling.

4.1   Objective Functions

The tabu search technique is employed in two phases: *minimising infeasibility* and *minimising cost*. In the phase of minimising infeasibility, the objective function is defined in Section 2 as $min(g(S))$ while in the phase of minimising cost, as $min(f(S))$.

4.2 Multi-neighbourhood Structures

Tabu Search may be conveniently characterized as a form of neighbourhood search (see Glover and Laguna (1993)), where each solution S has an associated set of neighbours $N(S)$ called the neighbourhood of S. Any solution $S' \in N(S)$ can be reached directly from S to S' by an operation called a *move*. The neighbourhoods in HACS are defined by the following move operations: *swapping-two-links, swapping-two-spells, inserting-one-spell*, and *recutting-block*.

In the following discussion, links and AROs are denoted by symbols with double subscripts, where the first subscript identifies the duty it belongs to. For example, d_{2j} is a link j belonging to duty D_2.

4.2.1 Swapping Two Links

Given two links d_{1i} and d_{2j}, this operation will remove these links and replace them with two new links d'_{1i} and d'_{2j} as illustrated in Figure 2.

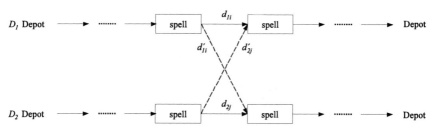

Figure 2. Illustration of swapping two links.

4.2.2 Swapping Two Spells

Given two links d_{1i} and d_{2j}, if none of them is a sign-off activity, this operation will first find the following links $d_{1,i+1}$ and $d_{2,j+1}$, and then remove these links and replace them with four new links d'_{1i}, d'_{2j}, $d'_{1,i+1}$ and $d'_{2,j+1}$ as illustrated in Figure 3.

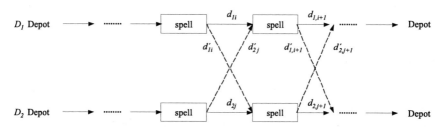

Figure 3. Illustration of swapping two spells.

4.2.3   Inserting One Spell

Given two links d_{1i} and d_{2j}, if d_{1i} is not a sign-off activity, this operation will first find the following links $d_{1,i+1}$, and then remove these links and replace them with three new links d'_{1i}, d'_{2j} and $d'_{1,i+1}$ as illustrated in Figure 4.

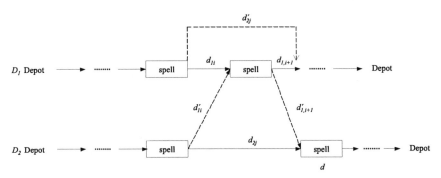

Figure 4. Illustration of inserting one spell.

4.2.4   Recutting-block

Given a link $d_{1i} = (p_{1i}, q_{1i})$, where p_{1i} and q_{1i} are AROs, if there exists another link $d_{2j} = (p_{2j}, q_{2j})$ such that $p_{1i} = q_{2j}$ or $p_{2j} = q_{1i}$, the link d_{2j} is called a *relevant link* of d_{1i}, we define the operation *recutting-block* as follows:

Two cases have to be considered. In case 1: $p_{1i} = q_{2j}$, if p_{1i} is not the first RO in the block, this operation will first find its previous RO r, and then replace the link $d_{1i} = (p_{1i}, q_{1i})$ with a new link $d'_{1i} = (r, q_{1i})$. Meanwhile, the relevant link $d_{2j} = (p_{2j}, q_{2j})$ should be replaced with a new link $d'_{2j} = (p_{2j}, r)$ in order to cover the piece of work between r and p_{1i}. Figure 5 shows an illustration. Figure 5(a) shows two original duties while Figure 5(b) shows the corresponding new duties. Similarly, in case 2: $q_{1i} = p_{2j}$, this operation will replace $d_{1i} = (p_{1i}, q_{1i})$ and $d_{2j} = (p_{2j}, q_{2j})$ with $d'_{1i} = (p_{1i}, r)$ and $d'_{2j} = (r, q_{2j})$, respectively.

4.2.5   Alternatives to the Recutting-block Operation

In the foregoing operations, the first three operations can be categorised as *swapping-links* (*swapping* for short) operations because they involve swapping several links while the set of AROs remains unchanged. The *recutting-block* operation is different from the swapping operations, which is used to revise the selection of AROs. In these operations, the spells are treated as basic units. In addition to swapping links, recutting-blocks play an important role to refine the current schedule because the initial set of spells and corresponding AROs were determined quite roughly. However, the recutting function is limited because only the ROs adjacent to the current AROs are considered as potential AROs. If this restriction were relaxed, the process would be impractical because there would be too many combinations to consider.

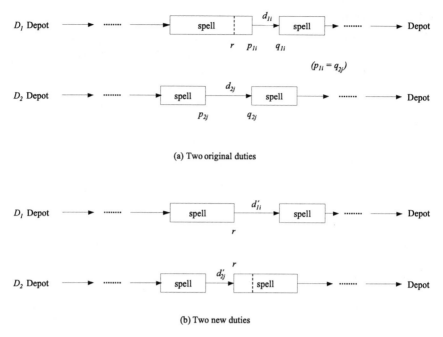

(a) Two original duties

(b) Two new duties

Figure 5. Illustration of recutting-block.

In the above swapping operations, *spells* are the basic units of driver work. By considering the *piece of work* (*piece* for short) covered between two successive ROs as a more basic unit, a spell consists of a number of pieces contiguously linked together. The three swapping operations could be applied to the links between the pieces instead of between spells, and the piece-based *swapping-two-links* operation would be too time-consuming, and the resulting duties would have too many fragmented pieces of work. A compromise is to retain the above spell-based swapping operations, and to introduce two piece-based swapping operations to replace the recutting operation.

The new piece-based swapping operations are *swapping-two-pieces* and *inserting-one-piece*, which are analogous to the spell-based swapping-two-spells and inserting-one-spell, respectively. The new operations can be similarly illustrated by Figures 3 and 4 by replacing "spell" with "piece."

4.3   Move Value

The foregoing operations decide moves. The value of a move (d_{1i}, d_{2j}) is defined as $f(S) - f(S')$ or $g(S) - g(S')$ depending on the phase of the algorithm. Obviously, only the duties altered by the move need to be evaluated. Since we have defined the cost and penalty of a link to be those of the duty it belongs to, the move value $f(S) - f(S')$ can be equivalently represented as $f(d_{1i}) + f(d_{2j}) - f(d'_{1i}) - f(d'_{2j})$ while $g(S) - g(S')$ can be similarly represented.

4.4   Tabu Tenure

Tabu tenure decides the number of iterations for which the current selected move must remain tabu. Glover and Laguna (1993) illustrate some rules to determine tabu tenure. They also indicate that values between 7 and 20 appear to work well for a variety of problem classes, while values between $0.5\sqrt{n}$ and $2\sqrt{n}$ appear to work well for other classes, where n is a measure of problem dimension. In our approach, we use n to denote the number of links, which equals the sum of the number of spells and the number of duties, in the current schedule. Considering practical problems with about 100 duties, which are mostly consisting of two or three spells (i.e., 3 or 4 links per duty), $\min(\sqrt{n}, 20)$ is defined as the tabu tenure.

4.5   Memory Scheme

It is generally recognised that finding a best move in the neighbourhood $N(S, H)$ (which depends on the current solution S and the search history H) may often be too time-consuming and it is usually crucial to determine a *candidate set* as a subset of $N(S, H)$ in order to restrict the number of solutions examined on a given iteration. The importance of determining a candidate set has been described a great deal in the literature (see, e.g., Glover and Laguna (1993)). However, as described later an efficient memory structure has been designed for this application such that updating after each move does not require much computation and, therefore, we use the entire neighbourhood as the candidate set. Swapping moves are different from recutting moves, and different memory structures are, therefore, designed.

In the Swapping Process: The candidate set (as large as the entire neighbourhood) is denoted as a lower triangular matrix $C = [c_{ij}]$, which records all the move values, where $i > j$; $i = 2, 3, \ldots, n$; $j = 1, 2, \ldots, n - 1$; and n is the number of links in the current schedule. However, it would be very time-consuming to update every value in C after a move. Since in driver scheduling the average number of links in a duty is usually very small, only several links contained in two duties change after a move. Every move is assigned a *swapping-status*, which indicates whether the link is in a currently operated duty. Only the moves in C containing the recently visited links need to be costed again while other values in C remain the same. Each move in C corresponds to a *tabu status*, which can be recorded in the upper triangle of C. However, most elements in the upper triangle of C will be useless because most of the moves are not recorded tabu at a given iteration. In order to use the memory efficiently, we design a *tabu list* to record the *tabu status* of each tabu-active move.

The tabu list is designed in HACS as a matrix $T = [t_{ij}]$, where $j = 1, 2$; $i = 1, 2, \ldots, m$; m equals the tabu tenure and the value of i denotes the tabu status of a swapping move stored in the ith row of the matrix. After a move, the first row in T is removed while the contents in the other rows are transferred into their previous rows. The currently selected move is put in the last row. T is initialised by $t_{ij} = -1$, where $t_{ij} \in T$. Any move not recorded in T is tabu-inactive.

In the Recutting Process: Any link contains two AROs; a vector $T = (t_1, t_2, \ldots, t_{2n})$, where n is the number of links in D, can, therefore, record all necessary information for each move. Any t_i in T denotes an object of *Recut*, which is a class that contains the items to be recorded for a move, e.g., a *link* and the *relevant link, move value*, and *tabu status*, etc. The vector T presents an entire neighbourhood. The tabu status is initialised to 0 and used to identify the ending iteration of the tabu tenure for the corresponding move.

5   Experiments and Computational Results

The HACS approach has been implemented using the Borland C++ Development Suite (Version 5.02) within an object-oriented framework. The quality of solutions is the major concern while the computational time is displayed. The experiments presented in this section are first carried out on the spell-based operations in order to evaluate the neighbourhood structures and to compare the tabu search technique with a steepest descent method; they then are carried out on the piece-based operations and the combination of the spell-based and piece-based operations.

5.1   Test Problems

The test problems selected are real-life problems accumulated at Leeds University, UK. Table 1 shows the properties and the best known solutions to the problems, which were generated by TRACS II, one of the most successful driver scheduling systems, which first generates a large set of legal potential duties, and then selects the best solution using ILP.

The elapsed times recorded in this paper were actual clock times to obtain the solutions while the problems were run on a 333 MHz personal computer (difference will be indicated if applicable). Since TRACS II does not report on elapsed times, and the duty generation and the ILP selection process were run separately, the elapsed times were estimated. In D9 and D10, slack parameters were used in the TRACS II runs (on a Pentium III personal computer), and much (over 85%) of the elapsed time was spent on generating the large number of potential duties.

5.2   Experiments on Spell-based Operations

Given an objective function, different neighbourhood structures may give rise to different "landscapes" of local optima. In Section 4, three swapping

Data	Vehicle	No.	No.	Best known solutions			
Code	Hours	ROs	Blocks	Potential Duties	No. Duties	Cost (hours)	Elapsed Time
D1	165.56	130	10	8090	24	180.18	3 mins
D2	178.46	200	13	58136	25	192.52	25 mins
D3	216.28	248	16	32904	31	235.05	10 mins
D4	227.52	175	21	11817	34	289.32	5 mins
D5	294.10	289	23	27068	42	327.48	10 mins
D6	332.06	356	30	32000	39	363.26	10 mins
D7	349.37	574	21	29500	49	408.47	10 mins
D8	404.46	646	30	29021	54	427.04	10 mins
D9	552.07	799	41	671541	80	632.32	over 5 hours
D10	786.47	859	63	999101	108	862.29	over 15 hours

Table 1. Properties and best known solutions of test problems.

neighbourhood structures have been defined, each of which has different local optima and leads to a different solution. Table 2 compares the solutions obtained using each swapping neighbourhood structure separately while the recutting neighbourhood structure is always applied. It also displays the results obtained by the HACS approach with all the neighbourhood structures compounded, in which each neighbourhood structure is applied in turn until no further improvement can be achieved. The table also presents the results obtained by a *steepest descent method* (i.e., from a given solution the move that leads to the best improved solution amongst the neighbourhood of the given solution is selected). The steepest descent method presented here is adapted from the HACS approach (with all the neighbourhood structures) but turning off the uphill move operation. The last row in the table shows the average relative percentage deviation (RPD) over the best known solutions. The elapsed times for all the methods presented in the table are less than 6 minutes while they are less than one minute for the problems D1 to D7 and are about 2 minutes for D8 and D9.

From Table 2 we can see that HACS with all the neighbourhood structures compounded produced better solutions than with only a single swapping neighbourhood structure. The application of multi-swapping neighbourhoods diversified the search and led to much better results. The average relative percentage deviation of the HACS solutions over the best known solutions was 3.24% in terms of number of duties and 1.27% in terms of cost. For the largest problem (D10) HACS produced a marginally better solution than TRACS II.

| Data | Single swapping neighbourhood | | | | | | Multi-neighbourhood | | | |
| | Swapping links | | Swapping spells | | Inserting spell | | HACS (Tabu Search) | | Steepest Descent | |
	No. Duties	Cost (hours)	No. Duties	Cost (hours)	No. Duties	Cost (hours)	No. Duties	Cost (hours)	No. Duties	Cost (hours)
D1	28	199.24	26	191.02	25	188.03	25	182.26	26	198.49
D2	26	192.37	26	199.21	26	200.41	25	196.50	27	192.15
D3	33	244.06	34	239.20	34	242.13	33	239.33	33	244.28
D4	36	285.57	36	289.16	37	292.37	35	287.48	35	295.04
D5	45	330.32	46	333.13	45	333.25	44	331.05	45	334.31
D6	44	388.13	44	384.00	46	393.12	42	371.15	43	373.39
D7	53	419.25	51	416.08	52	415.18	50	411.18	51	415.46
D8	58	457.10	58	457.22	59	470.33	55	438.52	58	453.14
D9	87	685.26	83	654.12	87	688.27	82	647.13	83	650.27
D10	113	889.11	110	865.26	118	927.02	108	856.11	110	864.00
RPD	8.19%		6.73%		8.52%		3.24%	1.27%	6.02%	3.16%

Table 2. Results obtained by the tabu search based approach with different neighbourhoods and by a steepest descent method.

The above experiments show that the steepest descent method is inferior to the HACS approach based on the tabu search technique. However, the steepest descent method produced better solutions than the HACS approach with a single neighbourhood structure.

5.3   Experiments on Piece-based Operations and the Combination of Piece-based and Spell-based Operations

Despite the use of an efficient memory structure and multi-neighbourhoods in HACS, the above-shown results are unsatisfactory. It was anticipated that the piece-based operations would improve the results because more ROs would be considered than the restrictive recutting operation. A number of experiments have been done, in which the same memory structure was used since links are still the basis of the memory structure.

As discussed in § 4.2.5, the recutting operation is replaced with the swapping operations: *swapping-two-pieces* and *inserting-one-piece* leading to the HACS approach with the spell-based and piece-based operations combined, in which the five swapping operations are applied in turn. The computational results (presented in Table 3) show:

- The HACS approach with the spell-based and piece-based operations combined, produced better solutions than the other variants of the HACS approach.
- In most cases, the improved HACS solutions to the test problems are close to the best known solutions. The average RPD over the best known solutions is 1.36% in term of number of duties and 1.30% in terms of total cost.
- For the largest problem in the table, HACS produced a better solution than TRACS II.

Data	No. Duties	RPD (%)	Cost (hours)	RPD (%)	Elapsed time (min:sec)
D1	24	0.00	182.29	1.21	00:24
D2	25	0.00	196.14	1.74	00:54
D3	32	3.22	239.15	1.77	01:02
D4	34	0.00	291.56	0.83	00:36
D5	44	4.76	330.18	0.76	00.59
D6	40	2.56	362.48	-0.17	01:40
D7	49	0.00	414.05	1.30	02:53
D8	56	3.70	450.34	5.50	09:17
D9	81	1.25	643.41	1.76	03:33
D10	106	-1.85	847.55	-1.69	17:54
	1.36		1.30		

Table 3. Results obtained by the HACS approach with the spell-based and piece-based operations combined.

6   Conclusions

This paper has presented HACS, a tabu search approach to the driver scheduling problem. HACS can produce solutions considerably quicker than TRACS II, an ILP-based generate-and-select approach. Theoretically, using slack labour rule parameters enlarges the search space providing more chance to produce a good solution. However, the duty generation process could take a very long time when the labour rule parameters are slack. Hence, TRACS II uses some soft filtering rules to prevent too many potential duties from being generated. The HACS approach does not need any filtering rules because it only refines the small solution set of duties constructed. Although the current HACS solutions are generally only getting close to as good as the TRACS

II solutions, HACS has the potential of being developed to yield solutions better than TRACS II because its search space is not limited by the soft rules.

This paper also presented the tabu search technique tailored for the driver scheduling problem, in which an efficient memory structure was devised with multi-neighbourhoods to diversify the search space. Experiments have shown that the multi-neighbourhood structures together are superior to a single neighbourhood structure, and the tabu search technique is superior to the steepest descent method. By combining spell-based and piece-based operations, HACS has allowed all ROs chance to be used while the driver work would not become too fragmented and the computation is efficient.

The research on the HACS approach is ongoing. Further improvements are expected in future. For example, it is anticipated that HACS could be easily extended to cater for the driver scheduling problem with time windows, which no existing driver scheduling systems can practically handle.

Acknowledgement

We would like to thank Dr. Ann Kwan for providing us with the test problems.

Bibliography

Aarts, E. and J.K. Lenstra (Eds.) (1997). *Local Search in Combinatorial Optimization*. Wiley, Chichester.

Fores, S., L. Proll, and A. Wren (1999). An improved ILP system for driver scheduling. In N.H.M. Wilson (Ed.), *Computer-Aided Transit Scheduling, Lecture Notes in Economics and Mathematical Systems*, 471, Springer, Berlin, 43–61.

Glover, F. (1989). Tabu search: Part I. *ORSA Journal on Computing 1*, 190–206.

Glover, F. (1990). Tabu search: Part II. *ORSA Journal on Computing 2*, 4–32.

Glover, F. and M. Laguna (1993). Tabu search. In C.R. Reeves (Ed.), *Modern Heuristic Techniques for Combinatorial Problems*, Blackwell, Oxford, 70–150.

Glover, F. and M. Laguna (1997). *Tabu Search*. Kluwer, Norwell.

Kwan, A.S.K., R.S.K. Kwan, M.E. Parker, and A. Wren (1999). Producing train driver schedules under different operating strategies. In N.H.M. Wilson (Ed.), *Computer-Aided Transit Scheduling, Lecture Notes in Economics and Mathematical Systems*, 471, Springer, Berlin, 129–154.

Kwan, R.S.K., A. Wren, and R.S.K. Kwan (2000). Hybrid genetic algorithms for scheduling bus and train drivers. In *Proceedings of the 2000 Congress on Evolutionary Computation*, La Jolla, San Diego, 285–292.

Reeves, C.R. (Ed.) (1993). *Modern Heuristic Techniques for Combinatorial Problems*. Blackwell, Oxford.

Rousseau, J.-M. and J. Desrosiers (1995). Results obtained with Crew-Opt: A column generation method for transit crew scheduling. In J.R. Daduna, I. Branco, and J.M.P. Paixão (Eds.), *Computer-Aided Transit Scheduling, Lecture Notes in Economics and Mathematical Systems*, 430, Springer, Berlin, 349–358.

Smith, B.M. and A. Wren (1988). A bus crew scheduling system using a set covering formulation. *Transportation Research A 22*, 97–108.

Wilson, N.H.M. (Ed.) (1999). *Computer-Aided Transit Scheduling, Lecture Notes in Economics and Mathematical Systems*, 471. Springer, Berlin.

Wren, A. and J.-M. Rousseau (1995). Bus driver scheduling – An overview. In J.R. Daduna, I. Branco, and J.M.P. Paixão (Eds.), *Computer-Aided Transit Scheduling, Lecture Notes in Economics and Mathematical Systems*, 430, Springer, Berlin, 173–187.

Experiences with a Flexible Driver Scheduler

Sarah Fores, Les Proll, and Anthony Wren

Constraint Programming and OR Group,
School of Computing, University of Leeds, Leeds, LS2 9JT, UK
{sarah, lgp, wren}@comp.leeds.ac.uk

Abstract. We present a flexible user-driven ILP tool for the optimisation component of the TRACS II driver scheduling system. The system allows the user to select
from a number of objective functions and to drive the LP relaxation through one of
a range of optimisation processes. As a default we provide a Sherali objective which
minimises the number of shifts, and within that yields the least cost. The default
method of solving the LP relaxation is by specialised primal column generation for
larger problems, and by dual steepest edge for smaller ones. The LP is capable of
working with over 200,000 previously generated shifts. Once the relaxed LP has
been solved, we reduce the size of the problem, and enter a specialised branch and
bound procedure. Using real data we show how, by being able to handle larger
problems, our system reduces the need for problem decomposition and can produce
better schedules.

1   Introduction

Transport operators typically require an allocation of vehicles to predetermined journeys, an allocation of drivers to vehicles, and rostering of people to a sequence of shifts. Previous workshops (see, e.g., Desrochers and
Rousseau (1992); Daduna et al. (1995); Wilson (1999)) have included attempts to combine vehicle and driver scheduling (see Freling et al. (1999)),
but tend to favour approaches which schedule vehicles in advance (see, e.g.,
Caprara et al. (1999); Rousseau and Desrosiers (1995)). TRACS II, developed at the University of Leeds, is a mathematical programming based bus
and train driver scheduling system. The system is based upon the IMPACS
system (see Smith and Wren (1988); Wren and Smith (1988)) which first generates valid shifts, and then selects a subset which covers all of the vehicle
work and minimises an objective function. The main enhancements made to
the Integer Linear Programming (ILP) component relate to the implementation of a choice of solution methods, and to improvements in the size of
problem that can be dealt with. Thus, whilst TRACS II has been shown to
successfully solve real scheduling problems, the ability to solve ever larger
problems enables the user to generate more and better shifts from which a
selection can be made, and reduces the need to decompose the problem into
smaller areas. This, in turn, is expected to yield better solutions.

2   Driver Scheduling

The main information received from the vehicle schedule is the time and location of convenient changeover points for drivers. These location/time pairs
are known as *relief opportunities* and include those at the beginning and end
of the vehicle work where a driver will travel with the vehicle to and from the
appropriate *depot*. The periods of time between the relief opportunities are
known as *pieces of work*. Before the driver scheduling can take place it may
also be necessary to note that certain vehicles or routes may only be driven
by a subset of drivers.

Driver scheduling consists of allocating all pieces of work to a set of drivers
in such a way that the work of a driver (*a shift*) is acceptable to union
regulations and the schedule is in some way efficient. The cost of the final
schedule may be determined by the number of shifts included, the actual wage
costs according to the hours worked, a penalty cost to deter features which
are 'legal' but undesirable, or any combination of all three. Side constraints
may be imposed to restrict the shift total or the combination of shift types
in the schedule.

The driver scheduling problem has been tackled by many different approaches. Most commercial software, including TRACS II, uses a mathematical programming approach which utilises a set covering or set partitioning
formulation to ensure that vehicle work is covered, with a suitable objective
function to differentiate between schedules. The problem with many such formulations is the large number of variables necessary to produce good schedules, which may lead to the formulation being intractable in a viable computing time. Genetic algorithms (see Kwan et al. (1999b)) and constraint
satisfaction (see Curtis et al. (1999)) have been investigated as techniques
which attempt to find good schedules using variants of local search. They
have the advantage that, for a set covering problem in the absence of side
constraints, they can always find *a* schedule where sometimes an 'optimisation' approach might not find a solution within the computing limits or with
an imposed target number of shifts. However, these techniques currently are
at an experimental stage and cannot, in general, match the performance of our
ILP approach. Nor currently can they handle side constraints. Algorithmic
and technological improvements continue to alleviate the inherent difficulties
that mathematical programming techniques have with large problems.

Recently there has been an increased interest in applying column generation techniques to driver scheduling (see, e.g., Caprara et al. (1999); Böhringer
et al. (2000); Freling et al. (2001); Kroon and Fischetti (2001)). Column generation approaches are particularly beneficial in solving problems with a large
number of variables but many practical applications require them to be integrated with heuristics. These approaches have predominantly been applied
to single operator intercity rail driver problems and, as yet, there is little
evidence that they can be readily adapted to different scenarios.

3   TRACS II

Fores et al. (1999) identified the principal three stages to the solution process of TRACS II :

- Generate a set of valid shifts
- Reduce the shift set (if required)
- Select a subset of shifts which cover the vehicle work and minimises an objective function

As well as improving the algorithms in these stages, there has been an emphasis on increasing their flexibility in terms of what they want to achieve, making experimentation a viable option.

3.1   Shift Generation

Parameter-driven heuristics are employed to first generate a set of shifts which are valid according to a set of labour agreement rules. The capabilities of these heuristics have been adapted and improved over the years so that many different user requirements can be satisfied within a unified framework. One of the advantages of generating shifts beforehand is that different shift sets can be formed using different agreements and maybe even merged before optimisation. Problems with the user-supplied parameters or data can be identified relatively quickly, possible evidence being that of forming a seemingly high or low number of shifts, or pieces of work remaining uncovered. Error correction or experimental changes can thus be carried out relatively quickly before searching for resulting schedules. Such features are offered so that experienced schedulers can 'share' their own knowledge rather than relying on a 'black box' to produce the schedules. The benefit is a more effective scheduling process.

3.2   Shift Reduction

The size of the generated shift set depends on the setting of a number of parameters and cannot be predicted. If the set is judged to be too large, it can be reduced according to a number of different scenarios which attempt to retain those shifts which are likely to be beneficial to a final schedule. The possibilities include discarding identical shifts, shifts wholly contained within other shifts, and shifts which are deemed inefficient according to a measure of cost effectiveness and coverage of the pieces of work it contains. This reduction process was particularly important when the optimisation component could not accept more than 30,000 shifts and 'inefficient' shifts had to be removed beforehand. The system can now accept much larger shift sets so the removal of duplicate shifts is often the only requirement before optimisation.

3.3 Optimisation

The selection of the subset of shifts used to form the final schedule from those previously generated is performed by means of the set covering model (1). The set covering model allows pieces of work to be allocated to more than one shift. It is not appropriate to use a set partitioning approach, which does not, since not all valid shifts have been previously generated, or shifts covering fewer pieces of work have been removed. Whilst we must ensure that all pieces of work are allocated to a unique driver, any overlap can be edited afterwards.

Given a problem with M pieces of work and N previously generated shifts, we can define the set covering model to be:

$$\text{Minimize} \quad \sum_{j=1}^{N} D_j \, x_j \tag{1}$$

$$\text{subject to} \quad \sum_{j=1}^{N} A_{ij} \, x_j \geq 1 \qquad \text{for } i = 1, \ldots, M$$

$$\{x_j : j = 1, \ldots, N\} \in S_c$$

$$x_j = \begin{cases} 1 & \text{if shift } j \text{ is used in the solution} \\ 0 & \text{otherwise} \end{cases}$$

$$A_{ij} = \begin{cases} 1 & \text{if shift } j \text{ covers piece of work } i \\ 0 & \text{otherwise} \end{cases}$$

D_j is determined by the objective function used

S_c is the set of side constraints

The model is solved as follows:

- Find an initial solution
- Relax the integrality conditions on x_j and solve
- If shift minimisation is required, add a side constraint to increase the shift total to the next highest integer (if necessary) and re-solve
- Reduce the size of the problem
- Find an integer solution by branch and bound

The user has control over a number of the specific parameters which determine the way in which the code runs, which are outlined below. Whilst a default route exists, it is sometimes appropriate to use different routines if the problem exhibits particular features or behaves differently in some circumstances. There are also parameters which could be changed which alter the sensitivity of the algorithms or allow different solution strategies, but these are often for experimental use, since they may not be effective when changed in isolation from other parameters.

3.3.1   Objective Function

Historically the main optimisation required was one which would prioritise the reduction of the number of shifts, and within that target reduce the total shift and penalty cost. The reason was that each shift in itself was seen to be costly to the company, incurring significant indirect labour costs. Fores et al. (1999) describes the implementation of a Sherali weighted function to accurately reflect this objective, which is appropriate in medium and long term planning situations in which the vehicle schedule is radically altered or a new one instituted. In situations where only minor changes are made to the vehicle schedule, it may be more appropriate to reduce total shift cost whilst using the same number of shifts as previously. Thus we see the main user objectives as being to minimise the number of shifts, the total cost, or a combination of both, and TRACS II has been extended to allow the user to select from these three different options. Users can supply penalty costs representing features which are in some way undesirable in the composition of a shift. For example, it may be preferable but not essential for a driver, after a break, to drive a bus travelling in the same direction as the bus he has left in order to reduce potential problems arising from traffic delays. However, the penalty costs are subjective values and could outweigh a more general goal of shift cost minimisation. For this reason, all three objectives can be selected to include or exclude the user-controlled penalty costs. The cost coefficient D_j of the objective function (1) is defined according to the objective function selected, where the default is to minimise the combination of shift total and total cost.

Combined shift/cost minimisation: Willers (see Willers et al. (1995)) introduced a Sherali weighted objective function to more satisfactorily achieve the desired overall combined objective.

$$D_j = W_1 + C_j \qquad \text{for } j = 1, .., N$$

C_j = cost (+penalty) of shift j
W_1 = 1 + sum of X largest C_j values
X = no. shifts (+ no. uncovered pieces) in initial solution

The large weight attached to each shift guarantees shift minimisation as a priority.

Shift minimisation:

$$D_j = 1 \qquad \text{for } j = 1, \ldots, N$$

Since the Sherali weighted objective function also guarantees the minimum number of shifts, the main use of this objective function without penalties is to determine the minimum number of shifts in the relaxed LP quickly. It is unlikely that any integer solution found would be useful because unit costs would not be sufficiently discriminating in finding good schedules. If the user requires penalty costs to be considered, a minimum shift total is first found, a side constraint is added to ensure this target is not exceeded and the objective changed to minimise the sum of penalty costs. In this way the integer solution would represent a minimum shift schedule containing the most acceptable duties.

Cost minimisation: Costs can consist of a pure shift cost or shift cost plus penalty.

$$D_j = \text{cost (+penalty) of shift } j$$

If this objective is chosen the integer solution indicates the cheapest schedule.

3.3.2   Constraints

There are necessary constraints on the model (1) which arise from the set covering nature of the problem. The user can also add side constraints which can relate to shift type, depot type, or the total number of shifts. The constraints can be expressed as follows:

$$\sum_{j=1}^{N} \delta_{kj} x_j \le U_k, \quad \sum_{j=1}^{N} \delta_{kj} x_j \ge L_k, \quad \sum_{j=1}^{N} \delta_{kj} x_j = T_k$$

where U_k is an upper limit on the number of shifts of type k, L_k is a lower limit on the number of shifts of type k, and T_k is a target number of shifts of type k. The δ_{kj} are defined as :

$$\delta_{kj} = \begin{cases} 1 & \text{if } x_j \text{ is a shift of type k} \\ 0 & \text{otherwise} \end{cases}$$

3.3.3   Optimisation

The two main choices of optimisation routine are that of a primal column generation approach and a dual steepest edge approach. Within each of these approaches several other algorithmic options are available. Both primal and dual methods are accelerated by providing an initial solution. Several heuristic methods of doing this are available within TRACS II. The default method constructs a schedule by sequentially looking at the currently uncovered piece of work having the least shifts available to cover it. From these, the shift covering the most currently uncovered work is selected, thus attempting to cover all work with a small number of shifts. Although any side constraints are considered during this process, it is not guaranteed that the initial solution will satisfy all the constraints. To provide an advanced start for the solution of the LP, the schedule has to be transformed into a basic solution to (1) which is either primal feasible or dual feasible, depending on which optimisation approach is being used.

It is recommended that dual steepest edge is executed on problems containing at most 30,000 shifts, where all shifts can be considered within limits imposed for storage and efficiency. However, the user can choose to run any size of problem using a column generation approach, which implicitly considers all shifts but retains only a working subset of at most 30,000 shifts. This approach is automatically executed if the shift size is greater than 30,000, regardless of user choice.

Primal Column Generation: An initial subset of shifts is selected from the submitted set to ensure a specified level of coverage for every piece of work. This is done whilst searching through shifts to find an initial solution. The LP relaxation of (1) is solved over the subset using primal steepest edge (see Forrest and Goldfarb (1992)). The simplex multipliers associated with any relaxed solution are used to identify pieces of work which may be better covered by available shifts not yet considered. A certain number of shifts can then be added to the subset if their reduced costs indicate an improvement to the objective function. More than one piece of work may be considered in each iteration. The number of shifts added per piece of work and the total number to be added in any major iteration are defined using parameters. Since the relaxed solution is always optimal over the subset of shifts currently defined, it is possible to proceed to the branch and bound at any stage. There is also an option which allows the column generation process to be halted if no significant progress is being made but, as yet, the number of major column generation iterations required has not hindered the running time enough to justify not implicitly considering ALL shifts provided beforehand.

In order to retain a working subset of at most 30,000 shifts, some deletion of shifts from the subset can be carried out if they are not currently in the optimal LP basis. The deleted shifts can return to the subset, if desirable, at a later stage.

Dual Steepest Edge: It has long been known that set covering problems are inherently degenerate and that primal simplex approaches to solving their LP relaxation spend many iterations making no reduction in the value of the objective function. The effect of degeneracy can be lessened by a dual simplex approach. This led Willers (see Willers et al. (1995)) to implement a dual steepest edge approach (see Forrest and Goldfarb (1992)) which showed a substantial reduction in solution time for the LP relaxation over the primal steepest edge approach used in pre-1995 versions of TRACS II.

3.3.4 Branch and Bound

The relief opportunities used in the relaxed solution are deemed to give an adequate representation of the relief opportunities which would be used in a schedule. Any shifts currently available to the model which do not use the relief opportunities are eliminated from the problem, and the constraints associated with eliminated relief opportunities are merged. If the column generation approach is being used, it is possible at this stage to reselect all shifts from the superset which use the relief opportunities selected, although adequate solutions are normally found from the smaller remaining set of shifts.

Searching the branch and bound tree requires a branching strategy, i.e., rules to select the node to explore next and to construct the branches. There are a number of node selection rules available in TRACS II which consider how infeasible the node solution is and how much the objective function has degraded at this node. A hierarchy of branching rules exist, viz relief opportunity branching, constraint branching and shift branching (see Smith (1986)). Within the branch and bound phase dual steepest edge is used to solve the LP at each node.

The user can define the number of nodes in the branch and bound tree that they wish to search to find a schedule. The default is 500 nodes but users may be happy to let the program run for a longer time in order to find a schedule for a difficult problem. With larger problems, and using the default branching strategy, it has been proved that increasing the node limit has allowed better solutions to be found.

Although it is usually successful, TRACS II does not guarantee finding a schedule at the first attempt, particularly when being introduced in a new organisation. This is partly because of the target number of shifts which may be imposed and partly because it performs a limited search of the branch and bound tree. If this is the case, the inherent flexibility of TRACS II can be exploited to find a schedule by:

- increasing the processing limits or using different algorithmic options
- introducing a side constraint to increase the target and re-solving with the same shift set
- performing a less (or more) severe reduction in the generated shift set prior to optimisation
- tightening (or relaxing) the operational parameters which control the generation of shifts

Which of these options to choose depends on the particular circumstances and that choice benefits from skill and experience.

4   Data Sets

Since its original inception as IMPACS, when it was limited to bus driver problems of at most 350 pieces of work and 5000 shifts, TRACS has undergone continuous development. It can now handle both bus and train driver problems and is considerably more flexible and robust, typically accepting problems with up to 2000 pieces of work and 100,000 shifts. Problems with 200,000 shifts and with 2600 pieces of work have recently been solved. The evolution has occurred in part because of increased computing power and in part because of the introduction and improvement of the solution algorithms such as column generation. The acceptance of larger shift sets in the optimisation component has allowed a more flexible shift generation module. In the past large problems have needed decomposition in order to fit into the limits imposed by TRACS II. The need for decomposition is lessened, and indeed our perception of what is a large problem has been substantially altered, by the improved capability of the system.

We identify two problems which were previously divided and show any differences in the approach and solution of them with the improvements previously mentioned.

Decomposition: One of the main problems of dividing a large problem is the overhead involved in schedulers discussing and swapping pieces of work when it appears that they cannot be easily covered in one subdivision. Certainly any decomposition can only be performed by schedulers who have some knowledge of possible scenarios of area splitting. However, there may be advantages when distinct areas respond to different constraints or can be built from different rules.

On the whole, removing the need for decomposition removes the possibility of inefficient splits, and allowing more shifts to be available to the ILP should increase the possibility of covering the complete problem with a better solution.

The problems presented were originally tackled by subdividing the network into (overlapping) geographical areas. Since the ILP system has improved in the intervening period, we have reoptimised to see if better solutions can be found without subdividing.

ScotRail: One of the larger driver scheduling problems conducted by our Group was for ScotRail. This company covers the whole Scottish rail network (see Figure 1), with (in 1998) over 350 drivers' shifts daily. There were 19 drivers' depots, ranging in size from 2 shifts daily up to 73. Drivers at any depot were qualified to work certain sections of track only, most sections being covered by between two and four depots. There were also restrictions on the types of traction unit that could be covered by any depot.

The operations ranged from very intensive commuter systems around Glasgow to remote areas of the Scottish Highlands. For example, the 138 miles from Helensburgh to Mallaig were used by only four trains daily in each direction in winter (three on the remotest section). The exercise undertaken in 1998 is described elsewhere (see Kwan et al. (2000)). We describe it here only in sufficient detail to show the consequences of applying the improved 2000 version of our ILP system.

In 1998, we were unable to tackle the whole of ScotRail as a single problem. Fortunately, the Scottish operations could be divided into three virtually self-contained areas. North Clyde suburban services running East-West through Glasgow were operated by about 90 shifts out of two depots. Services to the South and West of Glasgow were operated by about 100 shifts out of seven depots (one depot shared with the North Clyde services).

The remainder of Scotland was served by about 170 shifts from eleven depots. This area included the Highland areas from Mallaig, Oban and Kyle

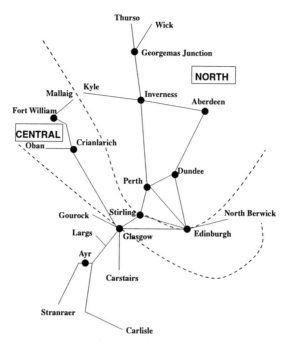

Figure 1. Subdivisions of the ScotRail Network.

of Lochalsh in the west and Thurso and Wick in the north, as well as the major cities of Aberdeen, Dundee, Edinburgh and Perth, and the services from Glasgow Queen Street to the north and east. Unfortunately, the trains in this area included over 1000 relief opportunities, too many for the program's capabilities at that time. It was, therefore, necessary to split these operations.

After some careful consideration, it was determined that two approximately equal sub-problems could be formed by considering separately the central Scottish area, and the rest of the country. For ease of reference we called these the Central and North areas, respectively. The Central area included all services running out of Glasgow Queen Street, and many of the services around Edinburgh and further east. The North area included services running north from Edinburgh across the Forth Bridge, and everything north of Dundee and Perth.

This division caused some problems relating to services to Stirling and Perth from both Edinburgh and Glasgow. Local services to Stirling and Dunblane were frequently interworked with other Central services out of Edinburgh and Glasgow, while services covering the same tracks, but continuing to Perth and beyond were in general best placed in the North area. There was also some difficulty with services operating across the Forth Bridge into Fife, as several of them were driven by crews who also worked on Central area services. We could of course have ensured that all the work of any existing shift was placed in the same subdivision, but this policy would have restricted us to forming shifts similar to the existing ones, and seemed equally arbitrary to the boundaries chosen.

The allocation of train work to subdivisions was an arduous task involving a careful search through all the train workings; this took several days, and was exacerbated by errors in the coding of trains which had been supplied to us. TRACS II was then run separately for each of the subdivisions.

The first runs of the system showed that portions of some trains might have been better served had they been placed in another division. For example, Aberdeen or Inverness to Glasgow trains were initially assigned to the North area, and were often covered efficiently by either Aberdeen, Inverness or Perth drivers. However, the last southbound trains of the day proved difficult to cover in efficient shifts. In practice they were best driven by Glasgow drivers who spent the first parts of their shifts working Central trains, and finished by driving an Aberdeen or Inverness train as far as Perth and then driving a southbound train back to Glasgow.

The initial solutions for the two areas were, therefore, inspected carefully, and the work of less efficient shifts was transferred to the other area if it was felt that this might enhance the quality of the overall solution. In addition to the Glasgow to Perth section, some work between Edinburgh and Stirling was moved in this way. This process of solving the problems separately, inspecting the results and transferring work between areas was carried out iteratively until we felt that no further improvements could be made. It involved skilled personnel over a period of about three person-weeks. The process was made particularly difficult by restrictions on the numbers of shifts that could be worked from certain depots. The results obtained in 1998 used 88 shifts in the North area and 82 in the Central area. This was a reduction of two shifts from the numbers in operation, despite our being required to allow 15 minutes for drivers to change trains at major stations compared with the 10 minutes often allowed in practice. Over the whole ScotRail weekday operations, our solutions indicated a saving of 15 shifts, most of this coming from the Clyde area, where the 15 minute minimum did not apply.

We have recently used the data from the North and Central areas to evaluate the effects of the improvements made to the set covering algorithms in the intervening period. In 1998, the relaxed LP solutions for the subproblems showed 88.00 and 77.39 shifts, respectively, implying that no integer solutions would exist with fewer than 88 and 78 shifts. The North subproblem quickly yielded an integer solution with 88 shifts. However, no solution could be found for the Central problem using a target of 78 shifts, within the limits of the branch and bound search process then being used. The model was then constrained to increase the target to 79, and still no solution was found. The target was increased in steps of one until eventually an integer solution was found with 82 shifts.

The questions to be addressed in the recent investigation were then:

- Could a solution be found for the Central area with fewer than 82 shifts, using in the first instance only those potential shifts generated for the 1998 exercise?
- Could a solution be found for the two areas together with fewer than 170 shifts?
- By allowing more potential shifts to be generated, taking advantage of the increased problem size that can now be handled, could a solution be found using fewer shifts for either area?

To resolve the first question, a set of potential shifts was generated, using the same parameters as in 1998. The set covering process again started to look for a solution with a target of 78 shifts, and progressively increased the target, until this time a solution was reached with only 81 shifts.

The second question has been more difficult to address. The parameter sets used in the two subproblems in 1998 were very different, reflecting the contrast between the generally sparse North area and the intensive commuter traffic of some of the Central area. In 1998, about 78k potential shifts had been generated for the North area. These had been reduced to 66k by removing redundant shifts. Applying the North area constraints to the combined problem in 2000 resulted in far too many potential shifts being generated (the computer run was abandoned with 400k potential shifts generated and many more to come). A compromise in the parameter set was, therefore, reached, and 184k shifts were generated, reduced to 80k. A relaxed LP solution was then found with 169 shifts, implying that with this set of generated shifts, no solution could be obtained with fewer shifts than the 88 and 81 found for the separate subproblems in 2000.

While these results may seem disappointing, yielding an improvement of one shift in one of the subproblems only, and no further improvement from combining the subproblems, there are good reasons for this. Our system in 1998 was already very powerful for problems of the size tackled then. We had not expected an improvement in either subproblem solution, so we were particularly pleased by the Central area solution.

Our solution for the combined problem uses one fewer shift than the separate solutions found in 1998, but does not yield any improvement over the separate solutions of the current year. This is perhaps unsurprising, because a great deal of effort was expended in deciding how the two subproblems should best be configured. A significant advantage gained from the ability of the system now to solve larger problems is of course, that it is no longer necessary to expend such effort.

The third question above remains to be answered. The research described here has set the foundations, but there has not yet been sufficient time to conduct proper experiments.

Regional Railways North East: Regional Railways North East operated a wide range of services over a complex network (see Figure 2). These included inter-city routes such as Newcastle to Liverpool, taking about four hours, and many local inter-urban and rural services.

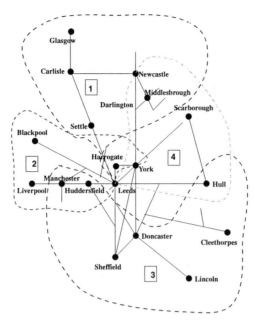

Figure 2. Subdivisions of the Regional Railways North East Network.

In 1996 our Group was commissioned by RRNE to apply alternative scenarios to their whole Monday to Friday operation. At that time, this was by far the largest and most complicated problem we had ever tackled, with around twenty depots or groups of drivers, and over 400 daily driver shifts. There were restrictions on route and traction knowledge for each depot, so that any individual portion of train work between relief opportunities could only be driven from a small number of depots. The route knowledge of neighbouring depots did, however, overlap considerably, so that the whole operation was a single interacting process rather than a set of individual depot-based problems. The numbers of shifts to be assigned to certain depots were also limited. For further details see Kwan et al. (1999a).

Again TRACS II was not, at that time, capable of dealing with such a large problem and it was necessary to decompose it. We subdivided the whole operation into five subproblems based on suitable combinations of depots, route and traction knowledge. One subdivision represented the electric powered services and was essentially self-contained. The remaining four subproblems resulted from the decomposition of the network shown in Figure 2. In order to achieve this, trains which worked across the whole area had to be broken into several sections. For example, the Newcastle to Liverpool train was split into Newcastle to York, allocated to subdivision 4, and York to Liverpool, allocated to subdivision 2. This was necessary in order to form subproblems with at most 500 pieces of work, the then limit for the optimisation component of TRACS II. As the column generation algorithm had not then been implemented, there was also a limit of 30,000 generated shifts in force. Because of the time constraints on the investigation, we were unable to devote the same level of effort to the decomposition as was the case for the ScotRail study. This, together with the substantial overlap between the network subdivisions, suggested that the improved capabilities of the current version of TRACS II ought to yield an improved solution.

The four subproblems were re-run with the current ILP system, using the same shift generation parameters as were used in 1996. However, the implementation of the column generation algorithm in the intervening period meant that a less severe reduction in the generated shift set was required prior to optimisation. Some gains may be expected simply from using a larger generated shift set than in 1996 (see Fores and Proll (1998)). We ignore these and concentrate on comparing the results obtained on the four subproblems with those on the combined problem, using the default routes through the current ILP system.

A generated set of 34,477 shifts for subdivision 1 yielded an integer solution with 70 shifts against the 69 shifts indicated by the LP solution; a target which the branch and bound search showed to be impossible to achieve. For subdivision 2, 82,443 generated shifts were submitted to the ILP system, the relaxed LP indicated at least 94 shifts. However, the branch and bound search quickly showed that there was no integer solution with this number of shifts and one with 96 shifts was then found. 41,631 generated shifts were submitted to the ILP system for subdivision 3 and an integer solution with 114 shifts was found from a relaxed solution indicating 109 shifts. Finally an integer solution of 46 shifts from a relaxed solution indicating 46 shifts from 1613 generated shifts was found for subdivision 4. The combined problem comprised 1459 workpieces for which in excess of 195k shifts were generated and reduced to 99k for submission to the ILP system. An integer solution with 318 shifts was found. Note that the lowest possible theoretical solutions of the separate subproblems total to 320, after allowing for infeasible targets. Thus the result for the combined problem is better than any result achievable for the separate subproblems.

The above results were achieved with the branch and bound search limited to 500 nodes which was the limit imposed in TRACS II until the 2000 version. Increasing the limit to 1000 nodes resulted in a decrease of 1 shift in the solution to both the combined problem and subproblem 2, with no change in the solution to the other subproblems.

5 Conclusions

We have shown that enhancements to TRACS II and, in particular, to its algorithmic kernel, based on ILP, have allowed us to handle driver scheduling problems of increased size and complexity. This reduces the need for scheduling scenarios to be decomposed into separate subproblems. In consequence better schedules may be found than previously and the need for substantial expert effort in performing the decomposition is decreased. It is likely, however, that there will always be problems for which decomposition is necessary on size grounds or for which it is beneficial because an expert scheduler can identify a set of self-contained work.

Acknowledgements

We are grateful to the UK Engineering and Physical Sciences Research Council for financial support under grant GR/M23205.

Bibliography

Böhringer, B., R. Borndörfer, M. Kammler, and A. Löbel (2000). *Scheduling duties by adaptive column generation*. Presented at: CASPT 2000 – The 8th International Conference on Computer-Aided Transit Scheduling, Berlin, June 2000.

Caprara, A., M. Fischetti, P. L. Guida, P. Toth, and D. Vigo (1999). Solution of large-scale railway crew planning problems: The Italian experience. In N.H.M. Wilson (Ed.), *Computer-Aided Transit Scheduling, Lecture Notes in Economics and Mathematical Systems*, 471, Springer, Berlin, 1–18.

Curtis, S.D., B.M. Smith, and A. Wren (1999). Forming bus driver schedules using constraint programming. In *Proceedings of the 1st International Conference on the Practical Application of Constraint Technologies and Logic Programming 1999 (PACLP99)*, The Practical Application Company Ltd., 239–254.

Daduna, J.R., I. Branco, and J.M.P. Paixão (Eds.) (1995). *Computer-Aided Transit Scheduling, Lecture Notes in Economics and Mathematical Systems*, 430. Springer, Berlin.

Desrochers, M. and J.-M. Rousseau (Eds.) (1992). *Computer-Aided Transit Scheduling, Lecture Notes in Economics and Mathematical Systems*, 386. Springer, Berlin.

Fores, S. and L. Proll (1998). Driver scheduling by integer linear programming: The TRACS II approach. In P. Borne, M. Ksouri, and A. El Kamel (Eds.), *Proceedings CESA'98 Computational Engineering in Systems Applications, Symposium on Industrial and Manufacturing Systems 3*, L'Union des Chercheurs et Ingenieurs Scientifique (UCIS), Villeneuve d'Ascq, 213–218.

Fores, S., L. Proll, and A. Wren (1999). An improved ILP system for driver scheduling. In N.H.M. Wilson (Ed.), *Computer-Aided Transit Scheduling, Lecture Notes in Economics and Mathematical Systems*, 471, Springer, Berlin, 43–61.

Forrest, J.J. and D. Goldfarb (1992). Steepest-edge simplex algorithms for linear programming. *Mathematical Programming 57*, 341–374.

Freling, R., R. Lentink, and M. Odijk (2001). *Scheduling train crews: A case study for the Dutch railways*. This Volume.

Freling, R., A.P.M. Wagelmans, and J.M.P. Paixão (1999). An overview of models and techniques for integrating vehicle and crew scheduling. In N.H.M. Wilson (Ed.), *Computer-Aided Transit Scheduling, Lecture Notes in Economics and Mathematical Systems*, 471, Springer, Berlin, 441–460.

Kroon, L.G. and M. Fischetti (2001). *Crew scheduling for Netherlands railways: "Destination: Customer"*. This volume.

Kwan, A.S.K., R.S.K. Kwan, M.E. Parker, and A. Wren (1999a). Producing train driver schedules under different operating strategies. In N.H.M. Wilson (Ed.), *Computer-Aided Transit Scheduling, Lecture Notes in Economics and Mathematical Systems*, 471, Springer, Berlin, 129–154.

Kwan, A.S.K., R.S.K. Kwan, M.E. Parker, and A. Wren (2000). *Proving the versatility of automatic driver scheduling on difficult train & bus problems*. Presented at CASPT 2000: The 8th International Conference on Computer-Aided Transit Scheduling, Berlin, June 2000.

Kwan, A.S.K., R.S.K. Kwan, and A. Wren (1999b). Driver scheduling using genetic algorithms with embedded combinatorial traits. In N.H.M. Wilson (Ed.), *Computer-Aided Transit Scheduling, Lecture Notes in Economics and Mathematical Systems*, 471, Springer, Berlin, 81–102.

Rousseau, J.-M. and J. Desrosiers (1995). Results obtained with Crew-Opt: A column generation method for transit crew scheduling. In J.R. Daduna, I. Branco, and J.M.P. Paixão (Eds.), *Computer-Aided Transit Scheduling, Lecture Notes in Economics and Mathematical Systems*, 430, Springer, Berlin, 349–358.

Smith, B.M. (1986). *Bus Crew Scheduling Using Mathematical Programming*. Ph.D. thesis, University of Leeds.

Smith, B.M. and A. Wren (1988). A bus crew scheduling system using a set covering formulation. *Transportation Research A 22*, 97–108.

Willers, W.P., L.G. Proll, and A. Wren (1995). A dual strategy for solving the linear programming relaxation of a driver scheduling system. *Annals of Operations Research 58*, 519–531.

Wilson, N.H.M. (Ed.) (1999). *Computer-Aided Transit Scheduling, Lecture Notes in Economics and Mathematical Systems*, 471. Springer, Berlin.

Wren, A. and B.M. Smith (1988). Experiences with a crew scheduling system based on set covering. In J.R. Daduna and A. Wren (Eds.), *Computer-Aided Transit Scheduling, Lecture Notes in Economics and Mathematical Systems*, 308, Springer, Berlin, 104–118.

Scheduling Train Crews: A Case Study for the Dutch Railways

Richard Freling[1,2], Ramon M. Lentink[2], and Michiel A. Odijk[2]

[1] Econometric Institute, Erasmus University Rotterdam,
P.O. Box 1738, NL-3000 DR Rotterdam, The Netherlands
`freling@few.eur.nl`
[2] ORTEC Consultants b.v., Gouda, The Netherlands

Abstract. In this paper, we consider a case study dealing with the scheduling of train crews at Dutch Railways (NS). A heuristic branch-and-price algorithm is used, which is suitable for large scale crew scheduling problems, such as the train guard scheduling problem at NS. Computational results show that our algorithm is capable of getting sub-optimal solutions within reasonable computation time.

1 Introduction

In this paper, we consider a case study dealing with the scheduling of train crews at Dutch Railways (NS). The Crew Scheduling Problem (CSP) is a well-known problem in the field of Operations Research and it consists of designing duties using pre-defined tasks with fixed starting and ending times and locations. Each duty must satisfy a set of work laws and agreements. The objective is to minimize a combination of fixed costs (the number of duties) and variable costs (e.g., penalties for less desired constructions).

For many years public transport companies have used planning systems which incorporate automatic crew scheduling. Although usually the CSP is solved after the vehicle schedules have been designed, it has recently also become possible to design integrated vehicle and crew schedules for small to medium-scale problems (see, e.g., Freling et al. (2001a)). After solving the CSP, the resulting duties must be assigned to individual crew members. This process is called crew rostering.

This paper deals with a case study, which has been carried out for NS by ORTEC Consultants in the Netherlands. NS currently plan their crew manually, but recently several researchers have done case studies to see whether automatic crew scheduling is an interesting alternative. Appropriate NS case studies are presented in Fischetti and Kroon (1999) and Kroon and Fischetti (2001). Both papers are more of a practical nature, and only very general information is provided about the solution procedure. In the first paper, a slightly different instance of the same case is studied as in this paper. The problem is formulated as a set covering model with additional constraints, and solved using a Lagrangian heuristic with implicit column generation.

In this paper, we propose a heuristic algorithm for the CSP, which uses column generation to solve linear programming problems and a branch-and-price heuristic to get integer solutions. Columns are generated implicitly using a dynamic programming algorithm. The heuristic is part of a general framework, which has been developed at ORTEC Consultants and is suitable for crew scheduling and rostering problems in different contexts. In fact, slightly different versions of this algorithm have been used for scheduling bus drivers and airline crew, and also for train and airline crew rostering. Furthermore, the approach allows for solving large scale CSPs within reasonable computation time, and is very robust in the sense that different scenarios with respect to, e.g., union and/or governmental regulations can easily be implemented. The mathematical techniques are not discussed in great detail here. A more comprehensive description of the subject can be found in Freling (1997); Lentink (1999); Freling et al. (2001b).

Most of the Operations Research literature on crew scheduling deals with bus driver scheduling and airline crew scheduling. From our experience, the main differences between scheduling train crew and bus drivers are that

- train crew travels more often as a passenger on a train to get to different locations,
- train crew can begin/end a duty at a relative longer distance from their home bases (sometimes spending the night at a hotel away from the home base),
- delays are more critical for trains because of the stronger interconnectivity of a train network.

Just like in the airline context, sometimes trains may be operated all round the clock, which makes it harder to schedule the crew on a daily basis. In the airline context, a schedule (also called a pairing) usually may cover two or three days.

This paper is organized as follows. In the next section, we introduce the NS case. One of our primary research objectives for this particular case was to adapt the algorithms in our framework to be able to deal with large-scale problems. In the next section, we present the branch-and-price heuristic, and we also discuss techniques for speeding up the algorithm. Finally, in the last section, we show computational results for the scheduling of train guards. For the NS instance, which consists of 1114 tasks, we are able to get a sub-optimal solution within reasonable computation time.

2 A Case Study for NS

In this section we discuss a case study carried out for NS. The case consists of four intercity lines. These lines connect the north east of the Netherlands with the urban part in the west, and, therefore, it is called the North-East

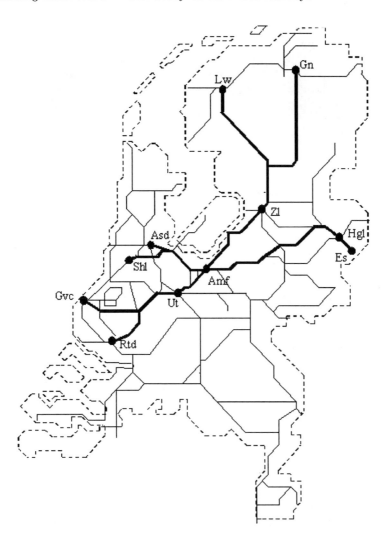

Figure 1. Lines that form the North-East case.

case. The network representing these lines is visualised in Figure 1, which is taken from Fischetti and Kroon (1999). In this paper a similar case is studied.

Train guards and drivers have to be assigned to the trains that operate on these lines. For this case study, we examined the generation of schedules for guards on a weekday for intercity lines. The generation of schedules for drivers is outside the scope of this case. It is assumed that a timetable has already been generated, and also that rolling stock has been assigned to operate the timetable.

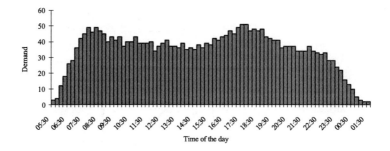

Figure 2. Demand for guards during a weekday.

A task is defined as the minimum portion of work that has to be carried out by one person. Here, this means a part of a line where it is not allowed for a guard to change trains in between. The data consisted of 1114 tasks that had to be assigned to the guards, which were not spread evenly across a day. In the morning and evening peaks more guards are necessary because more and longer trains are used. Moreover, when the length of a train exceeds certain thresholds more guards are necessary. Figure 2 shows the demand for guards during the whole day.

The scheduling solution must satisfy the following criteria:

- Efficient, i.e., the number of required guards is as low as possible.
- Robust, i.e., the duties must be of high quality with respect to delays of trains. This is taken care of by penalizing changes from one train to another within a duty.

The duties have to meet certain work laws and agreements in order to be feasible. These laws and agreements are described below. There are three levels of feasibility restrictions:

1. *High level restrictions.* Here we find rules that are called coupling restrictions because they relate to a set of duties. There are three of those rules in this case:
   (a) At most 5% of all selected duties have duration shorter than 5 hours.
   (b) At most 5% of all selected duties have duration longer than 9 hours.
   (c) The average length of the selected duties is at most 8 hours.
2. *Medium level restrictions.* Rules at this level determine the feasibility of a duty. Most restrictions in the case are at this level:
   (a) The maximum length of a duty is 9.5 hours.
   (b) The minimum length of a duty is 4 hours.
   (c) A duty with a length of at least 5.5 hours should contain a meal break of at least 30 minutes.
   (d) The meal break should not start more than 5.5 hours after the start of the duty.
   (e) The meal break should not end more than 5.5 hours before the end of the duty.

    (f) Each duty must start with a sign-in time of 20 minutes.

    (g) Each duty must end with a sing-off time of 15 minutes.

    (h) A duty has to start and finish at the same location.

3. *Low level restrictions.* A restriction on this level determines if two tasks can be performed consecutively:

    (a) The end location of the first task and the start location of the second task should be identical.

    (b) The second task can only start after the first task has finished.

    (c) In case of changing of trains a buffer of 16 minutes should be taken into account.

In the next section we discuss how these restrictions are incorporated in our solution approach.

3   A Heuristic Algorithm for the Railway Crew Scheduling Problem

The algorithm we propose for the railway CSP is part of a larger framework, developed at ORTEC Consultants, a medium sized company located in the Netherlands. Since its foundation in 1981, ORTEC has worked in close collaboration with the Erasmus University Rotterdam. The company is specialized in decision support systems that contain (advanced) Operations Research techniques. More information about some systems that are focused on railway planning can be found in Lentink et al. (2000). Several personnel planning systems are developed at ORTEC for various fields of industry. One such system is CDR-Lite (Crew Duty Rostering), a crew planning system for the airline and railway industry. The module for automatic crew scheduling and rostering is based on a branch-and-price algorithm for a generalized set covering or partitioning model. The general framework is very robust in the sense that both the constraints defining the feasibility of each duty, and the constraints defining the feasibility of a set of columns can be easily incorporated without changing the structure of the algorithm. In addition, for large scale problems several exact and heuristic techniques can be used in a straightforward manner to speed-up the algorithm. Here, we discuss the mathematical formulation for the CSP, and present a special version of the algorithm in the general framework. That is, a heuristic branch-and-price algorithm for the railway CSP.

3.1   Mathematical Formulation

The railway CSP is formulated as a generalized set covering model. Before presenting the model, we need the following definitions:

    R the set of all feasible duties

    I the set of tasks which need to be covered

$R(i)$ the set of duties covering task $i \in I$

d_r the cost of duty $r \in R$

b_i number of duties which need to cover task $i \in I$

X_r 1, if duty r is selected, 0, otherwise

R_k the set of duties with maximum length of k hours

R^k the set of duties with minimum length of k hours

t_r the length of duty $r \in R$ in minutes

The CSP is then formulated as follows:

$$\text{Minimize} \quad \sum_{r \in R} d_r X_r$$

$$\text{subject to} \quad \sum_{r \in R(i)} X_r \geq b_i \qquad\qquad \forall\, i \in I \qquad (1)$$

$$95 \sum_{r \in R_5} X_r \leq 5 \sum_{r \in R \setminus R_5} X_r \qquad\qquad (2)$$

$$95 \sum_{r \in R^9} X_r \leq 5 \sum_{r \in R \setminus R^9} X_r \qquad\qquad (3)$$

$$\sum_{r \in R} (t_r - 480) X_r \leq 0 \qquad\qquad (4)$$

$$X_r \in \{0, 1\} \qquad\qquad \forall\, r \in R \qquad (5)$$

The objective is to minimise the total costs of the cover. Constraints (1) are generalised set covering constraints, which state that at least b_i duties of the set R_i are in the solution. Constraints (2)-(4) are the high level restrictions as defined in the previous section. For example, constraints (2) state that for every 5 duties with a length shorter than 5 hours, we need 95 duties with a length longer than 5 hours. This implements restriction (1a) of the previous section. We need to use set covering because over-covers have a meaning in this context. Whenever a train crew needs to be *positioned* it is travelling as a passenger on a scheduled train. Thus, if a trip needs to be covered once but is covered twice, this means that one of the duties covering the trip contains a *positioning*. The right-hand-side of constraints (1) may be larger than one, because more than one crew member may need to cover a trip.

3.2   Heuristic Branch-and-Price

Branch-and-price is a special application of branch-and-bound, where column generation is used to solve linear programming (LP) relaxations with a huge number of variables. The generalised set covering model presented above also usually contains a huge number of variables (i.e., duties). These types of formulations can be used to model a wide variety of crew scheduling and rostering problems. For similar work on a general framework in the context of time constrained routing and scheduling problems we refer to Desrosiers et al. (1995). See also Barnhart et al. (1998) for a general discussion of column generation in the context of integer programming (IP). In the last decade many papers deal with column generation approaches for public transport scheduling problems. Here, we only briefly discuss the general outline of our approach, and go deeper into some particular details such as how to deal with the coupling constraints in a column generation context.

The column generation algorithm to solve the LP relaxation in the root node of the branch-and-bound tree starts with an initial selection of columns. Then, additional columns are generated implicitly while needed until the LP relaxation is solved to optimality. The medium level restrictions are checked during the generation of columns. The procedure to solve the IP problem uses the same column generation algorithm to solve the LP relaxations in each other node of the branch-and-bound tree. Here, the column generation procedure is started with the columns in the final LP problem of the parent node.

Two aspects are of major importance in this solution approach: how columns are generated, and which branching rule is used. The generation of columns is discussed in more detail in the next subsection. For the branch-and-price procedure, we branch on the X_r variables, where we only consider the 1-branch, i.e., we only solve the LP relaxation after fixing a variable X_r to one. Normally, for this type of branch-and-price algorithms one needs to use a special branching rule which is valid for the column generation procedure. The simple branching rule can be used here with column generation because we only solve the 1-branch and stop once an integer solution is found. It would not be possible to solve the 0-branch because we can not forbid a column to be generated in our column generation algorithm.

3.3   Generating Duties using Dynamic Programming

The underlying network structure of the CSP is exploited in the column generation procedure. The network is defined such that each node corresponds to a task and each arc corresponds to a feasible sequence of two tasks in one duty, where the low level restrictions have been checked. In addition a source and sink arc are added to the network, denoting the start and end of a duty. A path in the network corresponds to a feasible duty if the duty constraints are satisfied. See also Desrochers et al. (1992) for an example of a similar network for the CSP. Paths in the networks are constructed by solving a resource constraint shortest path algorithm using dynamic programming (see e.g., Desrosiers et al. (1995) and Freling (1997)).

For the column generation algorithm we need to find one or more paths that correspond to one ore more columns with negative reduced cost. The dual variables of the high level restrictions (2)-(4) need to be incorporated in the column generation algorithm. Let m be the number of constraints of type (1), let y_i be the dual variable corresponding to constraint (1), and let y_{m+1}, y_{m+2} and y_{m+3} be the dual variable corresponding to the coupling constraints defined as high level restrictions (1a), (1b) and (1c), respectively. Furthermore, let R^{5-9} be the set of duties with length between 5 and 9 hours. Then, the reduced costs of a column r in the set covering model is defined as follows:

$$
\bar{c}_r = \begin{cases}
c_r - \sum_{j \in J_i} y_j + 95 y_{m+1} - 5 y_{m+2} + (t_r - 480)\, y_{m+3} & \forall\, r \in R_5 \\[2mm]
c_r - \sum_{j \in J_i} y_j - 5 y_{m+1} + 95 y_{m+2} + (t_r - 480)\, y_{m+3} & \forall\, r \in R^9 \\[2mm]
c_r - \sum_{j \in J_i} y_j - 5 y_{m+1} - 5 y_{m+2} + (t_r - 480)\, y_{m+3} & \forall\, r \in R^{5-9}
\end{cases}
$$

Reduced costs need to be negative because we are minimizing. A feasible path up to a node in the network (except for the sink) is called a *partial path*. We can define the cost of arcs and a cost function for modifying the cost when adding an arc to a partial path. In this way the cost of a path from the source to the sink is such that it corresponds to the reduced cost of the corresponding column (duty) in the model. Similar, each constraint that defines the feasibility of a column can be translated to consumption on the arcs and a function for defining this constraint's consumption of a partial path when adding an arc to it. The reduced costs and the constraints are called *resources*, and the problem of finding the minimum cost path satisfying the constraints is called the resource constraint shortest path problem.

For the dynamic programming algorithm, dominance tests are carried out at each node in the network in order to reduce the state space. That is, keeping only those paths that can possibly be part of an optimal solution reduces the number of feasible paths up to a node. For this case three resources can establish dominance. The first resource is the duration of a duty after the meal break. The smaller the duration of the partial path, the more time is left for tasks that can be performed from here (according to rule (2e)). The second resource is the reduced costs of the partial path, i.e., the lower the reduced costs of a partial path the better. The last resource is more complicated. It corresponds to the duration of (the part of) the duty corresponding to the partial path, where shorter duties are better. Because a duty should have duration of at least 4 hours (see rule (2b)), this resource should only be used if the length of both paths is at least 4 hours. However, the coupling constraint (rule (1a)) plays a role in this as well. Let partial paths p_1 and p_2 have the same value for the first two resources. Path p_1 has a duration of 4.5 hours and path p_2 has a duration of 7 hours, so p_1 would dominate p_2. But because at most 5% of the selected duties may have duration of less than 5 hours, it may be that one or more complete paths constructed from partial path p_1 will be dominated by complete paths constructed from partial path p_2 due to a change in reduced costs. This because the dual variables with respect to this coupling constraint can only be taken into account once a path is completed. Note that once the partial path has duration of at least 5 hours we know that this dual variable need not be taken into account. The

Rule	Nodes	Arcs
Original network without reduction	1114	84497
Removing duplicate arcs and nodes in the network	820	38910
Further network reduction	820	26513

Table 1. Results of network reduction.

other two coupling constraints do not influence this criterion since the path on the 'safe side' is preferred.

3.4   Acceleration Techniques

In order to be able to solve the large scale scheduling of train guards, we experimented with several acceleration techniques to speed up our algorithm. In the last section we present some computational results which will show effects of these techniques on the computation time.

Reduced network size: We implemented several techniques for reducing the size of the network. As noted before, the number of guards required on a train depends on the length of the train among others. When more than one guard is necessary for a task, there are two possibilities with respect to the mathematical model. It is possible to see the tasks as different and insert them separately in the network. This is a straightforward manner but is undesired because tasks that have to be performed by more than one guard can be found regularly. This means an extra side constraint in the formulation but also extra nodes in the network with extra incoming and outgoing arcs. Therefore, another approach was implemented, where each identical task has to be covered more than once (i.e., $b_i > 1$ in the model for each such task i). As a result, the number of nodes as well as the number of arcs in the network decreases significantly because 'duplicates' are moved.

We briefly discuss two standard network reduction techniques we used:

- Remove arcs between the source/sink and nodes for which it is not allowed to start/end a duty. This is possible due to restriction on the location or on the start/end time of a duty.
- Remove arcs for which corresponding duration exceeds a value, which is based on the maximum duty duration.

The results of these network reductions are shown in Table 1. The last row contains the size of the final network that is used in the algorithm. We see that it is possible to reduce the network considerably, i.e., the number of nodes can be reduced by more than 25%, while the number of arcs by nearly 70%. These types of techniques are very important in order to solve large scale CSPs with our algorithm within reasonable computation time.

Dynamic network size: Especially in the first iterations of the column generation procedure a lot of paths can be generated with negative reduced cost, although we are only interested in few of them. Therefore, we can use a heuristic technique to speed up the procedure. This technique is focused on restricting the number of outgoing arcs at a node that will be considered. This number is updated dynamically during the column generation procedure. When tailing off, only small improvements in the objective function, is detected the number of outgoing arcs is increased. This continues until all outgoing arcs at all nodes are considered, so that optimality of solving the LP relaxation can still be guaranteed. The tailing off detection and the increase in number of outgoing arcs are the most important parameters for this technique. The strategy we employed is initializing the number of outgoing arcs to 5 and increasing it with 100 when in two consecutive iterations the value of the objective function decreases with less than 5%. Of course, we increase the number of outgoing arcs as long as not all outgoing arcs at each node are considered.

Multiple variable fixing: An acceleration technique for the branch-and-price part of our algorithm is multiple variable fixing. Instead of fixing only one variable X_r to 1 in each branch, we fix all variables with fractional value above 0.6 to 1 (see also Gamache et al. (1999)).

4 Computational Results

In Table 2 we summarize the computational results for the case described in this paper. This case consists of 1114 tasks that need to be assigned to duties for train guards. We tested the algorithm presented in the previous section and some of its variations. The second to fourth columns in the table show the results for the basic algorithm including the network reductions, and the fifth column without network reductions. The third and fourth columns show the results for multiple variable fixing, and dynamic network size, respectively. The results shown in each column are, respectively, the CPU runtime for solving the LP relaxation in the root node (in seconds on a Pentium II/400MHz/128Mb), the total CPU runtime of the algorithm, the objective value of the LP relaxation in the root node, the objective value of the final integer solution, the gap between these two, the number of branches needed to find the integer solution, and information about the solution: the number of duties, the average duration of the duties, the number of positionings, and the percentage of short and long duties.

The cost function is a combination of fixed cost per duty, and variable costs for positionings and change of trains. We have imposed the condition that at most one change of trains is allowed in a duty. As can be seen from Table 2, the basic algorithm (including network reductions) is able to find a good solution with 2.32% gap within 32 minutes. Multiple variable fixing greatly improves the runtime to slightly more than 11 minutes but the gap

Method	Basic	Multiple variable fixing	Dynamic network size	No network reduction
CPU LP	225	225	228	1193
CPU total	1935	687	1978	11473
LP value	31064	31064	31064	31086
best integer	31801	32255	31801	34427
gap %	2.32	3.69	2.32	9.71
Number of branches	102	14	102	102
Duties	132	134	132	143
Average duration	7:58	7:54	7:58	7:55
Positionings	73	81	73	154
short duties %	1.52	2.24	1.52	3.50
long duties %	3.79	2.99	3.79	4.20

Table 2. Computational results for the large dataset.

increases to no more than 3.69%. This means two extra duties and 8 extra positionings, but a decrease in average duty duration of 4 minutes. The large reduction in computation time is caused by the large decrease in the number of branches in our branch-and-bound algorithm. The dynamic network size did not perform well because of an extra runtime of 43 seconds. In previous research we noted that the dynamic network size can significantly improve the runtime when the network is larger. Finally, in the last column we can see that the runtime explodes to more than 3 hours when the network reductions are not used.

From this large data set we were able to derive a smaller data set, consisting of 216 tasks. Some results of the described methods on this case are condensed in Table 3. Here we see again that the multiple variable fixing strategy performs very well. Only 3 branches are needed to generate an integer solution, which of course greatly influences the CPU time of the algorithm. Moreover, we see that the dynamic network size strategy does not perform that well. Finally, the network reduction techniques of our algorithm are of great use for this data set as well.

5   Conclusion

Naturally, although the results are very promising, we need to be careful when drawing conclusions based on two instances only. Therefore, future research is necessary to test the algorithm and its variations on other instances as well. Although we know of other research on similar cases provided by NS, we

Method	Basic	Multiple variable fixing	Dynamic network size	No network reduction
CPU total	4.69	3.86	7.33	26.81
gap %	0.43	0.43	0.43	6.81
number of branches	17	3	17	22

Table 3. Computational results for the small dataset.

do not have access to information to provide a good comparison. In Fischetti and Kroon (1999) results are presented using the TURNI system for the same case but with slightly different constraints and objectives. Their approach is based on column generation as well, but using Lagrangian relaxations and heuristics instead of LP relaxations and branch-and-price (see also Caprara et al. (1999)). Adding different level constraints does have little impact on our solution approach. As mentioned before, the main advantage of our solution approach is its robustness. If some rules change this can be incorporated very easily in our approach and the algorithm needs very little tuning.

Bibliography

Barnhart, C., E.L. Johnson, G.L. Nemhauser, M.W.P. Savelsbergh, and P.H. Vance (1998). Branch-and-price: Column generation for solving huge integer programs. *Operations Research 46*, 316–329.

Caprara, A., M. Fischetti, P.L. Guida, P. Toth, and D. Vigo (1999). Solution to large scale railway crew planning problems: The Italian experience. In N.H.M. Wilson (Ed.), *Computer-Aided Transit Scheduling, Lecture Notes in Economics and Mathematical Systems*, 471, Springer, Berlin, 1–18.

Desrochers, M., J. Gilbert, M. Sauve, and F. Soumis (1992). CREW-OPT: Subproblem modeling in a column generation approach to urban crew scheduling problem. In M. Desrochers and J.-M. Rousseau (Eds.), *Computer-Aided Transit Scheduling, Lecture Notes in Economics and Mathematical Systems*, 386, Springer, Berlin, 395–406.

Desrosiers, J., Y. Dumas, M.M. Solomon, and F. Soumis (1995). Time constrained routing and scheduling. In M.O. Ball, T.L. Magnanti, C.L. Monma, and G.L. Nemhauser (Eds.), *Network Routing, Handbooks in Operations Research and Management Science*, 8, Elsevier, Amsterdam, 35–139.

Fischetti, M. and L. Kroon (1999). *Scheduling train drivers and guards: The Dutch "Noord-Oost" case.* Technical Report Management Rep. no. 25-1999, Erasmus University, Rotterdam.

Freling, R. (1997). *Models and Techniques for Integrating Vehicle and Crew Scheduling.* Ph.D. thesis, Tinbergen Institute, Erasmus University, Rotterdam.

Freling, R., D. Huisman, and A.P.M. Wagelmans (2001a). Applying an integrated approach to vehicle and crew scheduling in practice. This volume.

Freling, R., R.M. Lentink, and A.P.M. Wagelmans (2001b). *Crew planning for passenger transportation: A branch-and-price framework*. Technical report, Erasmus University, Rotterdam. In preparation.

Gamache, M., F. Soumis, G. Marquis, and J. Desrosiers (1999). A column generation approach for large-scale aircrew rostering problems. *Operations Research 47*, 247–263.

Kroon, L. and M. Fischetti (2001). Crew scheduling for Netherlands railways, "Destination: Customer". This volume.

Lentink, R.M. (1999). Crew Scheduling voor NS Reizigers (in Dutch). Master's thesis, Free University Amsterdam, Amsterdam.

Lentink, R.M., M.A. Odijk, R. Freling, and J.S. de Wit (2000). Use of operations research to facilitate and improve railway planning. In J. Allen, R.J. Hill, C.A. Brebbia, G. Sciutto, and S. Sone (Eds.), *Computers in Railways VII*, WIT Press, Southampton, 231–239.

Evaluating a DSS for Operational Planning in Public Transport Systems: Ten Years of Experience with the GIST System

Teresa Galvão Dias[1], José Vasconcelos Ferreira[2], and João Falcão e Cunha[1]

[1] GEIN-DEMEGI / INEGI, Faculdade de Engenharia da Universidade do Porto,
R. Dr. Roberto Frias, 4200465 Porto, Portugal
{tgalvao, jfcunha}@fe.up.pt
[2] SAGEI / INEGI, Universidade de Aveiro, Campus Universitário de Santiago,
3810193 Aveiro, Portugal
jvasferr@fe.up.pt

Abstract. The GIST Decision Support System (DSS) has been specified and developed by a consortium of two University R&D groups and five public transport companies, aiming to support the operational planning processes of the latter. It has so far been implemented with unquestionable success in several companies, although it has been difficult to measure objectively the gains it directly introduced into their daily operations.

This paper introduces the GIST DSS, describing its initial architecture and functionality, together with its more recent evolutions. It also presents the methods used in its specification, development and implementation, in particular the software process. In order to measure the success of GIST, this paper proposes a general DSS multidimensional impact evaluation model, and reports on the results of an experiment where this model is validated in five of the companies that are currently using GIST. Results of the validation process confirm the overall success of GIST, but we also highlight several areas where companies need more support and are not entirely happy. The experiment also seems to indicate that the impact evaluation model proposed is an appropriate tool to measure the success of a DSS.

1 Introduction

A consortium of two research institutions and five public transport companies has been working together over the past 10 years in one ambitious Operations Research (OR) project in Portugal. The GIST Decision Support System, resulting from such collaboration, is today in current use in six companies and their 25 operational planning centers.

The objective of the GIST DSS, in terms of its software tool, is to produce improved schedules of vehicles and crews. Therefore, it was initially expected that quantitative measures of its impact on operations could be easily made. For instance, it was expected that the number of buses and drivers required to support some predefined level of operation could be reduced with the help of the system, or that, with the same resources, more trips could be offered to the public. In fact, company users agree that this result has been achieved, but are unable to quantify it objectively. It has been difficult, if not impossible, either to get information from the companies stating objective gains (e.g., number of buses that are no longer required), or to involve the companies in controlled real scale experiments that provide quantitative results.

Despite these difficulties, the 10 years of experience with the GIST system already justify an assessment of the whole project. In particular, we feel that there were some lessons to be gained from a global evaluation to be carried out in the companies where GIST is used daily. We think that this evaluation process must include not only the DSS itself, but also the processes that contributed to its specification and development in the first place, and the processes that currently support its operation.

We, therefore, propose a model for the evaluation of a DSS for operational planning, based on its actual impacts on the companies. This proposed evaluation model, described in detail in Section 3, has been configured to the GIST system through a series of questionnaires that were then applied to the companies and users that actually have been using it. The results of such an experiment are discussed in Section 4. Our final conclusions are presented in Section 5.

The following Section 2 introduces the GIST system and presents the methods used to develop it, in particular the software development method that was used. We argue that the close involvement of endusers of GIST and their direct managers from the transport companies, with researchers, software designers, and programmers from the University R&D groups, contributed to the success of the project. This close involvement started in the early phase of specification, and has continued throughout the project. The results of the experimental evaluation of GIST confirm this, but also raise some questions regarding the success of the assistance phase, which follows GIST's implementation in the companies.

2 The GIST System

2.1 General Presentation

The GIST System is a DSS for the operational planning at mass transit companies resulting from a collaborative programme. This collaboration was formally established through the GIST consortium, involving two university OR units and five Portuguese mass transit companies. The university groups were INEGI, at the Faculty of Engineering of the University of Porto, and ICAT, at the Faculty of Sciences of the University of Lisbon. The mass transit companies involved in the consortium were Carris, STCP, Horários do Funchal, Vimeca and Empresa Barraqueiro.

At the beginning of this project almost all of these companies used manual procedures, occasionally supported by simple worksheets or customized software tools. The perceived high costs involved in the acquisition of one of the existing commercial systems, as well as the difficulties in the adaptation to each particular real-world setting, delayed the transition to automatic procedures in those companies. Besides, it was impossible for each company to afford the development of its own system.

Considering the reasons mentioned above, the main objective of the GIST project was to specify, develop and implement a new decision support system for operational planning flexible enough to handle all the different realities presented by a wide range of companies.

The architecture model and evolution of the GIST system, presented in Figure 1, were designed as a modular system, which provides flexibility and adaptability. The first version, implemented in 1994, addressed the following processes:

- Basic information management, which includes modelling of the network and routes, and the definition of trips.
- Vehicle scheduling, basically the process of assignment of trips to vehicles, mostly using algorithms or the user interface.
- Crew scheduling, involving the assignment of crews to vehicles, problem supported in algorithms but in practice solved using the user interface interactive tools.

The success of this version led to architectural changes and the further development of new modules, namely the Crew Rostering module that was implemented in 1999. More recently, a third version of GIST includes the INFOBUS and PIB modules, which are currently under development and are briefly described below.

The modules supporting basic information management allow the definition and maintenance of the data for the whole network, which comprises the depots, paths, routes, relief points, distances, travel times, the definition of the trips, and the specification of the public timetables.

The Vehicle Scheduling and the Crew Scheduling modules concern the assignment of vehicles and crews to the trips previously defined. Both modules use algorithms and heuristics of considerable complexity, together with graphical representations of solutions.

The Crew Rostering module concerns the actual assignment of duties to specific crews, taking into account, e.g., personnel breaks or holidays. This module was itself divided into three components:

- Roster templating, that creates a roster model that supports the generation of all the daily rosters within a given time horizon.
- Daily rostering, allowing the manipulation of the daily rosters, regarding the particular daily situation of the drivers.
- Roster control or post-rostering, that updates the information in the system with what actually has happened.

The INFOBUS module makes available the planning information produced with GIST, such as the timetables of each route, directly to the public.

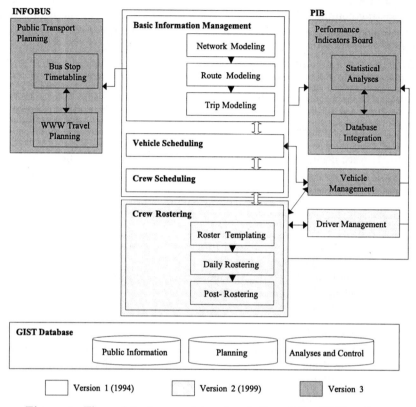

Figure 1. The architecture model and evolution of the GIST system.

INFOBUS can use a Geographical Information System and make its information available through the Internet. Beyond the production of personalized timetables for each station/bus stop, this module also allows interactive travel planning (e.g., answering questions such as "which is the nearest bus stop to some point in the city," or "which is the fastest path between two points using the transportation network").

The PIB, or Performance Indicators Board module, is based on a Data Warehousing system and aims at providing aggregate statistical information and key operational indicators useful for management purposes.

The first version of the GIST system, although modular, was fully integrated, providing the user with some feedback support between modules. For instance, if a vehicle solution requires more buses than available, the users would need to go back to trip planning and change the timetable.

In this first version of the GIST system, the algorithms implemented for the vehicle scheduling and for the crew scheduling problems were traditional OR algorithms (Paixão (1983); Paixão and Pato (1989)). For the vehicle scheduling we used an assignment formulation of the problem with several variations, aiming at providing the users with tools that could solve the particular problems of the companies. For example, the user could fix the number of vehicles available and assign penalties to the trips according to their priority. The algorithms implemented also considered the case in which the company had several depots.

The crew scheduling algorithms were based on a set covering formulation, but the experience showed that this approach was not the most appropriate for the companies in practice (Paixão and Pato (1989)). In the new versions of GIST we tried several other approaches, including hybrid genetic algorithms based on a set partitioning formulation (Galvão Dias (1995)) and other metaheuristics (Portugal (1998)).

The second version of GIST was fully modular, e.g., in the sense that the user who only needs to schedule drivers can input vehicle data directly into the module and evaluate different crew solutions. Obviously, he can also use a vehicle solution obtained from the previous module. All this is supported by a standard GIST database interface (as shown in Figure 1).

The third version of GIST basically extends the previous ones and corresponds to the development of the functionalities that an improved understanding of the needs of the companies allows. In particular, the third version will have improved algorithmic support in the crew scheduling process.

2.2   The Development Methodology Adoption

The development of GIST followed an Incremental Prototyping Model (IPM) methodology, which is based on the continuous evolution of the following three processes: design, development, and testing (DDT). This method aims at using an evolutionary project management style as explained in Woodward (1999). The IPM is an innovative method of software development that has emerged with the rapid evolution of software technology, advanced graphical user interfaces (GUI), and quality-oriented organizations in a very competitive business environment. Traditional approaches (Boehm (1981)) were based on sequential DDT processes, each one with a clear beginning and ending point. On the contrary, IPM requires an ongoing process, in which all the DDT processes are active throughout the whole project (see Figure 2).

This methodology is user-centered, employs rapid prototyping, and iterative usability testing (Boehm (2000); Nunes and Cunha (2000)). This approach promotes an orderly process with predictable progress among stages and requires selected enduser involvement. Usable prototypes, with increasing levels of functionality and robustness, must be regularly provided to the users. At some stage, when the prototypes are satisfactorily mature and agreed objectives have been attained, we must proceed and start training all users. However, this process does not stop the continuing DDT process. Instead, training often gives rise to important contributions to the evolving design and development processes.

The implementation process is carried out when the users are able to use the system autonomously. Usually, in this process, the old procedures are not

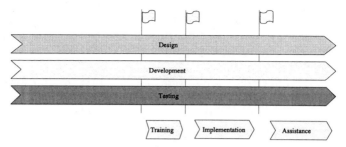

Figure 2. The GIST development methodology.

substituted by the system all at once. They are performed in parallel with the new system and the results and performance are compared. Finally, the company and the project team agree to a deadline for the complete substitution of the old procedures. Afterwards, an assistance process will keep the system correctly operating. Figure 2 summarizes the complete GIST development methodology, which includes, beyond the continuing DDT process, the Training process, the Implementation process and the Assistance process.

The continuing DDT process requires a significant effort of all parties involved, including the users. During the initial stages, the activities related to the Design process, such as the specification of requirements and the database design, demand the primary effort. At the final stages, the main activities are those related to the Testing process. The global effort involved in the DDT during the IPM process, and particularly the corresponding effort required from the users, is presented in Figure 3.

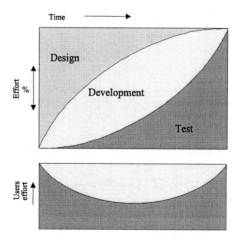

Figure 3. The global effort involved in the DDT process and the corresponding effort required from the users.

To support the application of the IPM we used easily maintainable specifications, based on object-oriented design, Object Modeling Technique (OMT) and Unified Modeling Language (UML) (Rumbaugh et al. (1991); Booch et al. (1999)), an open system architecture enabling existing components to be easily rearranged and new components to be incorporated, and friendly GUI.

One of the most important contributions to the success of an IPM is effective communication. The GIST project involved several companies, each with different technical languages, experiences, and rules. It was, therefore, crucial to homogenize the concepts and the vocabulary in order to allow effective communication. Another important issue was the promotion of back channel communication, in particular for the DDT, which was accomplished through email, chat rooms, groupware and customer relationship management (CRM). This was particularly important as the DDT was partitioned into two groups (Lisboa and Porto), having to deal with users in five companies, one of them on Madeira Island.

3 A Model to Evaluate a DSS for Operational Planning

The evaluation of a DSS for operational planning can be carried out in several different ways. For example, it can involve a group of experts simulating the use of the DSS in a wide range of scenarios. However, we believe that the evaluation of a DSS of this type must also be based on its actual *impacts* on the company operation, not only on its potential impact.

The mentioned *impacts* can originate directly from the DSS itself, but also from the processes that support it: the design, implementation, and assistance processes. Usually, the DSS development team has the main responsibility for such processes. In Figure 4 we present the most relevant impacts that can be achieved with the overall process, and propose a three level classification: *(1)* direct impact, *(2)* intermediate impact, and *(3)* final impact.

As *direct impact* we consider the effects originating directly from the DSS and its supporting processes. The *intermediate impact* result from the exploitation made by the companies of the direct impact. The *final impact* reflects the companies' highlevel objectives. We must underline that the *know-how* and the *attitude* of the DSS users can be considered both direct and intermediate impact. In summary, we can say that:

1. Direct impacts are closely related to DSS tools, capabilities, or experience, know-how, and attitudes of key-users.
2. Intermediate impacts have to do with the changes originated by the DSS in the company or in the way it organizes itself and its business processes.
3. Final impacts can be measured by the market position of the company or by its competitiveness; such final impacts are also called *Critical Success Factors* (Rockart (1979)).

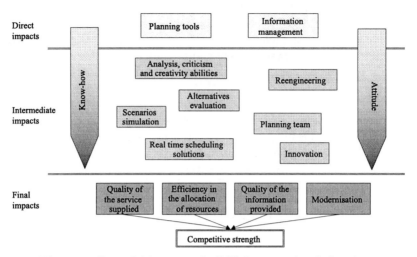

Figure 4. Potential impacts of a DSS for operational planning.

Measuring all these impacts objectively, even the ones with a quantitative nature, is a very difficult task. For example, how many companies are prepared to quantify the savings obtained with a new crew scheduling solution? We must remember that one of the major difficulties in formulating and solving the crew scheduling problem is the acceptance by the companies' staff of a concrete cost function to be optimized. The evaluation is even more complex for the qualitative impacts, such as the *users' attitude* or the *company level of modernization.*

Considering the difficulties in evaluating the impacts, a possible way of conducting the evaluation process is through a questionnaire answered by the companies' staff who are able to express valuable opinions about the DSS and its supporting processes. As these people are mainly technical staff, we should not expect them to be able to evaluate the final impacts, these will be noticed only by those with management responsibilities. In fact, the analysis of the inquiry results can give a precise idea of the direct and intermediate impacts, connecting them with their causes, i.e., the DSS and the three related processes of design, implementation, and assistance. However, this will not result in a cause and effect justification of the final impacts.

The final approach consists of complementing the responses to the inquiry performed with a set of interviews of the management staff, trying to establish a relationship between direct/intermediate impacts and final impacts (see Figure 5).

The evaluation process proposed can then be summarized by the following phases:

1. First, we need to carry out an inquiry among the DSS users in order to estimate the direct and intermediate impacts caused by the DSS and its supporting processes.

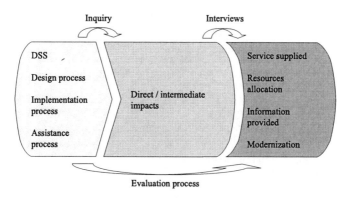

Figure 5. The evaluation process.

2. Next, we need to interview the management staff with the objective of extracting information about how the direct/intermediate impacts affect the final impacts.

3. Finally, we must bring together the results of the preceding phases in order to make a direct connection between the work developed by the DSS project team and the high-level results achieved by the companies.

The role of the analyst during the evaluation process is critical, particularly in the first two phases. The design of the questionnaire must take into account the cognitive profile of the potential respondents and should guarantee that the analysis of the results is conclusive. Besides, conducting the interviews and interpreting the answers received require appropriate experience and knowledge.

Figure 6 illustrates the model proposed to evaluate the performance of the DSS and the associated design, implementation, and assistance processes. For the sake of simplicity, the direct and intermediate impacts were grouped in four major categories: *(1)* information management, *(2)* procedures, *(3)* know-how and attitude, and *(4)* organization.

This model involves the construction of two basic matrices, a_{ij} summarizing the results of the inquiry process (phase 1), and b_{ij} showing the conclusions of the interview process (phase 2). Considered conjointly, these two matrices result in c_{ij}, which indicates the influence of the DSS and its supporting processes in the final impacts (phase 3).

We propose the following classification of the cells of the two basic matrices: 1 for null or weak connections, 2 for strong connections, and 3 for very strong connections. For the result matrix, we propose the following calculation of the percentages c_{ij}:

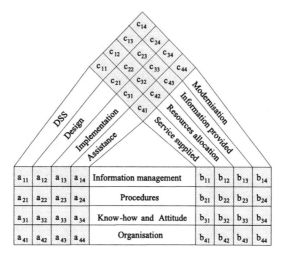

Figure 6. The evaluation model.

$$c_{ij} = 100 \cdot \frac{\sum_k a_{ki} \cdot b_{kj} - 1 \cdot \sum_k b_{kj}}{(3-1) \cdot \sum_k b_{kj}}$$

Note, that the denominator does not depend on the index i, remaining equal for all the c_{ij} in the same column. The value of this denominator lies between a minimum of 8 (i.e., $(3-1) \cdot (1+1+1+1)$) and a maximum of 24 (i.e., $(3-1) \cdot (3+3+3+3)$), and measures the potential of the impact caused in the objective associated with each column.

4   The Evaluation of the GIST System

The main motive in the evaluation model proposed in the previous section was the need to assess the GIST system performance, including the work developed by the project team.

The evaluation process involved five of the six companies that are currently using the system. To carry out the inquiry phase a specific questionnaire was designed and distributed to nearly fifty persons, mainly system users. The responses were anonymous and all the questionnaires that have been sent were returned. The interview phase was oriented to middle and top management staff, resulting in the opinion of ten other persons.

After performing the analysis and interpreting the information collected, the evaluation model was applied giving the results shown in Figure 7.

The most significant impacts were generated by the DSS itself, with the assistance process having the least impact. This last result was predictable considering the small size of the project team. All companies of the GIST

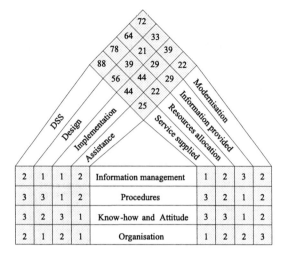

Figure 7. Applying the evaluation model to the GIST system.

consortium had the same expectations about the start of the system operation, but there was the need to distribute the implementation processes over time. As the project team was increasingly busy with new implementations, the assistance given to the companies that already used the system was not sufficient. This fault is being corrected by the execution of an assistance plan that includes, e.g., the distribution of a Newsletter and carrying out seminars promoting the exchange of experiences between the GIST users.

The quality of the service supplied is the most positively affected of the companies' highlevel objectives, which is mainly attributed to the characteristics of the DSS. At the other extreme, the quality of the information provided is not completely satisfactory, principally due to the design process. The reports specified did not correspond with the increase of the information available due to the use of the GIST system, and it is clearly urgent to correct this situation. Figure 8 shows in a more explicit way the conclusions drawn from the evaluation process.

5 Conclusions

The application of the model proposed by the authors to the evaluation of the GIST system proved to be very profitable and stimulating. Besides the encouraging indications obtained about the usefulness of the evaluation model, it assisted the project team in the analysis of its own performance and in the establishment of guidelines for future developments.

Globally, the work done was classified as very important and positive. The main merit goes to the system itself, which is considered to be responsible for significant improvements in all the companies' highlevel objectives, namely in the quality of the service supplied and in the resource allocation efficiency.

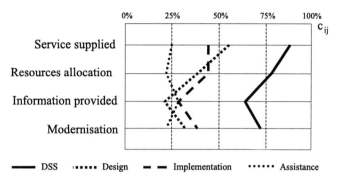

Figure 8. The results of the GIST system evaluation.

On the other hand, the expectations about the assistance process were not completely fulfilled and the reports available must be redefined.

The evaluation model could be applied separately to each company, department, or even to a specific type of user from one or more companies. The direct and intermediate impacts could also be grouped in more than the four categories just considered or even not be grouped at all. This could possibly facilitate the analysis of the results and enrich the conclusions derived. However, if we intend to generalize the results obtained, we must be aware of the statistical requirements that need to be guaranteed.

Despite the need for further validation with other decision support systems, in other contexts, the evaluation model we present has provided valuable experimental support, highlighting several characteristics of the success of GIST and allowing us to pinpoint areas that need special attention. Therefore, the evaluation model proposed seems to be fit for evaluation of important features of a DSS in a real context of use.

Bibliography

Boehm, B.W. (1981). *Software Engineering Economics.* Prentice-Hall, Englewood Cliffs.

Boehm, B.W. (2000). Unifying software engineering and systems engineering. *IEEE Computer 33*(3), 114–116.

Booch, G., J. Rumbaugh, and I. Jacobsen (1999). *The Unified Modeling Language User Guide.* Addison-Wesley, Reading.

Galvão Dias, T. (1995). Aplicação de Algoritmos Genéticos ao Problema da Geração de Serviços de Tripulações (Genetic Algorithms Applied to the Crew Scheduling Problem, in Portuguese). M.sc. thesis, Faculdade de Engenharia da Universidade do Porto, Porto, Portugal.

Nunes, N.J. and J.F. Cunha (2000). Whitewater: Interactive system design with object models. In M. Harmelen and S. Wilson (Eds.), *Object Oriented User Interface Design*, Addison-Wesley, Reading.

Paixão, J.M.P. and M.V. Pato (1989). A structural Lagrangean relaxation for two duty period bus driver scheduling problems. *European Journal of Operational Research 39*, 213–222.

Paixão, J.P. (1983). *Algorithms for Large Scale Set Covering Problems*. Ph.D. thesis, Imperial College, London, UK.

Portugal, R. (1998). Metaheuristicas para a geração de serviços para o pessoal tripulante (Metaheuristics for the Bus Driver Scheduling Problem, in Portuguese). M.Sc. thesis, Faculdade de Ciências da Universidade de Lisboa, Lisboa, Portugal.

Rockart, J.F. (1979). Chief executives define their own data needs. *Harvard Business Review 2 (March–April)*, 81–93.

Rumbaugh, J., M. Blaha, W. Premerlani, F. Eddy, and W. Lorensen (1991). *Object-Oriented Modeling and Design*. Prentice-Hall, Englewood Cliffs.

Woodward, S. (1999). Evolutionary project management. *IEEE Computer 32*(10), 49–57.

Crew Scheduling for Netherlands Railways "Destination: Customer"

Leo Kroon[1,2] and Matteo Fischetti[3,4]

[1] Erasmus University Rotterdam, Rotterdam School of Management,
  P.O. Box 1738, 3000 DR Rotterdam, Netherlands
  L.Kroon@fbk.eur.nl
[2] NS Reizigers, Department of Logistics, P.O. Box 2025, 3500 HA, Netherlands
[3] University of Padova, DEI, Via Gradenigo 6/A, 35100 Padova, Italy
  Fisch@dei.unipd.it
[4] Double-Click sas, Via Crescini 82/C, 35126 Padova, Italy

Abstract. In this paper we describe the use of a set covering model with additional constraints for scheduling train drivers and conductors for the Dutch railway operator NS Reizigers. The schedules were generated according to new rules originating from the project "Destination: Customer" ("Bestemming: Klant" in Dutch). This project is carried out by NS Reizigers in order to increase the quality and the punctuality of its train services. With respect to the scheduling of drivers and conductors, this project involves the generation of efficient and acceptable duties with a high robustness against the transfer of delays of trains. A key issue for the acceptability of the duties is the included amount of variation per duty. The applied set covering model is solved by dynamic column generation techniques, Lagrangean relaxation and powerful heuristics. The model and the solution techniques are part of TURNI, a commercial software package which is currently used by NS Reizigers for carrying out several analyses concerning the required capacities of the depots. The latter are strongly influenced by the "Destination: Customer" rules. We applied TURNI to crew scheduling instances of different sizes. These instances were all based on the 1999/2000 timetable of NS Reizigers.[1]

1 Introduction

Since a couple of years, there is a separation in the Dutch railway system between the capacity management and the maintenance of the railway infrastructure on one hand, and the exploitation of the railway infrastructure on the other. Capacity management and maintenance of the infrastructure are carried out by the government, and the railway operators are commercial organizations that are allowed to exploit the infrastructure by operating their trains. NS Reizigers is the largest Dutch railway operator, specialized in the railway transportation of passengers.

[1] This work was partially supported by the Human Potential Programme of the European Union under contract no. HPRN-CT-1999-00104 (AMORE).

Improving the quality and punctuality of the train services is currently one of the major issues in the Dutch railway system. Therefore, many projects are being carried out both by the government and by the railway operators. Projects carried out by the government involve the capacity and the dependability of the infrastructure. NS Reizigers currently carries out the project "Destination: Customer" ("Bestemming: Klant" in Dutch). This project aims at increasing the quality and the punctuality of its train services. Improving the quality of the train services involves, amongst others, a redesign of the circulation of the rolling stock in order to increase the seating probability, improved travel information (both before and during traveling), more safety measures in stations, and better facilities at stations for supporting the complete door-to-door movement of passengers (cycles, parking lots, train-taxis, buses, etc.). Improving the punctuality of the train services involves an improvement of the robustness and stability of the timetable, a decrease of the failure rate of the rolling stock by intensified maintenance, and a redesign of the duties for the drivers and conductors in order to increase their robustness against the transfers of delays. A consequence of the latter will be less transfers of drivers and conductors from one train to another, since such transfers may lead to transfers of delays from one train to another. Further details of these issues are provided in Section 3.

A commercially operating railway operator has to be able to react quickly to changes in the transportation market and to be able to adapt his operating plans accordingly. Hence, the railway operator should be able to modify his timetable quickly, and he should be able to modify the corresponding schedules for rolling stock and personnel as quickly as well. Furthermore, he should be able to carry out *what if* analyses easily, in order to study the long term consequences of certain future scenarios. In this situation, information systems that provide insight into the transportation market, as well as *intelligent* information systems that provide active support in the processes of timetabling and scheduling of rolling stock and personnel are indispensable tools. Indeed, timetabling and scheduling are highly complex tasks that are very time consuming when they are carried out manually.

In this paper we describe how we used TURNI for estimating the consequences of the new rules originating from the project "Destination: Customer" for the required capacities of the depots for train drivers and conductors. TURNI is a commercial software package which provides real active support to the planners in building up efficient and acceptable duties for the drivers and conductors. The "Destination: Customer" rules involve *(1)* the application of train teams consisting of a driver and a conductor, *(2)* the restricted assignment of train series to depots, and *(3)* the required amount of variation per duty. Especially the amount of variation per duty is a key issue for the acceptability of the duties.

The remainder of this paper is structured as follows. In Section 2 we describe some basic principles of the scheduling processes within the Logistics department of NS Reizigers. Section 3 describes the new rules originating from the project "Destination: Customer". In Section 4 we describe the TURNI system, focusing on the underlying set covering model and the corresponding solution techniques. Section 5 describes how we modeled the "Destination: Customer" rules in TURNI. In Section 6 we present computational results, based on the 1999/2000 timetable of NS Reizigers. The paper is finished in Section 7 with some conclusions.

2 Scheduling Train Drivers and Conductors

2.1 Depots

The numbers of drivers and conductors of NS Reizigers are about 3000 and 3500, respectively. Each driver and each conductor belongs to one of the depots. There are 29 depots, mainly located at the main nodes of the Dutch railway network (see Figure 1). However, also a number of smaller depots are located near the borders of the country. These depots are necessary because of the trains starting in these locations early in the morning or ending there late in the evening. Each driver or conductor operates from his own depot. That is, each of his duties starts and ends in his own depot and there are no overnight duties. For the drivers there are the additional rules that they are allowed to work on certain types of rolling stock and on certain parts of the infrastructure only, depending on their knowledge of the different types of rolling stock and the infrastructure.

2.2 Train Series, Trips and Duties

Each duty consists of a sequence of trips to be carried out by one single driver or conductor. Each trip is mainly characterized by a train number, a start and end location, and a start and end time. Also a number of other aspects of the trips are relevant, such as the involved train type (Intercity or Stoptrain) and train series. Here a train series is a set of trains that provide a regular direct connection between two end stations. The train series play an important role within the project "Destination: Customer," as will be described later. The following example describes a trip on the 833 Intercity train between Amsterdam (Asd) and Utrecht (Ut) from 9:28 to 10:01. The 833 train is a member of the 800 train series, which provides an hourly connection between Haarlem (Hlm) and Maastricht (Mt) (see also Figure 1).

| 833 | Amsterdam | Utrecht | 9:28 | 10:01 | IC | 800 |

The duties should be such that each trip is covered by one driver and a sufficient number of conductors. The number of conductors per trip mainly depends on the length of the corresponding train: the longer the train, the more conductors are required. The latter suggests that the process of scheduling drivers and conductors can start only after the timetable and the rolling

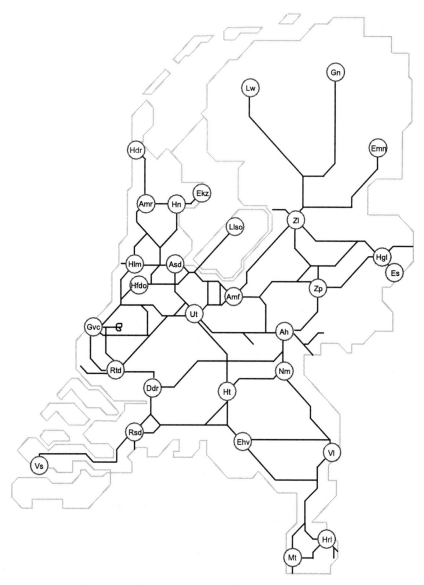

Figure 1. The crew depots of NS Reizigers.

stock schedule have been completed. This is also more or less what happens in practice, although in practice there are also some feedback iterations between scheduling personnel and scheduling rolling stock. Recently some research efforts were undertaken to integrate the scheduling of rolling stock and personnel (Freling (1997); Wren and Gualda (1999)).

Each duty has to satisfy certain constraints. To give a few examples (non-exhaustively): the maximum length of a duty is 9:30 hours. Each duty longer than 5:30 hours requires a meal break with a length of at least 30 minutes. The time between the start of the duty and the start of the meal break should be at most 5:30 hours and the same should hold for the time between the end of the meal break and the end of the duty. Another important constraint involves the minimum connection time: if a driver or conductor transfers from one train to another, then a minimum connection time of 25 minutes is to be taken into account. The mentioned constraints are hard constraints, but also several soft constraints are to be respected.

Besides the constraints that are to be satisfied by the individual duties, there are also several constraints that are not related to the individual duties but that are to be satisfied per depot or by the complete final schedule. For example, per depot at most 5% of the duties can be longer than 9:00 hours. Furthermore, per depot the average length of the duties should be less than 8:00 hours. There are also constraints on the percentage of night duties per depot. These constraints are to be satisfied during the scheduling process in order to facilitate the rostering process afterwards.

The main criteria in the scheduling process of drivers and conductors are *feasibility*, *efficiency*, and *acceptability*. Feasibility means that it should be possible to carry out the obtained schedules in practice, and that they are sufficiently robust for outside disruptions and for delays of trains. A key parameter here is the minimum connection time that was mentioned before. The larger the minimum connection time, the more robust the schedule will be. Efficiency means that the percentage of productive time in the duties is high. Non-productive time includes, besides the required pre-time at the start of a duty, the required post-time at the end of a duty and the meal break, also the gaps between the trips, and the so-called P-trips (passenger trips or positioning trips). A P-trip is a trip where a driver or conductor does not act as such, but moves as a passenger in a train from the end location of his previous trip to the start location of his next trip. Acceptability is a qualitative aspect of a schedule, referring to the probability that the obtained schedule will be accepted by the drivers and conductors. Within NS Reizigers, acceptability by the drivers and conductors is highly related to the level of variation in the duties (the more variation, the better).

2.3 Scheduling and Rostering

The duties to be carried out are created by the central Logistics department of NS Reizigers. After the duties have been generated, they are distributed over the depots. Usually, after some iterations of negotiation and adaptation, the duties are accepted by the depots. The generation of the rosters for the depots, based on the duties, is carried out locally in the depots. Also in the rostering process many constraints have to be satisfied. For example, after a couple of night shifts, a driver or conductor should have a certain number of days off before he may carry out his next duty. One reason for splitting the central scheduling process from the local rostering process is the fact that both processes are highly complex. Therefore, combining the scheduling and the rostering processes into one process would result in huge combinatorial problems that are practically insolvable. Note that, from a mathematical point of view, the scheduling problem and the rostering problem share a similar structure. Indeed, both problems involve the arrangement of objects into proper sequences. In the scheduling problem the trips have to be sequenced into duties and in the rostering problem the duties have to be sequenced into rosters. For more detailed descriptions of the problems crew scheduling and crew rostering we refer to Caprara et al. (1997, 1999a).

3   "Destination: Customer"

As was described earlier, the project "Destination: Customer" is being carried out by NS Reizigers with the objective to obtain an improved quality and punctuality of the train services provided to the customers. One of the aspects to be taken into account within this project is the generation of efficient and acceptable schedules for the drivers and conductors with a maximal robustness with respect to the transfer of delays of trains. For the duties, the following aspects originate from the project "Destination: Customer" which will be described in the next sections.

- Operating in train teams
- Variation per duty
- Assignment of train series to depots

3.1   Operating in Train Teams

According to the new rules originating from the project "Destination: Customer," the drivers and conductors will carry out their duties together in so-called train teams. This means that each day a driver and a conductor form a train team and that they carry out the same duty together. The composition of the train teams on a certain day will be different from the composition of the train teams on the previous day, depending on the rosters for drivers and conductors.

An advantage of the introduction of the train teams is that it will lead to an improved robustness of the schedules. Indeed, in the new situation, only two components will be required for operating a train, namely a train and a train team. In the traditional situation, three components were required, namely a train, a driver and a conductor. By reducing the number of components required for operating a train, the probability is increased that all components will be available in time. Furthermore, the rules for the variation in duties (see § 3.3) aim at keeping also the train and the train team together as much as possible. This will even further increase the mentioned probability.

Although for the majority of the trains one conductor is sufficient, a limited number of trains should be serviced by two conductors. In general, the second conductors cannot be part of a train team since there is no matching driver. Therefore, the introduction of the train teams will also require the introduction of special duties for second conductors.

3.2   Variation in Duties

For the drivers and conductors, the amount of variation in the duties is a key issue for the acceptability of the duties. In fact, in the current situation most duties are quite varied. The drivers and conductors prefer to keep it like this as much as possible. On the other hand, from the point of view of the robustness of the schedules, it would be better to keep the trains and the train teams as much as possible together. As was described above, this would even further increase the probability that all components required for operating a train are available in time. However, keeping a train and a train team together during a complete duty is impossible, because the train team needs a meal break after some time. Furthermore, the trains are usually scheduled in an up-and-down fashion between the two end stations of the involved train series. Thus, keeping the train and the train teams together for a longer period of time would result in duties without any variation.

Therefore, the following rules were introduced in order to create an acceptable compromise between the objective of keeping the rolling stock and the train teams as long as possible together and the objective of having an acceptable amount of variation in the duties.

Each duty longer than 5:30 hours not involving a long (> 150 km) Intercity train series should contain at least two out of the following three elements:

- Two train series
- Two train types (Intercity or Stoptrain)
- Two corridors of the infrastructure

These rules are considered as hard rules, which means that they must be satisfied by each duty. Note that the meal break is a natural split in a duty, which requires the train team to leave their train. Therefore, it is also a natural point for the train team to switch from one train to another. As a consequence, the variation rules allow for the generation of duties with the following simple structure: the first part of the duty before the meal break is carried out on one train servicing a certain train series, and the second part of the duty after the meal break is carried out on another train servicing another train series on another corridor. Together with the rules for the assignment of train series to depots, which will be described in the next section, the majority of the duties will have the meal break in the home depot, which has additional managerial advantages.

3.3   Assignment of Train Series to Depots

The "Destination: Customer" rules also describe that, in principle, each train series may be assigned only to those depots where the train series has a starting and/or ending train. For example, according to this rule the trips on the 800 train series between Haarlem (Hlm) and Maastricht (Mt) that was described earlier may be assigned to the depots Haarlem and Maastricht. Furthermore, in the early morning there are some trains in this train series which start in Utrecht (Ut) or Eindhoven (Ehv) and in the late evening there are some trains that end in these locations. Therefore, the trips on the 800 train series may be assigned to these depots as well.

The rationale behind this assignment rule is that each driver and conductor will operate more often on the same train series than in the traditional situation. Indeed, in the traditional situation each train series could basically be assigned to each of the 29 depots. The latter holds in particular for the conductors. The application of the assignment rule will lead to a higher exposure of the drivers and conductors to and thereby more experience with the peculiarities of the train series and the timetable. This will lead to an improved punctuality of the train services. Obviously, the assignment rule has a large impact on the duties for the drivers and conductors and on the required capacities of the depots. Hence the need for an effective automatic crew scheduling system for making a priori estimates of these consequences.

There are a few exceptions to the mentioned assignment rule. Indeed, for some of the depots a strict application of the rule would result in a number of allowable train series that is too small for satisfying the variation rules that were described in the previous section. In such a case, the depot is considered as a *satellite depot* of a nearby larger depot. This means that the satellite depot may be assigned also to (part of) the train series of the larger depot in order to be able to satisfy the variation rules. Satellite depots are mainly located near the borders of the country where only a limited number of train series is available.

4 TURNI

In order to be able to estimate the consequences of the "Destination: Customer" rules for the required capacities of the depots as good as possible, in particular the consequences of the new rules involving the assignment of train series to depots, the Logistics department of NS Reizigers decided to look for intelligent algorithmic approaches for supporting the crew scheduling process. One of the available alternatives turned out to be TURNI. This system has a relatively simple user-system interface, but a powerful scheduling module. So far, we have used TURNI mainly to carry out several *what if* analyses concerning the required capacities of the depots. For this purpose, only a few customizations of the system to the specific situation at NS Reizigers were necessary for obtaining useful results.

4.1 Model Description

Since it is difficult to describe the constraints to be satisfied by the individual duties explicitly in a mathematical programming model, the model underlying TURNI is a set covering model with a number of additional constraints. If a set covering model is applied, then the solution mechanism consists of a duty generation module and a duty selection module. After a certain set of feasible duties has been generated by the duty generation module, the duty selection module aims at selecting a subset of feasible duties in such a way that each of the trips is covered by at least one of the selected duties, that the relevant additional constraints are satisfied, and that the total involved costs are minimal. The set of feasible duties may be generated *a priori*, or it may be generated *on the fly* during the solution process.

Set covering models have been popular in the airline industry for many years (Barnhart et al. (1994); Desrosiers et al. (1995); Hoffman and Padberg (1993); Wedelin (1995)). However, in the railway industry the sizes of the crew scheduling instances are, in general, a magnitude larger than in the airline industry, which prohibited the application of these models in the railway industry until recently. But due to the increase in the computational power of nowadays' computers and algorithms, set covering models become more and more applicable in the railway industry as well (Caprara et al. (1997, 1999a); Kroon and Fischetti (2000); Kwan et al. (1999)). We mainly handled instances with up to 2500 trips. However, we also carried out a number of experiments with much larger instances (about 8000 trips corresponding to all trips of a single day), which were quite promising for the near future.

Using the notation $t = 1, \ldots, T$ for the trips to be covered, $d = 1, \ldots, D$ for the potential duties, and $c = 1, \ldots, C$ for the additional constraints to be satisfied, the set covering model with additional constraints can be formulated as follows:

$$\text{Minimize} \sum_{d=1}^{D} c_d \, x_d \qquad (1)$$

subject to

$$\sum_{d=1}^{D} a_{t,d} x_d \geq 1 \qquad \forall \, t = 1, \ldots, T \qquad (2)$$

$$\sum_{d=1}^{D} b_{c,d} x_d \leq u_c \qquad \forall \, c = 1, \ldots, C \qquad (3)$$

$$x_d \in \{\, 0, \ 1\} \qquad \forall \, d = 1, \ldots, D \qquad (4)$$

Here the meaning of the binary decision variables is as follows:

$$x_d = \begin{cases} 1 \text{ if duty } d \text{ is selected in the final solution,} \\ 0 \text{ otherwise.} \end{cases}$$

In the 0-1 matrix $a_{t,d}$, each row represents a trip and each column represents a feasible duty, and $a_{t,d} = 1$ if and only if trip t is covered by duty d. Besides the regular trips that *must* be covered, TURNI also allows one to include a number of *suggested* or *inadvisable* trips into the model. These trips may be covered by a duty, but they need not be covered. Adding such trips to an instance may be helpful for finding a feasible schedule or for improving the overall efficiency of the schedule. There is no constraint (2) corresponding to the inadvisable trips, which are only considered within the duty generation module. As to the suggested trips, for each such trip t we have a dummy duty that covers trip t only and has a cost equal to the user-defined penalty for leaving trip t uncovered. Examples of the use of these additional trips are given in § 5.4 and 6.2. The additional constraints (3) are not related to the individual duties (such constraints are handled at the duty generation level), but to certain forbidden combinations of duties. These constraints may be related to, e.g., the maximum number of duties per depot, or the maximum average length of the duties per depot.

As stated earlier, the cost coefficients in (1) represent the fact that the main objective is to minimize the number of duties required to cover all trips. However, also other cost aspects, such as additional costs for P-trips, uncovered suggested trips, covered inadvisable trips, or penalties for discouraging undesirable characteristics or violated soft constraints are handled by our model. In particular, the user is allowed to specify a penalty term for each unit of slack in constraints (2), so as to reduce (or forbid) trip over-covering.

4.2   Solution Technique

The resulting model is solved by applying dynamic column generation (Barnhart et al. (1994); Desrosiers et al. (1995)), Lagrangean relaxation and several heuristics. Dynamic column generation means that the duties are not generated a priori, but on the fly during the solution process. The application of this dynamic approach is required by the fact that the complete set of feasible duties is extremely large in general, hence generating all feasible duties a priori is not a feasible approach. We next give a brief sketch of the solution framework used by TURNI.

The overall execution is organized into a sequence of *passes*, in each of which the system tries to obtain better and better solutions. The initialization pass is aimed at determining an initial set of duties covering all the input trips. The first pass finds the first efficient solutions for the current problem and is designed to reach soon a feasible solution by using an aggressive solution strategy. The next passes improve upon the solutions found in previous passes, by using a more accurate solution strategy. Typically, the best solutions are determined already during the very early passes, but in some hard cases also later passes can find improved solutions.

Within each pass, the algorithm iterates between a *duty generation* module and a *duty selection module*. The duty generation module generates new prospective feasible duties based on a *dynamic programming* heuristic. The underlying network is a time-space network in which events are represented by nodes and trips are represented by arcs, as is customary in such applications (Barnhart et al. (1994)). Within the network, states that are dominated by other states are removed heuristically, in order to keep the size of the network manageable. The prospectiveness of a feasible duty is evaluated based on dual information related to the linear programming relaxation of the underlying set covering model. However, TURNI uses Lagrangean relaxation and subgradient optimization instead of Linear Programming for calculating the required dual information, in a vein similar to the one that was proposed in Caprara et al. (1999b) for the solution of pure set covering problems. Feasible duties generated in earlier stages whose prospectiveness turns out to be low during later stages of the algorithm may be deleted later in order to keep the number of active duties within certain bounds.

The duty selection module looks for a heuristic solution for the overall model, based on the currently available set of feasible duties. The applied heuristic is also driven by the Lagrangean dual information, following the approach proposed in Caprara et al. (1999b). According to this scheme, as soon as the subgradient optimization is close to convergence, a simple heuristic of greedy type is applied after each updating of the Lagrangean multipliers. The greedy heuristic iteratively evaluates and selects duties from the set of active duties, based on their current Lagrangean costs. After each selection, the current Lagrangean dual solution is locally improved by reducing as much as possible the multiplier of each constraint, which turns out to be saturated by the current partial solution. In particular, each time a duty d covering trip t is selected, the Lagrangean multiplier of the corresponding set covering constraint (2) is put to its lower bound (i.e., to 0 for the pure set covering model, or to $-p_t$ in case a positive over-covering penalty p_t is associated with trip t) and the duty Lagrangean costs are increased accordingly. An analogous mechanism is applied the first time the selection of the current duty d lets the left-hand side of a certain inequality in (3) reach (or exceed) the corresponding right-hand side value.

Within each pass, the generation and selection phases are cyclically applied for a certain number of iterations, hopefully updating the best solution found. Thereafter, a *fixing* procedure is activated to select some particularly efficient duties of a certain target solution, and to fix them as belonging to the final solution. The target solution is defined either as the best solution currently available (in the first passes), or as the best solution found since the last fixing (in the later passes, so as to diversify the search). As to the choice of the duties to be fixed, this depends on a duty score computed as a function of the current Lagrangean cost and of the current duty efficiency. The latter is defined as the ratio between the overall duration of the trips covered by the duty (skipping all trips covered by the previously fixed duties as well as the inadvisable trips) and the paid working time. The overall process is then repeated on the trips not covered by the fixed duties: the duty generation and selection phases are iterated for a while on the subproblem resulting from fixing, new duties are fixed, etc. In this way, millions of possible duties are typically generated within the duty generation module, and

thousands of alternative solutions are constructed and evaluated during the
selection phase.

Fixing typically leads to better and better solutions, up to a point where
the current solution cannot be improved any further without releasing some
of the fixed duties. In this situation, the program activates a backtracking
mechanism that consists of unfixing a certain number of fixed duties. These
duties are chosen among those that appear to be less efficient according to
the above definition of duty efficiency.

After a certain number of backtrackings (ranging from 5 to 10, depending
on the instance size), the program ends the current pass, unfixes all the currently fixed duties, and applies a refinement procedure to hopefully improve
the current best solution (as well as the minimum cost infeasible solution
found so far, if different), by means of trip exchanges among duties. The refinement procedure is based on the solution of a series of bipartite matching
problems that optimally reassign suitably-defined duty pieces. The refinement procedure turns out to be particularly effective on large size instances
where the set covering model may get into trouble in generating and selecting
small variations of duties.

After the completion of the refinement procedure, a new pass begins: the
overall procedure is reapplied on the current set of duties. As a heuristic rule,
after the second pass the dynamic programming duty generation is deactivated (during the whole pass) with a certain frequency, so as to intensify the
use of the selection procedures and to allow for early backtrackings based
on lower bound criteria. The elaboration continues until some stop condition
(time limit, maximum number of passes, etc.) is met.

Computational experience has shown that this solution approach has the
advantage of finding quickly good solutions for the problem at hand (already
after the first fixings, the solutions are typically comparable or better than the
best manual ones), while using the remaining processing time for improving
them by exploring different fixing patterns. For more details on the basic duty
selection mechanism, we refer the reader to Caprara et al. (1999b). Other
approaches for solving large set covering problems can be found in Balas and
Ho (1980); Balas and Carrera (1990); Barnhart et al. (1994); Beasley (1990);
Desrosiers et al. (1995); Wedelin (1995).

4.3 Output and Other Functions

The results of a run of the algorithm can be shown by TURNI in tabular
format, where the successive trips in each duty are listed along with some
duty statistics, or in a graphical format in the form of a Gantt chart or a time-
space diagram. An example of such a Gantt chart is shown in Figure 2 which
represents some stoptrain duties for train teams from the depot Maastricht
(Mt) in the southern part of the Netherlands. This Gantt chart clearly shows
that these duties indeed have the structure imposed by the "Destination:
Customer" rules. That is, each duty longer than 5:30 hours consists of one
part before and one part after the meal break, each one involving a different

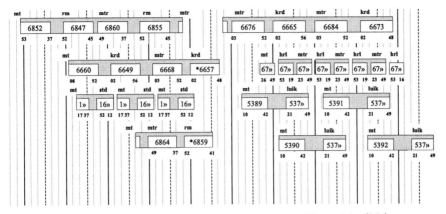

Figure 2. Duties with variation for the depot Maastricht (Mt).

train series. In this Gantt chart, the train series can be recognized from the first two digits of the train number.

An interesting feature of TURNI is the fact that it allows one to observe the best obtained solutions while the algorithm is still running. This is enabled by the multi-tasking facilities of the underlying operating system. At the same time, the system continuously provides an optimistic estimate (i.e., a Lagrangean lower bound) of the optimal solution costs. In this way, the user of the system may decide for himself whether or not he wants to interrupt the run of the algorithm.

Other useful functions provided by TURNI are the following. First, the system allows the generation of so-called deadheading trips. We will not go into the details of this deadheading mechanism here, but in general this mechanism automatically generates additional trips at appropriate positions in time and space in such a way that a feasible covering of all trips is obtained. This mechanism uses a complete distance matrix containing the time required to travel from one location to another. This mechanism is particularly important if one has only data available on the train movements of regular passenger trains and not on the additional train movements for positioning the rolling stock. A generated deadheading trip triggers the fact that some of these positioning trips are missing in the available data set. Also useful is the fact that the system allows one to check whether a certain manual solution satisfies all constraints. Here the obtained log file points at all the constraints that are violated by the given manual solution.

5   Modeling "Destination: Customer" and Other Rules

5.1   "Destination: Customer" Rules

TURNI contains a mechanism for specifying for each trip the depots that may be used for servicing the trip. This mechanism could be used to handle the "Destination: Customer" rules concerning the assignment of train series to depots in a straightforward way.

Handling the variation rules was less straightforward. First, we recoded the train series indicators of the trips in such a way that trips involving train series that should not appear in the same duty obtained the same train series indicator. For example, two stoptrain series covering (for a large part) the same corridor of the infrastructure obtained identical train series indicators. Furthermore, in order to guarantee that the train teams will not transfer too often from one train series to another, we used an upper bound on the number of transfers per duty from one train series to another. Finally, we used a mechanism for reducing the number of duties without the required amount of variation. That is, duties longer than 5:30 hours involving only one train series could be penalized or forbidden in several gradations.

Although this was not part of the official "Destination: Customer" rules, we also used a mechanism to prevent the generation of duties with so-called *artificial variation*. That is, duties for which the majority of the work is carried out on one train series and only a very minor part of it is carried out on another train series. This mechanism counts a train series in a duty only if the amount of work on this train series exceeds a certain lower bound.

Although from a technical point of view the mechanism for handling the variation rules operated as expected, it had to be handled with some care in practice. In particular, if the generation of duties without the required amount of variation is strictly forbidden, then the instances should be chosen carefully. If an instance does not contain a sufficiently large number of different train series, or if the amounts of work on the different train series do not sufficiently match with each other, then it is obviously impossible to generate duties satisfying the variation rules. Although a careful selection of the instances to be handled indeed usually leads to useful results, we are currently improving the mechanism for handling the variation rules by making it somewhat more subtle. We want to achieve this by making the variation constraints dependent on the involved depots and the involved train series. This will allow one to focus on the generation of duties satisfying the variation rules initially for only a subset of the depots, while satisfying the variation rules for the other depots can be postponed to later instances.

5.2   Underway Train Changes

From the point of view of the robustness of the duties, it is desirable that a train team remains as long as possible on the same train and does not get off the train somewhere underway. In order to achieve this, we used the following approach: each trip record was extended with an additional attribute describing whether the trip is the First (F) or the Last (L) trip of the train. Now an underway train change is defined as a transfer of a driver or conductor from trip t_1 to trip t_2, where

$\mathrm{TRAIN}(t_1) \neq \mathrm{TRAIN}(t_2)$ and

      t_1 is NOT the Last trip of $\mathrm{TRAIN}(t_1)$, or

      t_2 is NOT the First trip of $\mathrm{TRAIN}(t_2)$.

The duty generation module of TURNI was modified in such a way that the number of underway train changes per duty can be bounded or penalized. In the above definition, the first condition implies that the trips t_1 and t_2 are related to trains with different train numbers. In general, this implies that the trains are different. The first part of the second condition means that the train corresponding to trip t_1 has not yet come to its end station. Hence, if the driver or conductor leaves this train, then the train has to be supplied with a different driver or conductor on its remaining route to its end station. This is undesirable, since this may lead to a transfer of a delay. The second part of the second condition means that, if the driver or conductor boards the train corresponding to trip t_2, then it is somewhere underway the route of this train, which is again undesirable. This mechanism for structuring the duties turned out to be quite useful, in particular when dealing with long Intercity train series. The mechanism could be used to make a trade-off between maximum robustness of the duties (by not allowing any underway train changes) and maximum efficiency (by allowing underway train changes).

5.3 Additional Rules for Train Teams

The fact that drivers and conductors will operate in train teams implies that for the train teams at least the traditional rules for the drivers have to be applied. These traditional rules for the drivers are somewhat more restrictive than the traditional rules for the conductors. In most cases, these more restrictive rules for the drivers are related to the combining and splitting of trains, and then they are due to the fact that a driver has a fixed position in front of the train. Therefore, a driver is less flexible in his movements than a conductor who can move more or less freely throughout the train. Splitting and combining of trains occurs in several train series of NS Reizigers. Although the additionally required synchronization of trains may have a bad impact on the punctuality, it also allows for more direct connections for passengers and at the same time an efficient use of the infrastructure.

    The following example involves the combining and splitting of trains in the 6000 train series (Utrecht-Tiel) and the 16000 train series (Geldermalsen-'s Hertogenbosch) in Geldermalsen. A train in the 6000 train series arriving from Utrecht (Ut) in Geldermalsen (Gdm) is split into two parts: the front part of the split train continues to Tiel (Tl) as a 6000 train and the rear part of the split train continues to 's Hertogenbosch (Ht) as a 16000 train. As a consequence, the train team of the arriving train should continue to Tiel and the part of the 16000 train bound for 's Hertogenbosch should be supplied with another train team. In the reverse direction, the 6000 train from Tiel and the 16000 train from 's Hertogenbosch are combined into one single 6000 train from Geldermalsen to Utrecht. The train from 's Hertogenbosch arrives

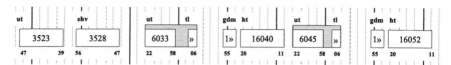

Figure 3. Duty involving the 6000/16000 train series for depot Utrecht (Ut).

first in Geldermalsen and becomes the front part of the combined train. The train from Tiel arrives second and becomes the rear part of the combined train. Thus the train team arriving from 's Hertogenbosch may continue on the combined train, and the train team from Tiel has to leave the train.

We modeled these additional rules by recoding the train numbers of the trips where necessary. For example, for the trains 6040 and 16040, there are three trips, namely

6040	Tiel	Geldermalsen	12:36	12:46	6040
6040	Geldermalsen	Utrecht	12:49	13:11	16040
16040	's Hertogenbosch	Geldermalsen	12:20	12:38	16040

Here the first column represents the original train number and the last column represents the recoded train number that we really used. The fact that the recoded train number of the Tl-Gdm trip is different from the recoded train number of the Gdm-Ut trip, together with the minimum connection time of 25 minutes that is to be respected between two trips with different train numbers, prohibits the train team on the 6040 Tl-Gdm trip to continue on the 16040 Gdm-Ut trip. This train team may continue, e.g., on the next 16000 trip Gdm-Ht. Furthermore, due to the recoding of the trips, the train team on the 16040 Ht-Gdm trip may continue on the 16040 Gdm-Ut trip.

That this recoding of the trips indeed has the required effect, is shown by the duty in the Gantt chart in Figure 3. This duty includes two times the route Utrecht-Tiel-Geldermalsen-'s Hertogenbosch-Utrecht, which is correct according to the given description. Note that the Gantt chart shows the recoded train numbers. The variation in this duty is provided by the trips from Utrecht (Ut) to Eindhoven (Ehv) and vice versa on the Intercity trains 3523 and 3528 of the train series 3500.

5.4 Duties for Second Conductors

As was described earlier, some trains must be serviced by two conductors. Thus the introduction of the train teams requires the generation of separate duties for these second conductors since for the second conductors there is no matching driver. There are some exceptions from this rule, since for the duties on a couple of long Intercity lines, train teams consisting of one driver and two conductors will be used.

A problem that we encountered here is the fact that the involved set of trains is only a relatively small subset of the set of all trains. This implies that the trips for the second conductors do not fit nicely together, as the trips for the train teams do. This problem is solved partly by the fact that it was decided that most of the "Destination: Customer" rules need not be applied to the duties for the second conductors. Hence, the trips for the second conductors may be assigned to all depots and the variation rules need not be satisfied here: only the traditional rules, involving aspects such as the average duty length per depot and the meal break, need to be satisfied. Nevertheless, the instances of the scheduling problem containing only the trips for the second conductors turned out to be infeasible. In particular, it turned out to be impossible to cover each of these trips by a feasible duty.

We solved this problem by first using the deadheading mechanism of TURNI. After having run the instance with the deadheading mechanism activated, which resulted in a number of deadheading trips, we added to the instance a set of additional timetabled inadvisable trips matching the generated deadheading trips as good as possible. By rerunning the extended instance with the deadheading mechanism deactivated, we finally obtained a feasible solution covering all trips and containing timetabled trips only.

Nevertheless, it turned out that in general the duties for the second conductors will be intrinsically less efficient than the duties for the train teams. However, within NS Reizigers the advantages of the introduction of the train teams that were described earlier are considered as more important than the rather low efficiency of the duties for the second conductors.

6   Computational Results

In this section we present computational results based on the experiments that we carried out. All experiments were performed on a PC with a Pentium II processor with 450 MHz and an internal memory of 128 Mbytes. Note that currently much faster PC's are available.

6.1   Instances

All instances that we studied were based on the 1999/2000 timetable of NS Reizigers. Since this timetable and its circulation plan for the rolling stock are both more or less cyclic with a cycle length of one week, the crew schedules can be cyclic with a cycle length of one week as well. Furthermore, although there are some differences between the timetables and the circulation plans for the different workdays of the week, these are in general so small that, from a crew scheduling point of view, the workdays can be considered as identical. Only during the weekends the timetables and the circulation plans are significantly different. In particular, during the weekends there are less early trains, there are no additional trains for rush hours, and for some train series the frequency is reduced. Traditionally, each workday counts for about 15% of the work, Saturday counts for about 13% of the work and Sunday counts for about 12% of the work.

In accordance with the foregoing, we split our set of trips into instances for a single workday (in particular, Tuesday), instances for Saturday, and instances for Sunday. Furthermore, we split these instances further into instances for train teams and instances for second conductors. The numbers of trips per day for train teams varied between about 7500 for Sunday and 9500 for Tuesday.

6.2   Train Teams/Separate Train Series

We started the analysis of each individual day by handling the involved train series separately from each other. Obviously, this will prohibit the generation of duties satisfying the variation rules. Therefore, the results of the analysis of the separate train series were not used directly as part of the final results of our analysis. However, as a starting point for obtaining the final results, it learned us how easy it would be to move parts of the involved work from one depot to another without increasing the total number of duties. Furthermore, it gave a lot of insight into the peculiarities of the different train series. This proved to be quite useful in the analysis of the combined instances.

For example, some train series have a complicated circulation of the rolling stock. An example is given by the 6000/16000 train series that was described in § 5.3. If we would not have prepared the data of this train series based on an instance containing no other train series, then in a larger instance it would have become impossible. Another example is the 5500 train series Utrecht-Baarn. In principle this train series should be serviced from the depot Utrecht (Ut), since this is the only depot along this train series. However, in the early morning, the first train from Baarn to Utrecht is supplied with a unit of rolling stock from Amersfoort (Amf). This implies that there is an early duty on this train series for the depot Amersfoort. In order to get a feasible solution for this case, a set of inadvisable trips Utrecht-Amersfoort on another train series (from which only one trip is to be selected) will be required for getting the crew from Amersfoort back home by the end of its duty. Similarly, there is also a late duty on this train series for the depot Amersfoort. This duty requires a set of inadvisable trips Amersfoort-Utrecht on another train series for getting the crew from Amersfoort to Utrecht by the start of its duty.

In general, each of the separate train series is handled very quickly, since the corresponding numbers of trips per train series were relatively small. The largest of these instances contained about 400 trips. The computing times for solving these instances varied from seconds for the smaller instances to minutes for the larger ones. According to the planners, the obtained solutions for these instances were of very good quality. Anyway, the solutions produced by TURNI usually had the same numbers of duties as the solutions that were created manually by the planners.

6.3   Train Teams/Combined Train Series

For solving the real scheduling problem, we combined the instances for the separate train series in such a way that the final duties could also satisfy the variation rules. This sometimes required a careful composition of the instances. The latter was particularly true for the areas of the satellite depots, since in these areas with only few train series it is hard to satisfy the variation rules. Here we handled instances with up to 1000 trips. We limited the size of these instances in order to have a better control over the final result. In the areas of the large depots with many train series (Amsterdam, Rotterdam, The Hague, Utrecht, Eindhoven), we generated instances that were much larger. In these areas it is relatively easy to satisfy the variation rules, and, therefore, the obtained duties were almost always immediately acceptable. The largest instances that we handled here contained about 2500 trips.

We pre-processed these instances as much as possible by applying the available forcing mechanism, which can be used to force certain pairs of trips to be scheduled directly after another in one duty. Applying this mechanism may improve the quality of the solution, since certain required combinations of trips can not be broken anymore. At the same time it may lead to a reduced computing time.

In almost all of these experiments it appeared that after the second pass of the algorithm (see § 4.2) only marginal improvements of the quality of the solution were obtained. In particular, the number of duties required to cover all trips hardly ever decreased after the second pass. Obviously, the time required to get to this point depends on the number of trips in the involved instance. For an instance with about 1000 trips it required usually about 15 to 20 minutes on our hardware. For an instance with up to 2500 trips some more time would be necessary, but in all cases this point was reached within one hour of computing time. In all cases the gap between the costs of the obtained solution and the Lagrangean lower bound was quite small, usually less than 1%. This gap is computed based on the actual set of duties in the set covering model. Therefore, the real gap may be somewhat larger. However, taking into account also the comments of the planners, we believe that there is not much room for improvement of the solutions.

In these experiments, we also had to manipulate the numbers of duties for the depots in order to obtain an optimal match of capacity and demand in the depots. In order to achieve this, we used both the mechanism for bounding the number of duties in a certain depot and the mechanism for attracting some extra duties to another depot by applying a cost reduction for the duties there. The first mechanism is handled by the additional constraints (3) and the second mechanism is handled by modifying the objective function (1) appropriately. Unfortunately, it turned out that a perfect match could not be obtained, due to the rather strict rules for the assignment of train series to depots.

6.4   Duties for Second Conductors

The numbers of trips per day for second conductors varied between 1200 for Sunday and 1500 for Tuesday. For each day, these numbers of trips could be handled in one instance. Although these instances were not extremely large, it took somewhat more time to obtain nearly optimal solutions here. This was mainly due to the fact that we had to use the deadheading mechanism both for obtaining feasible solutions and for obtaining better ones. The heuristic deadheading mechanism sometimes succeeded only after a couple of passes to find a crucial deadheading trip, thereby improving the quality of the solution. Nevertheless, the final results after a couple of hours of computing time for these instances were quite in accordance with the results that were obtained manually by the planners in parallel.

7   Conclusions

Within NS Reizigers, it is strongly believed that a proper structuring of the duties for the drivers and the conductors will lead to an improvement of the punctuality of the train services. The new rules originating from the project "Destination: Customer" aim at the specification of proper structures for the duties. On one hand, the duties should be as robust a possible against the delays of trains, and, on the other hand, they should allow for an acceptable amount of variation per duty. The new rules describe an effective compromise between these aspects. The new rules also describe that each train series may be assigned to a limited number of depots only.

In this paper we described how we used TURNI to estimate the consequences of these new rules for the required capacities of the depots. Whether the "Destination: Customer" rules will indeed lead to an improvement of the robustness of the duties for drivers and conductors will have to be proved in practice. However, we also intend to use a simulation model for assessing the effects of the new rules a priori.

For the drivers, the practical conclusion of the experiments is that the total number of available drivers will be sufficient for covering all the work. However, the drivers are sometimes wrongly distributed over the depots. For the conductors, the practical conclusion is that an additional number of conductors will have to be hired for covering all the work. This is first due to the fact that the more restrictive rules for the drivers will also be applied for the conductors, because drivers and conductors will operate in train teams. It is also due to the intrinsically inefficient duties for the second conductors. Also the conductors are sometimes wrongly distributed over the depots. Nevertheless, in practice no drivers nor conductors will be obliged to move to other depots. In order to obtain schedules that will be feasible in practice, the "Destination: Customer" rules will initially be relaxed to some extent in order to guarantee that all the work can be covered by the available drivers and conductors. For example, one option for dealing with this is to apply the new rules initially only on weekdays and not during the weekends.

Bibliography

Balas, E. and M.C. Carrera (1990). A dynamic subgradient-based branch-and-bound procedure for set covering. *Operations Research 44*, 25–42.

Balas, E. and A. Ho (1980). Set covering algorithms using cutting planes, heuristics and subgradient optimization: A computational study. *Mathematical Programming Study 12*, 37–60.

Barnhart, C., E.L. Johnson, G.L. Nemhauser, M.W.P. Savelsbergh, and P.H. Vance (1994). Branch-and-price: Column generation for solving huge integer programs. In J.R. Birge and K.G. Murty (Eds.), *Mathematical Programming: State of the Art 1994*, University of Michigan, 186–207.

Beasley, J.E. (1990). A Lagrangian heuristic for set covering problems. *Naval Research Logistics 31*, 151–164.

Caprara, A., M. Fischetti, P.L. Guida, P. Toth, and D. Vigo (1999a). Solution of large scale railway crew planning problems: The Italian experience. In N.H.M. Wilson (Ed.), *Computer-Aided Transit Scheduling, Lecture Notes in Economics and Mathematical Systems*, 471, Springer, Berlin, 1–18.

Caprara, A., M. Fischetti, and P. Toth (1999b). A heuristic method for the set covering problem. *Operations Research 47*, 730–743.

Caprara, A., M. Fischetti, P. Toth, D. Vigo, and P.L. Guida (1997). Algorithms for railway crew management. *Mathematical Programming 79*, 125–141.

Desrosiers, J., Y. Dumas, M.M. Solomon, and F. Soumis (1995). Time constrained routing and scheduling. In M.O. Ball, T.L. Magnanti, C.L. Monma, and G.L. Nemhauser (Eds.), *Network Routing, Handbooks in Operations Research and Management Science*, 8, Elsevier, Amsterdam, 35–139.

Freling, R. (1997). *Models and techniques for integrating vehicle and crew scheduling*. Ph.D. thesis, Erasmus University, Rotterdam, The Netherlands.

Hoffman, K.L. and M.W. Padberg (1993). Solving airline crew scheduling problems by branch-and-cut. *Management Science 39*, 657–682.

Kroon, L.G. and M. Fischetti (2000). Scheduling drivers and guards: The Dutch Noord-Oost case. In R.H. Sprague (Ed.), *Proceedings of the 34th HICSS Conference*, IEEE, Piscataway.

Kwan, A.S.K., R.S.K. Kwan, M.E. Parker, and A. Wren (1999). Producing train driver schedules under different operating strategies. In N.H.M. Wilson (Ed.), *Computer-Aided Transit Scheduling, Lecture Notes in Economics and Mathematical Systems*, 471, Springer, Berlin, 129–154.

Wedelin, D. (1995). An algorithm for large scale 0-1 integer programming with application to airline crew scheduling. *Annals of Operations Research 57*, 283–301.

Wren, A. and N.D. Gualda (1999). Integrated scheduling of buses and drivers. In N.H.M. Wilson (Ed.), *Computer-Aided Transit Scheduling, Lecture Notes in Economics and Mathematical Systems*, 471, Springer, Berlin, 155–176.

Selecting and Implementing a Computer Aided Scheduling System for a Large Bus Company

Michael Meilton

Schedules Development Manager, FirstGroup plc, Macmillan House, Paddington Station, W2 1TY London, UK
mmeilton@skomer.freeserve.co.uk

Abstract. This paper outlines the procedure undertaken by FirstGroup to select a company wide computer aided scheduling system for their UK bus operations. The project started with a review of the scheduling functions and requirements across FirstGroup in 1998. Selection of a system suitable for all 24 operating companies took place in 1999. Implementation began in April 2000, and is expected to take two years. By December 2000, three companies had installed the system.

1 Introduction

FirstGroup is a company involved with transportation on a global basis with bus, tram, train and airport interests in the UK and operations in the United States of America. The UK bus operation has grown by acquisition to approximately 9000 vehicles, with companies being progressively added to the portfolio. This has led to the knitting together of companies with many different systems and differing operating agreements.

Scheduling has seen little development in the 24 UK bus companies within FirstGroup in recent years. Some companies continue to compile schedules manually while others use one of seven different commercially available computer aided scheduling systems. It was felt appropriate to consider a standardised approach to computer aided bus scheduling for a number of reasons.

- Driver wages is the largest cost element in bus operation (45%) and effective control of this is essential. While it is often possible to produce solutions to problems that comply with the relevant labour constraints, it is important that this is achieved within minimum cost.
- Increased requirement to source computerised schedule information for third parties. With the introduction of other common systems across FirstGroup there is good commercial sense in feeding these directly from one common system. The changes in information technology with the introduction of journey planning software and other features via the Internet etc. require information to be readily available in a standard computerised format. Within the UK the timetable for every bus service provided is required to be registered with the Government appointed Traffic

Commissioners. It is proposed that the submission of such registrations will in the near future be undertaken electronically.

- Concern over the age / experience profile of current schedulers. A survey undertaken in 1998 of schedulers within FirstGroup revealed that over 70% of them were aged over 40 and a similar percentage had over 10 years experience of compiling schedules, which importantly included experience of producing schedules manually prior to the introduction of computer aided systems. Only 6% of schedulers were aged under 30! Scheduling is seen by many managers as a "black art" and schedulers are considered as a not easily replaceable resource.

- A multitude of systems within FirstGroup represents a high level of risk. With the intimate knowledge of the systems restricted to a relatively small number of users, in the case of some systems only one or two people within FirstGroup, this placed an unacceptable level of risk upon the loss of those individuals. Experience has shown this to have been the case at some locations where schedulers have left or unfortunately died. A standard system across FirstGroup would remove such problems, enable assistance to be provided between companies as required and provide a common system for such issues as training.

2 Initial Review

To undertake an initial review of the scheduling function across FirstGroup a working party was established in January 1998. This consisted of eight people; two Senior Directors, one Company Director, the IT manager responsible for bus operation, two managers of scheduling functions and two schedulers. It was felt essential that the working party should contain a number of staff who regularly compiled schedules and it was fortunate that the IT manager also had previous experience within a scheduling environment. These schedulers represented a broad cross section of companies and had experience in the use of different computer aided systems. One scheduler had no experience of any other method than manual compilation.

It was felt appropriate that a survey should be undertaken to establish exactly what the arrangements were for the production of schedules within each company and what developments were planned or felt desirable for the future. Two questionnaires were produced and circulated to all companies during February 1998. Whilst covering similar topics one was designed for those responsible for the management of the scheduling function, the other was directed at the individual schedulers. The aim, which was broadly achieved, was to involve and tap the experience of as many schedulers within FirstGroup as possible.

The results of the questionnaires provided a very comprehensive picture of how scheduling was undertaken within FirstGroup, the staff who undertook it and how the various Managements viewed the function. They showed varied arrangements with some companies retaining a centralised scheduling function and others having schedulers located at individual depots. The survey also revealed a lack of any formal training at a number of locations. A detailed insight was provided of the many systems in use from the viewpoint of the experienced user. The general tone of most replies accepted that the schedules function within companies had lacked development in recent years and welcomed the prospect of an improved and enhanced system. There was a general wish for greater flexibility in being able to influence the results produced by systems with the ability to indicate desired features prior to optimisation. The degree of interest in their chosen discipline from a large number of schedulers was evident from the number of constructive and detailed comments volunteered in the questionnaires returned.

The review of answers confirmed the initial perception of the working party that there was merit in looking at those Computer Aided Scheduling Systems currently available and ascertain their suitability as a potential Group wide system. It was appreciated that within the operations of the UK bus division of FirstGroup there were varying needs and types of services ranging from frequent intensive urban operation to rural once a week market day services.

3   Requirements Definition and Trial Exercise

To obtain a direct comparison between the commercially available systems it was proposed to produce a detailed Requirements Definition (FirstGroup (2000)) which outlined the functionality we would expect or wish to see developed within any scheduling system. This was drawn from experience with existing systems and the comments received in reply to the future developments identified within the questionnaires to staff. The Requirements Definition went beyond the capabilities of existing systems, and sought an indication from suppliers of their planned future developments. A draft Requirements Definition was circulated to each scheduler / manager who had responded to the initial questionnaires for their comments. Their responses were then incorporated into a final draft that was again circulated for any final comments.

It was felt important that, once selected, the system should be implemented across FirstGroup in as short a period of time as was practical. It was essential that any supplier should have experience of the British bus market and accordingly all those with such experience were invited to participate in an evaluation trial. Every potential supplier was required to engage at their own expense in a comparative trial to evaluate the effectiveness of each system in dealing with complex scheduling problems. This would provide a fair evaluation of the abilities of each system to minimise cost. It was felt that in selecting a problem, a balance had to be struck between ensuring the problem was non-trivial and not placing an unreasonable burden upon suppliers. It was also very desirable that the exercise should be based on an actual operational example.

The ideal problem was felt to be one that, although in an urban environment, demonstrated many of the features present in rural operation. It was desirable to include vehicles and drivers from several depots on a complex set of services meeting restrictive labour agreements. A small self contained group of service operated by First Eastern Counties met these requirements admirably. First Eastern Counties is the major provider of bus services in East Anglia, with depots in all significant towns. The largest conurbation served is the City of Norwich, which has numerous bus routes, provided by either midi, or larger conventional vehicles. It is also served by numerous services that connect it with towns and villages outside the City.

The data for this trial was based upon the Monday to Friday Conventional and SLF (Super Low Floor) bus network provided within Norwich in October 1998. It was selected as it was a small, but fairly complex operation, with many features common throughout FirstGroup bus subsidiaries.

The network consisted of six main corridors including 14 routes, within which there were a number of small variations. Full timetable, mileage and vehicle workings for this network were provided to suppliers. The routes included many special features, such as variable running time, circular operations, loop and bifurcating journeys. Suppliers were required to provide a full set of accurate schedules replicating the current operation and including point to point mileages and vehicle statistics. The provision of statistics in comparison with the current operation served to confirm that data had been input correctly and that the system can accommodate complex timetables.

Suppliers were provided with as comprehensive details of the current operation as possible. In addition to the current timetables and vehicle workings full details were provided of the current duty schedule and the associated labour agreement. It was anticipated that further questions would arise whilst undertaking the trial and so a nominated co-ordinator from FirstGroup was made available to answer any queries. It was the desire of FirstGroup to be as interactive as possible with all suppliers during this process and welcome the clarification of any issues that were unclear in either the Trial Exercise or within the Requirements Definition.

4   Duty Schedule Comparison

The creation by each supplier of a set of vehicle workings that replicated the existing operation permitted the compilation of four duty scheduling scenarios to various labour conditions to be undertaken. The use of identical vehicle workings and labour constraints would allow a fair comparison to be made of the abilities of each system to produce cost effective answers. Four scenarios for driver scheduling were specified. Each scenario specification set out the applicable labour constraints, and how the payable hours should be calculated. The objective of the exercise was to produce a schedule with the minimum payable hours. It was accepted that in some instances this might not necessarily be within the minimum number of duties. The varied labour constraints applied to the four duty schedule scenarios were based around different agreements found within FirstGroup.

The intention was to assess the capabilities of each system by a direct comparison of results using identical data. This would allow a judgement to be made as to the quality of the system and the quality of those supporting it, as to achieve the best results would not only require an effective system but staff experienced and competent in their use and understanding of it. This was considered an important element of the trial exercise as a depth of knowledge of the product within any supplier was essential to achieving a swift implementation within FirstGroup. The knowledge and abilities of suppliers were further tested by the requirement to complete the exercise within six week of receiving the initial information. This was intended to be, as it proved, a tight deadline. This enabled a judgement to be made as to how suppliers were able to manage a situation that was felt to be comparable with that created by a swift implementation. It also ensured that to produce an effective answer suppliers had to be very familiar with their product.

The problems presented within the four scenarios represented complex problems that required systems to show a high degree of flexibility, or an ability to be easily amended to meet new requirements. An indication from the earlier questionnaires to staff was that those systems that had been tailored for individual companies' needs had stood the test of time far better than systems bought "off the shelf" with little or no adaptation. It was, therefore, essential that the system had either to be capable of being adapted easily during implementation to meet these needs or to be so flexible that this was no longer an issue. The complexity of the duty scheduling scenario was intended to evaluate the capabilities of suppliers and systems to accommodate complex and differing constraints.

5   Nature of the Problems

Scenario 1 was the most complex, replicating the labour agreement to which the original schedules had been compiled and falling within the UK Domestic regulations for drivers hours. The problem covered four depots (Woodcock Road, Vulcan Road, Roundtree Way and Surrey Street) and two classes of vehicle (Conventional and SLF). Drivers from any depot could drive vehicles from any other although there were, in some cases, restrictions as to which class of vehicle could be driven. The ability of drivers to work on vehicles regardless of their depot of origin meant that the duty scheduling problem had to be considered as a whole and not sub divided into four separate problems, one for each depot.

Two types of duty were required, 4-day week duties of around ten hours, and 5-day week duties of around eight hours. Some depots had only one type. Complex variable minimum break constraints were applicable to the longer 4-day week duties, based upon the total length of the duties' spreadovers. All breaks were to be taken in the City centre.

The fact of the maximum spell of work being 4 hours 8 minutes without a break, coupled with a maximum spreadover of 12 hours 59 minutes, meant that most long duties consisted of several spells of work, usually a minimum of 3, often 4 and at time 5 or more. This, coupled with the frequency of relief opportunities, (approximately every 40 minutes), presented a problem that could generate a large number of potential shifts.

The requirement for drivers to start and finish at the same location forced drivers to travel on scheduled services at times to their home depot. This traveling had to conform to the timetables, and in the case of one depot also required an allowance for a 15 minute walk from the bus stop nearest the depot, to the actual depot. Seven different relief points were used within the City centre at Norwich, which provided a complex web of potential travel opportunities for drivers, with the associated time allowances for doing so.

Standard allowances were incorporated for such activities as signing on and off, refueling buses and cashing in. There was a preference for maintaining the present number of duties at Woodcock Road and Vulcan Road, with any saving of duties to be made elsewhere. The existing manually produced duty schedule had been provided to suppliers to provide an indication of the current cost and distribution of duties, and acted as a benchmark to allow them to assess the effectiveness of any results produced.

Scenario 2 was developed from Scenario 1, with a relaxation of duty constraints. Drivers from each location were restricted to driving only one class of vehicle (regardless of which depot it was from), except for Surrey Street drivers who were able to drive any class. This again ensured that the entire operation had to be considered as one problem and not subdivided into smaller sub-problems. Within Scenario 2 duties could be allocated as desired to minimise payable hours, although duties continued to have to sign on and off at the same location.

Scenario 3 allowed drivers signing on at any of the depots to sign off in the City centre. Likewise duties starting in the City centre could sign off at any depot. It was not acceptable to sign on at one depot and off at another. The labour constraints applicable to drivers duties within this scenario were also amended, with a complex calculation for the individual cost of each duty. This replicated agreements found elsewhere within FirstGroup.

The significant restriction within this scenario was that the average work content for all duties at each separate location should be as near to eight hours as possible, without exceeding this. It was anticipated that solutions could easily have been found with fewer duties, but that these would exceed the eight hours maximum at some locations. The intention of this scenario was to evaluate how capable systems were in restricting the average amount of work that each location was able to undertake whilst still producing the most cost effective answer.

Scenario 4 evolved from Scenario 3 and reflected features similar to those operated by certain companies within FirstGroup.

6 Evaluation of Results

The responses to the Requirement Definition and the results of the Trial exercises were evaluated to determine the abilities of each system, with the intention of short listing those most effective for further consideration. An analysis of responses to the Requirements Definition revealed a significant variation in the capabilities between the systems evaluated. This was followed through into the results from the trial exercise where some suppliers were unable to produce acceptable (or in some cases any) solutions. There were, however, sufficient acceptable answers that a meaningful comparison was possible and from this a short list of preferred systems was produced.

The evaluation process so far had provided, through the Requirements Definition, a good indication of the capabilities of each system and suppliers' proposals for future developments. The trial exercises had given a comparison of the abilities of each system to solve an identical series of complex duty scheduling problems and, as importantly, a meaningful comparison of the cost for each acceptable answer. The desire of FirstGroup to talk to and support suppliers throughout the six week trial exercise process had provided an opportunity to obtain an insight into how each supplier was structured, operated and their capabilities.

Within FirstGroup there is a very varied level of computer literacy among the staff involved with the compilation of schedules for bus operation. The number of people employed within this specialised discipline is small and there is a relatively small turnover of staff. Therefore, it is essential that any system should be simple to use and easily understood by those existing schedulers with little previous computer knowledge. To assess this element each short listed supplier was allocated a full day to undertake a comprehensive demonstration of their system to the schedulers who had been involved with the original working party. The breadth of experience within this group stretched from some who had used computer aided scheduling for 15 or more years to one who still undertook the task manually. They also had experience of a wide range of differing systems which gave a broad perspective to the group.

The level of this group's experience led to some very detailed and specific questioning on various areas of functionality. This ensured that a full understanding of each system from the perspective of a user was achieved. At the conclusion of all the demonstrations the group of schedulers evaluated the systems from this perspective, with particular note being taken of the view of those schedulers with little or no experience of computer aided scheduling. It was essential that any system was simple and easy to use. The short listed systems were ranked in order of preference from a user's perspective.

The exercise until this point had been undertaken without reference to the costs of purchasing any particular system. The evaluation was to determine which systems were the most appropriate for use within FirstGroup. Following short listing the selected companies were invited to submit a commercial proposal for discussion. Suppliers were expected to give consideration to the issues raised within the Requirements Definition. These included the FirstGroup requirement that the product met individual company needs and suppliers' proposals to address the elements within the Requirements Definition not presently met. The trial exercise had provided an indication of the potential difference in the costs of results produced by each system which was an element to be borne in mind when considering the proposals.

7 Implementation Proposals

Details of the commercial discussions remain confidential but were concluded with the selection of the "Schedules Office" suite of programs supplied by Omnibus Solutions and Leeds University, (e.g., Fores et al. (1999, 2001); Wren and Kwan (1999)) the implementation of the system beginning in April 2000. All the elements mentioned earlier were considered in selecting the successful system. An important element of the process has been the involvement as far as practical of every scheduler within FirstGroup. The support of the schedulers is of paramount importance for the introduction of the new system to be effective and it is hoped that their earlier involvement will have stimulated interest. It is essential that schedulers can "take ownership" of the system when it is introduced to their company and want to work on the new system. The willingness of suppliers to tailor the software to suit the needs of each individual company as necessary was considered an essential element. This was to create the feeling, and in some cases the reality, that the product has been designed for them.

To confirm the abilities of the selected system to Senior Directors further trials were undertaken using data from other companies, coupled with a further examination of the complete operations of First Eastern Counties in the Norwich area. These results produced savings of up to 4% on payable hours at locations where scheduling had been done without computer assistance, and saving of over 1% in some locations where it had been used. In each trial the local schedulers were involved within the exercise and gave a final acceptance to any answers. This was essential to ensure that the results produced were considered operable by staff with local knowledge. In all cases the local staff involved were surprised by the levels of saving achieved.

The significance of the proposed system being introduced to all companies within FirstGroup should not be underestimated. Each of the 24 constituent bus operating companies within FirstGroup functions with a significant degree of autonomy. The procurement of a scheduling system would, until recently, have been a matter for individual companies. The imposition of a "Group" scheduling system on companies has the potential to be viewed with scepticism and represented an interesting challenge. Whilst Managers and Schedulers had been aware of the proposals throughout the project, it still remained essential to convince each company of the benefits of the change and provide an example of how these benefits can be demonstrated upon schedules with which they are familiar. The full support of the company is essential for an effective implementation.

The implementation at each company was planned to involve several stages phased over a fourteen week period. An initial meeting between the project team and the company would take place during week 1. This would

- Obtain details of all the labour agreements and requirements within the company to ascertain what, if any, customisation of the system was required. Example of current operations would be provided to ensure labour agreements were interpreted correctly. These would be used to confirm that the requirements of each labour agreement are fully covered by the system.
- Obtain the information required to establish a current schedules database for use within the system. It was proposed to undertake the conversion of existing schedules into the correct format for use with the new system at companies which currently use a different computer aided scheduling system. This would ensure that once the implementation has been completed each company would have the system and a complete database of current schedules information.
- Obtain details of interfaces with other systems currently available. These would be replicated to ensure a seamless transition from the old to the new system, essential for maintaining confidence.
- Produce a statement of requirement which would be signed by both the project team and Company Director, outlining the proposals and responsibilities of each party. This specifies minimum levels of equipment, knowledge of staff and commitment to provide staff with sufficient time and resources to undertake the implementation effectively. It defines the involvement of the project team in respect of what will be provided and the proposed training program. The document would be signed again upon completion of implementation to confirm the implementation to the satisfaction of the company. The reasoning for such a formalised approach was to ensure that everyone was aware of what is required of them.
- Establish a "lead scheduler" with extensive knowledge of the company's operations to be part of the implementation team for that specific company to provide advise and information as required. They would also be involved with any trials undertaken and judge the acceptability of any schedule produced.

During weeks 2-8 the labour agreements and requirements of the company would be examined and any customisation required would be undertaken. If required a database of current schedules would be created for the current operations and this would be used to provide data to undertake alternative schedules on current operations. These alternative schedules were to ensure the system could produce schedules to the labour constraints of the company and to demonstrate to the company the potential of the system using data with which they were familiar. The aspiration would be to produce an alternative schedule demonstrating a potential saving for a current operation.

Training in the various modules of the system would be undertaken between weeks 8 and 12 and structured to meet the needs of individual companies. It was anticipated that schedulers with little or no computer aided scheduling experience would take longer to become familiar with the requirements of the system than those who currently use computer aided systems. The training would be split into several sections to ensure that schedulers are able to gain experience and confidence in certain areas before moving on to the next stage. The broad split of functions is

- Timetabling and vehicle scheduling
- Duty scheduling for simpler scenarios
- Duty scheduling for complex scenarios
- Provision of Rosters for staff
- Publicity, interfaces to other systems

A requirement upon the companies was to ensure staff have a basic knowledge of Windows before the commencement of training.

It was proposed to conclude the initial training program after approximately 14 weeks, by which time training would have been given in all aspects of the system and a period of time would have been allowed to gain familiarity. A general review day is provided at the end of the 14 weeks to cover any issues that might have arisen or clarify points of concern, At this point companies will countersign the Statement of Requirement, mentioned earlier, to confirm the requirements objectives outlined have been achieved.

It was accepted that there might be problems arising subsequent to completion of the initial implementation and, in addition to on screen help, a one point telephone support and help desk facility will be provided by the suppliers. This is also complemented by the establishment of an Intranet site upon which schedulers, project team or suppliers can post notices or questions. It is felt that an increased dialogue between schedulers from different companies will help to broaden their outlook in what can be an isolated occupation. For the first few months companies would be "hand held" with informal telephone calls and visits to see how the system was meeting the company needs and, as importantly, ensuring it is being used to its full potential.

8 Longer Term Developments

During the implementation programme, which it is anticipated will take two years to complete, there are expected to be significant developments within the system, prompted by the tailoring to meet company requirements, enhancements to meet the Requirements Definition and other developments initiated by the supplier. It is planned that companies completed within the implementation will continue to be updated on a regular basis. To further stimulate dialogue between schedulers it is proposed to establish a FirstGroup Users Group, similar to those that exist for current systems to complement the Intranet site.

The focus of this project has been to obtain a product that best suits FirstGroups' UK bus operations. With the successful selection of a system and the commencement of introduction into companies it is appropriate to consider extending the application to FirstGroups' rail businesses or overseas bus operations.

9   Schedulers

The system will be used by schedulers with vastly different experiences and capabilities. A concern with the introduction of any new system is to ensure it is used to its full potential. A user with a basic knowledge of scheduling will obtain a better result from the system more swiftly than one without. It is, therefore, proposed, in addition to the training required to introduce the system to companies, to initiate a series of Schedules Training Courses for existing Schedulers. Whilst for some this will be reiterating existing knowledge, it will allow others the chance to gain this. It will also provide a forum for an exchange of views and provide another avenue to establishing contact between schedulers. The experience from these courses will be used to formulate a strategy for the training needs of future schedulers.

The system has now been successfully installed within three companies and is on target to be introduced within the remainder within the next 18 months. Experience has shown that initially sceptic schedulers are persuaded on the merits of the new system by both its ease of use and quality of results achieved. Each company has required a degree of customisation to better match the constraints of the labour agreements and this willingness to adapt the system to suit individual needs has been a factor in building the confidence in new users. Initial indications suggest savings in excess of 2% scheduled payable hours at locations where scheduling was previously compiled manually and in excess of 0.5% where previously compiled using computer aided methods.

Software developments have also been made to provide the scheduler with tools to better influence the results. These include the provision of a complex variety of features that can be pre-specified, ranging from total pre specification of complete shifts to two separate individual journeys (such as a morning and afternoon school movement) to be incorporated within the same duty, with the system determining the remainder of the duty content.

The ability to obtain a duty schedule with the minimum amount of change from the existing operation has been developed. This is useful when there has been a minor change to the vehicle workings. This allows the scheduler to evaluate the cost implications of a minimum change option as opposed to a full recast of duties.

10   Conclusions

The initial consultation with schedulers and managers proved invaluable. It helped identify unmet needs with existing systems and confirmed the anticipated age and experience profile of current compiling staff.

The project team was keen from the outset to involve all companies within the group. Each company has been encouraged to share their unique experiences in an attempt to draw out the most comprehensive criteria for a system. Companies involved in the consultation process now have an understanding and interest in the system that was eventually chosen, this will be of great value as the implementation process progresses.

The team started with a wide range of personal experience. Our initial thoughts concerning the position of scheduling were recycled time and again during the consultation process. The main factor that we did not fully appreciate prior to the process, was that companies had a very varied range of constraints that were not effectively being met by their present system. This identified the need to ensure systems were capable of being tailored to meet each individual company needs.

The responses from suppliers to a comprehensive Requirements Definition and complex scheduling problem enabled an informed judgement to be made on the capabilities of each system. The most significant element of the selection process was to be able to compare the abilities of each system with a number of scenarios using identical data. This allowed a judgement to be made of their effectiveness against each other.

This user led approach has raised the profile of scheduling within FirstGroup and highlighted the potential savings that are achievable with an effective system. The involvement of schedulers from different companies in the project has already led to a broadening of their outlook and it is proposed to develop this further.

The success of the project can only be judged subsequent to implementation. Success should not only be judged by the cost savings achieved but also by the ease it is accepted by companies and their continued use of it.

Bibliography

FirstGroup (2000). *Computer aided scheduling system requirements definition.* FirstGroup private document.

Fores, S., L. Proll, and A. Wren (1999). An improved ILP system for driver scheduling. In N.H.M. Wilson (Ed.), *Computer-Aided Transit Scheduling, Lecture Notes in Economics and Mathematical Systems*, 471, Springer, Berlin, 43–61.

Fores, S., L. Proll, and A. Wren (2001). *Experiences with a flexible driver scheduler.* This Volume.

Wren, A. and R.S.K. Kwan (1999). Installing an urban transport scheduling system. *Journal of Scheduling 2*, 3–17.

Days-off Scheduling in Public Transport Companies

Dulce Pedrosa and Miguel Constantino

Faculdade de Ciências da Universidade de Lisboa, Portugal
dulce@icat.fc.ul.pt, miguel.constantino@fc.ul.pt

Abstract. The assignment of weekly rests to workers in large companies is, in general, conditioned by strict labor union rules. Some of these rules establish, e.g., the (average) number of rest days per week, a minimum and a maximum number of days for the length of a rest period or work period, the number of weekends or Sundays off each p weeks, etc. On the other hand, companies must have enough workers available each day, in order to satisfy the internal workforce demand. The solution approach adopted by some companies consists of assigning workers to cyclic schedules. All workers assigned to the same cyclic schedule have the same type of rest/work periods. Since the weekly workforce demand is not constant in general, these cyclic schedules have to be carefully planned in order to minimize the necessary resources (number of workers).

We propose an integer programming model for which the solutions consist of a set of cyclic schedules as well as the number of workers assigned to each schedule. Since the number of possible schedules may be very large, we use column generation to solve the linear programming (LP) relaxation, and to obtain a set of basic cyclic schedules. The number of workers assigned to each schedule is then obtained by using a simple LP heuristic. To solve the pricing subproblem within the column generation procedure we use a network in which paths correspond to schedules, and solve a shortest path problem. This algorithm has been integrated in the GIST Decision Support System, in use by several public transport companies in Portugal. We also present our computational experience with some large randomly generated instances.

1 Introduction

In this paper we describe an algorithm to solve the weekly days-off scheduling problem in large public transport companies. The algorithm has been integrated in the GIST Decision Support System, in use by several public transport companies in Portugal. GIST (*Gestão Integrada de Sistemas de Transportes*) is a transit scheduling system, which has been developed within the last decade by a consortium of two research groups at the University of Lisbon (ICAT) and University of Oporto (INEGI), and five Portuguese companies of public transport, see Cunha et al. (1993a,b); Cunha and Sousa (2000).

GIST is a modular system, which provides user-friendly solutions to scheduling problems in different stages of the planning procedure. One of these stages is the so-called *crew rostering*, where assignments of each duty to the crew are built such that all daily duties are covered, and assuring that all the rules in use in the company are satisfied. As a part of the crew rostering problem, a company has to distribute the duties over the workers, assuring weekly rest days according to the rules. The rules in use are usually established after hard negotiations with labor unions. They impose many constraints on the schedules such as bounds on the number of consecutive working days and rest days, and on the frequency of Sundays off.

On the other hand, the assignment of duties in most companies in the GIST consortium requires equity in the long run, i.e., "good" and "bad" schedules should be evenly distributed to the workers. Since the concept of "good" schedules is highly subjective, the solution adopted in most companies is to build *cyclic schedules* or *rotations*, where a group of workers has to rotate among a set of selected (balanced) schedules. This approach is considered here for the scheduling of weekly days off. Two factors make the building of these schedules non-trivial: *(1)* the rules in use in the companies mentioned above; *(2)* the non-constant distribution of the duties within the week, which is not the same on week-days, Saturdays, or Sundays for the companies in the GIST consortium.

The approach we propose considers solutions with more than one cyclic schedule. Using several cyclic schedules, instead of only one, allows the company to make better use of the resources, by minimizing the total number of workers needed. All these schedules satisfy the labor union rules, assuring in this way that no schedule is "too bad." Assignment of workers to different types of schedules may be decided by some priority criteria, such as seniority.

In this paper we present an integer programming formulation for the problem of days off scheduling, where the variables correspond to cyclic schedules. Since there may exist exponentially many variables, we use a column generation procedure to solve the linear relaxation of the integer program, and then obtain a (heuristic) feasible solution. The procedure is illustrated with a real life case taken from one of the companies in the GIST consortium. We also present computational results with randomly generated data following the rules in use in the companies.

Several authors have considered the problem of weekly days off scheduling before. Bartholdi III and Ratliff (1978); Bartholdi III et al. (1980); Bartholdi III (1981); Paixão and Pato (1984); Pato (1989) developed algorithms to find a minimum cost cyclic schedule with consecutive workdays alternating with consecutive rest days and with constraints on the number of days off in a week. Other authors propose lower bounds for the total workforce needed in a restricted problem where each worker receives at least A out of every B weekends off and the demand is fixed on weekdays and on weekend days (see Baker and Magazine (1977)) or can vary from day to day (see Burns and Carter (1985)).

In the next section we give the main definitions in order to formalize the problem, and present an integer programming model. In Section 3 we describe the column generation algorithm. We introduce a network model, which is used to formulate the pricing subproblems as shortest path problems. The heuristic procedures are described in Section 4. In Section 5 we report on the computational experience and in the last section we present our conclusions and possible extensions of this work.

2   Definitions and Formulation

In this section we describe and formalize the problem in study, considering the rules used in the companies of the GIST consortium. We define a cyclic schedule, which is an important component of the solutions we aim to obtain. Then we present an integer programming formulation and show why a column generation approach is desirable.

The problem presented is described in the following way: suppose that we have to cover a workforce demand on each day for the planning horizon with crew members that are grouped in teams. Consider also that all crew members in one team work on the same cyclic schedule defined as a sequence of rest days and working days satisfying the union rules imposed, but different teams can work on different cyclic schedules. The problem is to find a preferable small set of schedules and assign teams to those schedules, in order to minimize the number of crew members needed.

Now we formalize some concepts and give some formulations. Consider a company with a daily demand to satisfy, measured in number of workers, over a horizon divided in periods of equal length called *weeks*. We can think of a week as a seven day period from Sunday to Saturday, but in general we may consider weeks of m days where the days are called *weekdays* and represented by $i, i = 1, \ldots, m$. The demand within a week can vary, but it is the same for each weekday in different weeks, and is represented by d_i for weekday i.

Each day can be of one of two types: a *workday* or a *rest day*. A *cyclic schedule* consists of a sequence of work and rest days, such that the last day in the schedule is followed by the first day, and satisfying some given rules. The number of days in a cyclic schedule is assumed here to be a multiple of the week length m. The rules we consider reflect the practice in the companies in the GIST consortium.

A cyclic schedule with $m \cdot s$ days can be represented by a 0-1 matrix with m rows and s columns. Each row $i = 1, \ldots, m$ corresponds to a weekday, while each column $t = 1, \ldots, s$ corresponds to a week. The element a_{it} has the value 1 if the weekday i in week t is a workday and 0 otherwise. From now on we identify a cyclic schedule through the matrix representing it.

A work period is defined as a sequence of consecutive workdays and a rest period is a sequence of consecutive rest days.

Example: *The following example is taken from one of the companies in the GIST consortium. This cyclic schedule is in use in the company. The weeks have 7 days, from Monday ($i = 1$) to Sunday ($i = 7$). The sequence formed by Saturday and Sunday in week 6 and Monday in week 7 is a rest period; the sequence formed by Tuesday through Sunday in week 7 is a work period.*

	Week/Individual Schedule						
	1	*2*	*3*	*4*	*5*	*6*	*7*
Mon	0	1	1	1	1	1	0
Tue	0	0	1	1	1	1	1
Wed	1	0	0	1	1	1	1
Thu	1	1	0	0	1	1	1
Fri	1	1	1	0	0	1	1
Sat	1	1	1	1	0	0	1
Sun	1	1	1	1	0	0	1

The rules used to construct a (feasible) cyclic schedule are as follows:

1. The number of days in any work period must be greater or equal than some minimum value w_{min} and less or equal than a maximum of w_{max} days.
2. Similarly, the number of days in a rest period must be greater or equal than r_{min} and less or equal than r_{max}.
3. If the number of weeks in the schedule is s, then the schedule must have exactly $rs \cdot s$ rest days (and consequently $(m - rs) \cdot s$ workdays), which brings us an average number of rs rest days per week (and $m - rs$ workdays per week). Note that $(m - rs) \cdot s = \sum_{i=1}^{m} \sum_{t=1}^{s} a_{it}$.
4. In np consecutive weeks the schedule must have at least a fixed number p of rest days in a given weekday dp. This rule will be replaced by a slightly looser rule: in each np weeks, the schedule should have on *average* at least p rest days in weekday dp. Usually, the special day is Sunday as we show in the example.

Example: *For the company from which the example was taken, the rules are*

- $w_{min} = 1$ *(it is not specified)*, $w_{max} = 6$.
- $r_{min} = 2$, $r_{max} = 3$.
- $rs = 2$ *(in average two rest days per week).*
- *A Sunday off each 6 consecutive weeks.*

It is straightforward to check that these rules are satisfied by the schedule shown above.

A cyclic schedule can be seen as a *master schedule*, which can be decomposed in as many *individual schedules* as the number s of weeks. Note that all individual schedules within a master schedule have the same sequence of work and rest periods. Hence, these individual schedules satisfy the equity requirement. Workers can be assigned to any of these individual schedules. However, in order to distribute workforce among the days of the week, the number of workers assigned is the same for all individual schedules. A *team* is a group of s workers that are assigned to each of the s individual schedules within a master schedule of s weeks.

Example: *We can build seven different individual schedules. One individual schedule is given by the sequence of weeks (columns) 1,2,3,4,5,6,7 in this order. This means that the person assigned to the schedule works in the first week according to column 1, in the second week according to column 2, etc. Another individual schedule is given by the sequence of weeks (columns) 2,3,4,5,6,7,1 in this order. A person assigned to the schedule works in the first week according to column 2, in the second week according to column 3, etc.*

In what follows, we will use the term schedule to designate either cyclic or master schedule. Let S be a schedule with s weeks. We assume that there is a cost, real or fictitious, associated to S, which is proportional to the number of teams of s workers assigned to S. For the companies in the GIST consortium this cost is considered to be the total number of workers assigned to the schedule (which is s times the number of teams). In general, we assume the cost c of assigning one team to S depends on the number of times a given weekday is a workday. Let p_i be a cost to be considered each time weekday i is a workday in the schedule ($i = 1, \ldots, m$).

The number of times i is a workday is given by $a_i = \sum_{t=1}^{s} a_{it}$, where $a_{it} = 1$ if day i is a workday in week t, and so $c = \sum_{i=1}^{m} p_i a_i$ is the cost of assigning one team to S. If p_i is constant, say $p_i = 1/(m - rs)$ then

$$ c = \frac{1}{m - rs} \sum_{i=1}^{m} a_i = \frac{1}{m - rs} \sum_{i=1}^{m} \sum_{t=1}^{s} a_{it} = s $$

(see rule 3 above).

The problem we want to solve can be stated as follows: find a set of cyclic schedules satisfying all the rules and the number of teams assigned to each schedule, such that the demand (duties) can be satisfied, at minimum cost. The number of possible schedules is potentially infinite but, in general, companies consider an upper bound on the number of weeks in a schedule, so the number of schedules considered is finite (although, as we will see, can be exponentially large).

Example: *The duties in the company are organized according to their depots. For each depot the demand (the number of duties to be assured) follows the same pattern: it is constant from Monday to Friday, lower on Saturday, and the lowest is on Sunday. For a particular depot the data is $d_1 = d_2 = \cdots = d_5 = 240, d_6 = 135, d_7 = 125$*

One solution for this instance, that used to be adopted by the company, uses two schedules: the schedule presented before (denote it by S^2) and the schedule in which Saturday and Sunday are the rest days (S^1). S^1 is a schedule of one week only. If the minimization criterion is the total number of workers, the best solution achieved with these two schedules is to assign 105 teams of 1 worker to S^1 and 27 teams of 7 workers to S^2, yielding a total of 294 workers. However, there are 10 workers too much working on Sundays. We will see later that an optimal solution for this case has 292 workers and it is obtained by adding a third schedule.

Now we present an integer programming formulation for this problem. Let $\{S^j, j = 1, \ldots, n\}$ be the set of schedules to be considered, and let s^j and c^j be the number of weeks and the cost of schedule S^j, respectively. Let

$$a_{it}^j = \begin{cases} 1 & \text{if day } i \text{ in week } t \text{ of schedule } S^j \text{ is a working day} \\ 0 & \text{otherwise} \end{cases}$$

and let $a_i^j = \sum_{t=1}^{s^j} a_{it}^j$ be the number of workers in each team assigned to S^j working on weekday i. The array of a^j is called the *workforce array* of schedule S^j. Let d_i be the demand (measured in number of duties or workers) on weekday i, for $i = 1, \ldots, m$. Define the variables x_j as the number of teams of s^j workers assigned to schedule S^j $(j = 1, \ldots, n)$. The formulation of the problem is given by:

$$(F) \qquad \text{Minimize} \sum_{j=1}^{n} c^j x_j \qquad \qquad (1)$$

subject to

$$\sum_{j=1}^{n} a_i^j x_j \geq d_i \qquad i = 1, \ldots, m \qquad (2)$$

$$x_j \geq 0 \text{ and integer}, \quad j = 1, \ldots, n \qquad (3)$$

Constraints (2) guarantee that in each day of the week there are enough workers to satisfy the demand.

The number of feasible schedules can be very large. For example, given the rules on the length of rest and work periods, we can find a considerable growth in the number of schedules with only seven weeks as can be seen in Table 1. In this table it is also shown how the number of schedules depends on the parameters r_{min}, r_{max} (minimum and maximum consecutive rest days) and w_{min}, w_{max} (minimum and maximum consecutive working days). The number of schedules presented corresponds to schedules that guarantee the total number of 14 rest days on seven weeks, i.e., an average of two rest days per week. In order to find the schedules we use complete enumeration.

The number of schedules is potentially infinite if there is no limit on the number of weeks in a schedule. We will assume that the schedules have

r_{min}	r_{max}	w_{min}	w_{max}	# schedules
1	3	3	6	9773349
2	4	4	7	6075
2	4	4	8	25365
2	3	4	8	16669
1	3	4	8	527125
2	3	5	6	70
1	3	5	6	522
1	3	5	5	42

Table 1. Number of schedules for different rule parameters.

at most T weeks. As we saw in Table 1, it is not reasonable to generate all schedules if T is of moderate size (e.g., 7) and the rules are not tight. We propose to use a column generation procedure to solve problem (F). Our approach will be a heuristic approach: first, we will solve the linear programming relaxation of problem (F), and then we will use its solution and the schedules generated to obtain a feasible integer solution.

3   Column Generation Procedure

As we want to solve a generalized problem where the rules could be flexible, we will apply a column generation approach (see Dantzig and Wolfe (1960); Wolsey (1998); Winston (1993)) to the linear relaxation of problem (F), in order to generate master schedules, as they are needed. This approach has been used in Gilmore and Gomory (1961) and Desrochers and Soumis (1989) for similar problems. We describe here how the column generation works and how we can find master schedules by solving a shortest path problem in an appropriate network.

Let $(\overline{F})$ denote the linear relaxation of (F), i.e., $(\overline{F})$ is obtained from (F) by dropping the integrality constraints in (3). The problem corresponding to $(\overline{F})$ is called the *master problem*. Let N be a small set of columns (master schedules). The *restricted master problem* is defined as:

$$(\overline{Fr}) \qquad \text{Minimize} \sum_{S^j \in N} c^j x_j$$

subject to

$$\sum_{S^j \in N} a_i^j x_j \geq d_i \qquad i = 1, \dots, m$$

$$x_j \geq 0 \qquad S^j \in N$$

New columns are generated, as they are needed, when their reduced costs are negative and, therefore, they are candidates to improve the value of the solution. A solution x for $(\overline{Fr})$ can be extended to $(\overline{F})$ by setting to zero the variables x_j for $S_j \notin N$. This extended solution is optimal for $(\overline{F})$ if there is an optimal solution for the dual problem of $(\overline{Fr})$ which is feasible for the dual problem of $(\overline{F})$, see Wolsey (1998).

Let w_i, $i = 1, \ldots, m$ be the dual variables associated to constraints (2) in the master problem. We can formulate the dual problem of $(\overline{F})$ as follows.

$$(D\overline{F}) \qquad \text{Maximize} \sum_{i=1}^{m} d_i w_i$$

subject to

$$\sum_{i=1}^{m} a_i^j w_i \leq c^j \qquad j = 1, \ldots, n \qquad (4)$$

$$w_i \geq 0 \qquad i = 1, \ldots, m$$

The dual $(D\overline{Fr})$ of $(\overline{Fr})$ is as problem $(D\overline{F})$, with constraints (4) defined only for j such that $S^j \in N$. Suppose x and w are optimal basic solutions of $(\overline{Fr})$ and $(D\overline{Fr})$, respectively. If there is a schedule S^j such that the corresponding constraint (4) is violated by w, then x is not optimal for $(\overline{F})$ and x_j is a variable with negative reduced cost and a candidate to enter the basis in the next basic feasible solution. In this case the variable (column) x_j is added to the restricted master problem $(\overline{Fr})$.

The reduced cost $\bar{c}^j$ of variable x_j is given by $\bar{c}^j = c^j - \sum_{i=1}^{m} a_i^j w_i$. As assumed before, let p_i be the cost to pay if weekday i is a workday, for $i = 1, \ldots, m$. Hence $c^j = \sum_{i=1}^{m} a_i^j p_i$. The new master schedule or column to add to problem $(\overline{Fr})$ will be such that

$$\bar{c}^j = \sum_{i=1}^{m} (p_i - w_i) a_i^j < 0 \qquad (5)$$

In order to find such a column or to prove that all reduced costs are non-negative, we solve the *pricing subproblem*, which consists in finding a schedule minimizing the left hand side of (5).

It is the way a column is found that makes the distinction between some applications of the column generation procedure. In the Cutting Stock Problem of Gilmore and Gomory (1961), the subproblem is a Knapsack Problem, while in Desrochers and Soumis (1989) the authors identified the subproblem with a shortest path problem for the Crew Scheduling Problem.

Next we describe how, in this case, the pricing subproblem can also be formulated and solved as a shortest path in a special acyclic network. A similar network has been proposed by Balakrishnan and Wong (1990) for a problem of sequencing work and rest periods with more than one shift per day.

In the first network model considered we assume that rule number 4 for the feasibility of the schedules is not to be considered, i.e., there is no special day that must be a rest day from time to time. Afterwards, we will see how to incorporate this rule in the network model.

The nodes in the network are defined as follows. Let $l = 1, \ldots, T \cdot m$ represent the days in the horizon, and let $k = 0, \ldots, T \cdot rs$ represent the cumulative number of rest days. We associate two nodes to each pair (k, l): $W(k, l)$ and $R(k, l)$, indicating that k rest days have been considered up to day $l-1$ and day l is the beginning of a work period ($W(k, l)$) or the beginning of a rest period ($R(k, l)$). Moreover, node $W(0, 1)$ is the start node, while we consider a dummy end node. Such a network is shown in Figure 1. The arcs in the network correspond to a work period or to a rest period where:

1. A work period must start on a node $W(k, l)$ and end on a node $R(k, l + nw)$, where $w_{min} \leq nw \leq w_{max}$.
2. A rest period must start on a node $R(k, l)$ and end on a node $W(k + nr, l + nr)$, where $r_{min} \leq nr \leq r_{max}$.
3. Assuming that we begin the schedule with a work period (first node is $W(0, 1)$), we have arcs to the end node from each node of the form $W(n \cdot rs, n \cdot m)$, $1 \leq n < T$, and from $R(T \cdot rs - nr, T \cdot m - nr)$, $r_{min} \leq nr \leq r_{max}$.

Each path from the start node to the end node will define an alternate sequence of work and rest periods which corresponds to a feasible schedule with the total number of rest days and workdays assured.

Example: *Suppose the number of weeks is $T = 2$ and each week has seven days. The length of the work periods is between 4 and 6 days, and the length of the rest periods is between 1 and 3 days. The average number of rest days on each week is 2, which gives a total of 4 rest days in two weeks.*

We present the network model in Figure 2 with all feasible arcs. The path represented by the arcs in bold corresponds to a master schedule with a sequence of 6 workdays, 2 rest days, 4 workdays, and 2 rest days.

In matrix form this master schedule is given by:

$$\begin{bmatrix} 1 & 0 \\ 1 & 1 \\ 1 & 1 \\ 1 & 1 \\ 1 & 1 \\ 1 & 0 \\ 0 & 0 \end{bmatrix} \quad \textit{With workforce array:} \quad \begin{bmatrix} 1 \\ 2 \\ 2 \\ 2 \\ 2 \\ 1 \\ 0 \end{bmatrix}$$

We consider costs associated to the arcs as follows: The cost associated to an arc representing a work period is given by $\sum_{l \in D}(p_l - w_l)$, where D is the sequence of weekdays in that work period. The arcs representing rest periods have null costs, as well as the arcs leading to the dummy end node.

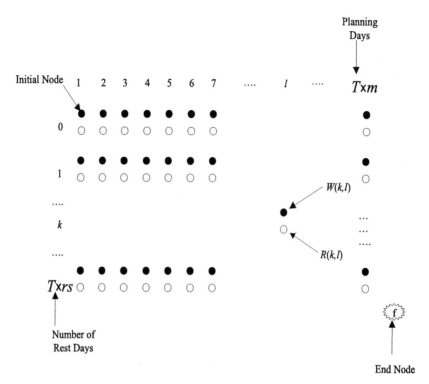

Figure 1. Network representation.

The shortest path in this network corresponds to the schedule with minimum (reduced) cost.

Note that the costs of the arcs will ultimately depend on the specific day of the week of the first day in the horizon. Suppose this day is weekday 1 (e.g., Monday). It is assumed that this day is a workday, and that the last day in the schedule (weekday m, Sunday in the 7 week example) is a rest day. Then, the schedules generated are characterized by having always one work period starting on Monday (and the rest period just before ending on Sunday). In order to consider all possible combinations, we should consider as many networks as days of the week. In each network the first day of the horizon will be a different weekday.

If the cost of the shortest path in one of these networks is negative, then the column representing the workforce array should be added to the restricted master problem. Otherwise, no other schedule will improve the solution and we have found the optimal solution of problem $(\overline{F})$.

We can observe in Figure 1 that many of the nodes are not needed to obtain the shortest path. The shortest path in one of those acyclic networks can be obtained using a simple Dynamic Programming recursion, with complexity bounded by $O(T^2 \cdot m^2)$ in the worst case.

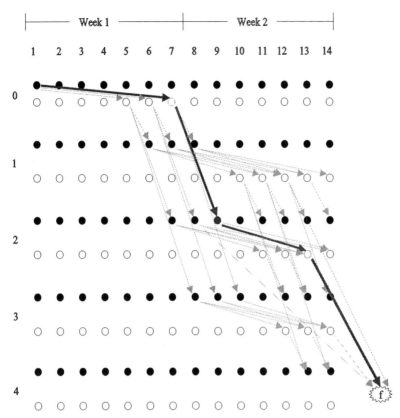

Figure 2. Network model with feasible arcs.

In order to find the shortest path we consider the labels $EW(k,l)$ and $ER(k,l)$ as the length of the path from the origin to node $W(k,l)$ and $R(k,l)$, respectively. Let

$$EW(1,0) = 0$$
$$EW(k,l) = +\infty, \quad ER(k,l) = +\infty, \quad k > 0, l > 1$$

Considering the feasible rest and work periods we get:

$$EW(k,l) = \min_{r_{min} \le nr \le r_{max}} \{EW(k - nr, l - nr)\}$$

$$ER(k,l) = \min_{w_{min} \le nw \le w_{max}} \{ER(k, l - nw) + c_{l,nw}\}$$

$$\text{with } c_{l,nw} = \sum_{lt=(l-nw)\bmod m}^{(l-1)\bmod m} (p_{lt} - w_{lt})$$

So, the value $EW(rs \cdot s + 1, s \cdot m + 1)$ associated to the dummy end node, represents the cost of the shortest path from the start node to the end node.

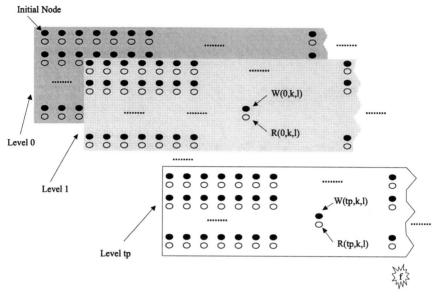

Figure 3. Multiple levels.

With this network model, rule number 4 required by feasible schedules is not guaranteed. This model is extended in order to assure that the average number of days that weekday dp is a rest day is at least p days out of each np weeks. In T weeks we need at least $tp = \lceil (T \cdot p)/(np) \rceil$ rest days on day dp. To do this, we add a new dimension to the network model. Each node is characterized by a new parameter $h = 0, \ldots, p$ representing the number of times the special day is a rest day. In this case we will have p levels in our network, as it can be seen in Figure 3. The nodes are represented by $W(h, k, l)$ and $R(h, k, l)$, indicating that k rest days have been considered up to day $l - 1$, h of them are in the special weekday, and day l is the beginning of a work period $(W(h, k, l))$ or the beginning of a rest period $(R(h, k, l))$.

The arcs in this network are defined in a similar way as before:

1. A work period must start on a node $W(h, k, l)$ and end on a node $R(h, k, l + nw)$, where $w_{min} \leq nw \leq w_{max}$.
2. A rest period must start on a rest node $R(h_1, k, l)$ and end on a work node $W(h_2, k + nr, l + nr)$, where $r_{min} \leq nr \leq r_{max}$ and $h_2 = min\{tp, h_1 + nh\}$ where nh is the number of times the special weekday appears in the interval $[l, l + nr]$.
3. Assuming that we begin the schedule with a work period (first node is $W(0, 0, 1)$), we have arcs to the final node from each node of the form $W(h, n \cdot rs, n \cdot m)$, such that $n/np \leq h/p$, $n = 1, \ldots, T - 1$, and from the nodes $R(h, T \cdot rs - nr, T \cdot m - nr)$, $r_{min} \leq nr \leq r_{max}$ yielding feasible schedules.

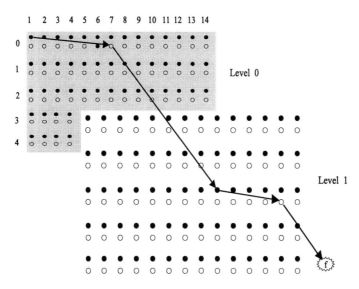

Figure 4. Example path.

Using the example, with the additional constraint of one rest day on Sunday (day 7), we have the path shown in Figure 4. In this case, the shortest path can be found in a way similar to the first network model.

4   Heuristic Algorithms

Here we present the heuristic algorithms to obtain an assignment of teams to schedules. These heuristics are based on the column generation procedure described just before, and they use the generated schedules as a basis to that assignment.

When the column generation procedure terminates we end up with a set of schedules, say N_g, and a solution x such that $x_j = 0$ for $j \notin Ng$, which is optimal for the LP relaxation $(\overline{F})$. If the components of x are integer, then x is also optimal for problem (F). Otherwise, N_g and x are used to obtain a hopefully good solution for (F).

The heuristics we consider are of two types: *(1)* rounding of the LP solution; *(2)* branch-and-bound restricted to the columns generated.

The rounding heuristics work as follows: *(1)* choose a variable x_{j*}, fractional in the LP solution, to be round up; *(2)* solve the Linear Program $(\overline{Fr})$, with $N = N_g$ and the value of x_{j*} fixed to $\lceil x_{j*} \rceil$, to obtain a new LP solution where x_{j*} is integer; *(3)* repeat steps *(1)* and *(2)* until there are no more fractional variables; *(4)* check if some variables can have its value reduced, while keeping the solution feasible.

Each time step *(2)* is performed, one variable is fixed to an integer larger than its value. Hence this procedure terminates after a finite number of steps and a feasible solution is obtained.

We consider two rounding heuristics, corresponding to two different ways to choose the variable x_{j*}. In heuristic 1, we consider x_{j*} maximizing $\{(x_j - \lfloor x_j \rfloor) \cdot s_j\}$ for $j \in S$, which aims to find the column where the round up value gives the least number of workers too much. In heuristic 2 we choose the variable maximizing $\{(\lceil x_j \rceil - x_j) \cdot s_j\}$, i.e., the variable which rounding up gives the possibility of decreasing the values of the others variables.

The other heuristic considered performs branch-and-bound restricted to the columns generated.

When the objective function (1) represents the total number of workers (W) needed to satisfy the demand, we use the information provided by three general lower bounds on the optimal solution value, in order to speed up the column generation procedure. These lower bounds are similar to the ones presented in Burns and Carter (1985).

The first lower bound is related to the number of rest days on a fixed day of the week. If we must have p rest days on N weeks for a fixed day dp with demand d_{dp}, then

$$LB_1 : W \geq \frac{d_{dp}N}{N - p}$$

The second lower bound is related to the number of working days on a week. If we have to assure a number of rs rest days on a week, then

$$LB_2 : W \geq \frac{1}{m - ds} \sum_{i=1}^{m} d_i$$

The third lower bound is related to the maximum demand on a week:

$$LB_3 : W \geq \max_{i=1,\ldots,m} \{d_i\}$$

It is straightforward to see that all these three lower bounds are dominated by the optimal LP value $v(\overline{F}) = \sum_{j \in N_g} s_j x_j$. Let $LB = \max\{LB_1, LB_2, LB_3\}$. If, in some iteration of the column generation procedure, $v(\overline{Fr}) = LB$, then the optimal solution has been attained and the procedure can stop.

Example: *For one particular depot in the company the demand is given by* $d_1 = \cdots = d_5 = 240$ *duties from Monday to Friday,* $d_6 = 135$ *on Saturdays and* $d_7 = 125$ *on Sundays. The initial set of schedules considered as initial solution to the algorithm is given by those used at this company, which are:*

$$S^1 = \begin{bmatrix} 1 \\ 1 \\ 1 \\ 1 \\ 1 \\ 0 \\ 0 \end{bmatrix} \qquad S^2 = \begin{bmatrix} 0 & 1 & 1 & 1 & 1 & 1 & 0 \\ 0 & 0 & 1 & 1 & 1 & 1 & 1 \\ 1 & 0 & 0 & 1 & 1 & 1 & 1 \\ 1 & 1 & 0 & 0 & 1 & 1 & 1 \\ 1 & 1 & 1 & 0 & 0 & 1 & 1 \\ 1 & 1 & 1 & 1 & 0 & 0 & 1 \\ 1 & 1 & 1 & 1 & 0 & 0 & 1 \end{bmatrix}$$

Using these master schedules, the initial solution uses 105 masters schedules S^1 and 27 masters schedules S^2, with a total of 294 workers and 10 at Sunday at work but with no duty to do. This is also the initial LP solution. The general lower bound for this problem is given by $LB = max\{150, 292, 240\} = 292$.

Several optimal solutions were found by the column generation procedure. One of those solutions uses the two master schedules described before and only one more schedule:

$$S^3 = \begin{bmatrix} 1 & 0 & 1 & 1 \\ 1 & 1 & 0 & 1 \\ 1 & 1 & 0 & 1 \\ 1 & 1 & 0 & 1 \\ 1 & 1 & 1 & 0 \\ 1 & 1 & 1 & 0 \\ 0 & 1 & 1 & 0 \end{bmatrix}$$

In this case the solution is $x_1 = 105, x_2 = 21$, and $x_3 = 10$.

5   Computational Experience

We present the results of computational tests with some randomly generated instances. We have considered 10 test instances, with the demand (number of daily duties) for each weekday randomly generated in the interval $[1000, 2500]$. We have considered an upper bound of $T = 26$ for the number of weeks in the schedules, and weeks of $m = 7$ days. The rules parameters are: $r_{min} = 2$, $r_{max} = 4$, $rs = 2$, $w_{min} = 4$, $w_{max} = 8$.

The LP solver callable Library CPLEX 4.0 (see CPLEX (1995)) was used on a Sun Sparc Station LX, and the algorithm was implemented in C programming language. In Table 2 we present some of the results obtained as the value of the lower bound (LB), the value of linear relaxation, the time used in the linear relaxation phase, and the value obtained for the three heuristics presented: heuristic 1 (H1), heuristic 2 (H2), and branch-and-bound (BB) restricted to the columns generated. The time used to reach an integer solution is not presented because the values were always below one second.

In problems 1, 2, 5, and 9, the value of the linear relaxation is larger than the general lower bound. For these instances the column generation procedure did not stop before finding schedules with 26 weeks.

Problem	LB	Linear Rel.	Time	Work Schedules Generated	Solution		
					H1	H2	BB
1	2464.6	2621.1	2271.21	17	2323	2625	2622
2	2220.0	2255.0	2283.73	14	2258	2260	2255
3	2759.0	2759.0	2.09	7	2761	2762	2759
4	2333.6	2333.6	9.72	16	2337	2338	2334
5	2636.2	2726.3	2289.53	13	2730	2730	2727
6	2680.4	2680.4	4.59	9	2686	2686	2681
7	2458.0	2458.0	2.89	6	2459	2459	2458
8	2822.2	2822.2	3.03	6	2828	2828	2823
9	2110.8	2122.3	2292.80	16	2123	2125	2123
10	2124.0	2124.0	4.59	11	2124	2124	2124

Table 2. Numerical results.

The heuristic solutions obtained by the two rounding heuristics and the branch-and-bound method show that the branch-and-bound obtains better solutions than the rounding heuristics. The branch-and-bound solutions are optimal for these problems. This fact is explained by the set of schedules used: branch-and-bound uses all the schedules presented in the linear relaxation while the rounding heuristics use only the schedules with positive value. The time used for the heuristics is always less than one second.

In the results presented in Table 2, we can see that the value of the linear relaxation is very close to the optimal solution. This is a result of the structure of the problem and probably of the nature of the problems tested: instances with a large randomly generated demand and flexible rule parameters.

6 Conclusion

The procedure presented here has been tested in the companies in the GIST consortium. As a result, new types of cyclic schedules have been obtained. Collated with the presented solutions, the companies have shown interest in this type of solutions, since they can satisfy the labor union rules and, at same time, guarantee an optimum use of the workforce resources.

Nowadays, the contractual rules in a company are always with changes on duties and on the scheduling of days off. The functionality of the presented algorithm has to be enhanced in this point, which has the potentiality of being adapted to other restrictions (rules) in the generation of feasible schedules. Moreover, the use of parameters to define periods of work and rest becomes fruitful, because it allows the scheduler to add alternative solutions to the ones that are currently accepted in the company.

This algorithm can be extended to more complex problems, as is the case of the work assured with shifts. We could easily adapt the column generation algorithm to this case by adding new levels to the network model for the pricing subproblem.

The main purpose of this study is not to present a fixed solution for a company, but to supply a tool to the scheduler, helping him in decisions concerning the assignment of days off to the drivers.

Bibliography

Baker, K.R. and M. Magazine (1977). Workforce scheduling with cyclic demands and day-off constraints. *Management Science 24*, 161–167.

Balakrishnan, N. and R.T. Wong (1990). A network model for the rotating workforce scheduling problem. *Networks 20*, 25–42.

Bartholdi III, J.J. (1981). A guaranteed-accuracy round-off algorithm for cyclic scheduling and set covering. *Operations Research 29*, 501–510.

Bartholdi III, J.J., J.B. Orlin, and D. Ratliff (1980). Cyclic scheduling via integer programs with circular ones. *Operations Research 28*, 1074–1085.

Bartholdi III, J.J. and D. Ratliff (1978). Unnetworks, with applications to idle time scheduling. *Management Science 24*, 850–858.

Burns, R.N. and W.W. Carter (1985). Work force size and single shift schedules with variable demands. *Management Science 31*, 599–607.

CPLEX (1995). *Using the CPLEX Callable Library, Version 4.0.* CPLEX Optimization, Inc., Incline Village, USA.

Cunha, J.F. and J.P. Sousa (2000). The bus stops here – GIST: Decision-support system for public transport planning in Portugal. *OR/MS Today 27*(2), 48–53.

Cunha, J.F., J.P. Sousa, T. Galvão Dias, and J.L. Borges (1993a). *GIST – A Decision Support System for the Operational Planning at Mass Transport Companies.* Presented at 6th International Workshop on Computer-Aided Scheduling of Public Transport.

Cunha, J.F., J.P. Sousa, T. Galvão Dias, and J.L. Borges (1993b). *GIST: LINE Concept in a DSS for Operational Planning at Public Transport Companies.* Technical report, Faculdade de Engenharia da Universidade do Porto, INEGI, Porto.

Dantzig, G.B. and P. Wolfe (1960). Decomposition principle for linear programs. *Operations Research 8*, 101–111.

Desrochers, M. and F. Soumis (1989). A column generation approach to the urban transit crew scheduling problem. *Transportation Science 23*, 1–13.

Gilmore, P.C. and R.E. Gomory (1961). A linear programming approach to the cutting stock problem. *Operations Research 9*, 849–859.

Paixão, J.M.P. and M. Pato (1984). *Escalonamento Óptimo de Pessoal Sujeito a Horários Cíclicos.* 4, CEAUL, Lisboa, Portugal.

Pato, M. (1989). *Algoritmos para Problemas de Cobertura Generalizados.* Ph.D. thesis, Faculty of Sciences-University of Lisbon.

Winston, W.L. (1993). *Operations Research: Applications and Algorithms* (3 ed.). Duxbury, Belmont.

Wolsey, L.A. (1998). *Integer Programming.* Wiley, New York.

Modeling Cost and Passenger Level of Service for Integrated Transit Service

Mark Hickman[1] and Kelly Blume[2]

[1] Department of Civil Engineering and Engineering Mechanics
The University of Arizona, P.O. Box 210072, Tucson AZ 85721-0072, USA
mhickman@engr.arizona.edu
[2] Kittelson and Associates, 610 SW Adler, Suite 700, Portland, OR 97205, USA
kblume@kittelson.com

Abstract. In the United States, many transit agencies are considering integrating their demand-responsive service with traditional fixed-route service. In some cases, it may be advantageous to the transit agency or to the passenger to coordinate traditional demand-responsive transit service with fixed-route service. The demand-responsive service connects passengers from their origin to the fixed route service and (or) from the fixed route service to their final destination. Such a service is expected to reduce the cost of transit service, but also will affect the level of service experienced by passengers. The integrated transit service problem is to schedule both passenger trips (or *itineraries*) and vehicle trips for this service. In considering the literature, this research proposes a scheduling method that explicitly incorporates both transit agency cost and passenger level of service. More specifically, the model assumes: *(i)* a fixed-route bus schedule; *(ii)* desired passenger pick-up and drop-off points; *(iii)* time window constraints for passenger pick-ups, drop-offs, and transfers; and *(iv)* passenger level of service constraints, including maximum travel times and number of transfers. Using this information, the proposed technique determines which trips are eligible for integrated service using the passenger level of service constraints. A schedule is then created for both the passenger trips and the vehicle trips, so that the total cost of service is minimized. The method is illustrated using a case study of transit service in Houston, Texas, showing the possible cost advantages and changes in passenger level of service with integrated service. The contributions of the research include: *(i)* a new heuristic for scheduling integrated transit trips that accommodates both passenger and vehicle scheduling objectives; and, *(ii)* an illustrated method for evaluating the operating cost and passenger level of service implications of integrated transit service.

1 Introduction

In the United States, many transit agencies have been considering integrating their demand-responsive service with traditional fixed-route service. In some cases, it may be advantageous to the transit agency or to the passenger to coordinate traditional demand-responsive transit service with fixed-route service. The demand-responsive service connects passengers from their origin to the fixed route service and (or) from the fixed route service to their final destination. Using this concept, transit agencies can extend demand-responsive service into low-density markets or may substitute demand-responsive service for fixed-route service. In these cases, operating costs may be reduced, and the level of service to passengers may increase by providing door-to-door service. In other situations, longer trip lengths and growing patronage for demand-responsive service may lead a transit agency to consider providing at least part of the trip on fixed-route service, thereby reducing operating costs.

From a scheduling perspective, the integrated transit service problem is to schedule transit trips that may be carried by some combination of demand-responsive and fixed-route transit service. Both passenger trips (or *itineraries*) and vehicle trips must be scheduled. Past research on this specific problem includes the work of Wilson et al. (1976) and Liaw et al. (1996). The first work examines scheduling of integrated service where several demand-responsive services operate in different geographic zones that are connected by a fixed-route service. The problem is formulated with a passenger utility function as its objective, subject to various level of service constraints. Operator costs are not included directly in the model. To schedule passenger and vehicle trips, a trip insertion heuristic is used. Somewhat in contrast, the second work examines scheduling of integrated service using operating cost as the objective function. The problem is formulated using hard time window constraints, but no other passenger level of service measures are included in the model. An on-line heuristic is used to generate passenger itineraries, and the passenger and vehicle trips are further refined using simulated annealing.

A more general approach to multi-modal and flexible transit scheduling is discussed in several other references, including Gerland (1991); Crainic et al. (1998); Horn (1999); Malucelli et al. (2001). These works have made important contributions in scheduling and routing integrated transit service. These approaches are more general than the methodology described here, and they have great advantage in modeling new transit service options. In contrast, the model described in this paper is specific to the integration of fixed-route and demand-responsive systems, constrained to the context of current transit service options in the US.

What is still lacking in this context is a scheduling method for integrated service that includes both the passenger and operator objectives. To this end, this research explicitly incorporates both transit agency cost and passenger level of service directly in the model. From the transit agency's perspective, the goal in scheduling vehicle trips is to minimize the total cost of service. On the other hand, passengers desire a high level of service; e.g., minimizing travel time, transfer time, and the number of transfers. To balance agency and passenger objectives, this research introduces a heuristic to schedule integrated trips that minimizes transit agency cost, subject to passenger level of service constraints. In the model formulation, the following elements are given: *(i)* a fixed-route bus schedule; *(ii)* desired passenger pick-up and drop-off points; *(iii)* time window constraints for passenger pick-ups, drop-offs, and transfers; and *(iv)* passenger level of service constraints, including maximum travel times and number of transfers. Using this information, the proposed technique determines which trips are eligible for integrated service using the passenger level of service constraints. A schedule is then created for both the passenger trips and the vehicle trips in the integrated service, so that the total cost of service is minimized.

This paper describes and illustrates a proposed two-stage heuristic to solve the integrated service scheduling problem. In the following section, this two-stage heuristic is described. In the third section, the proposed heuristic is used to illustrate the possible advantages of integrated service through an illustration of transit service in Houston, Texas. The case study is used to identify the potential cost and level of service implications for a transit agency considering shifting portions of some demand-responsive trips to the fixed-route service. The final section presents conclusions on the value of this scheduling method.

2   The Scheduling Heuristic

2.1   Overview

Development of an integrated transit service schedule comprises two main tasks: scheduling passenger trips and scheduling vehicle trips. In the scheduling of passenger trips, an itinerary is developed for each integrated service request in which:

1. A paratransit vehicle may pick up the passenger from his/her origin and "feed" him/her to an appropriate fixed-route stop.
2. A fixed-route vehicle will then pick the passenger up and transport him/ her to another fixed-route stop.
3. A second paratransit vehicle may carry the passenger from the second fixed-route stop to the door of his/her destination.

One or more of the paratransit "legs" may be excluded, and multiple itineraries are possible for a single request. Figure 1 conceptually illustrates the scheduling of a single passenger's request where two transfers must be made.

The proposed approach decomposes this problem into two parts. First, one must find a feasible passenger itinerary, connecting the passenger's origin with the passenger's destination with transit service that maximizes the passenger's level of service. If such a passenger itinerary can be found that meets these level of service requirements, the passenger's trip is scheduled. Second, the paratransit trip legs must be added to a vehicle's schedule. This is done through existing vehicle routing heuristics for paratransit service. Through this decomposition, it is believed that this technique improves upon that of Liaw et al. (1996) by explicitly considering the passenger's level of service. It also improves upon the technique of Wilson et al. (1976) by explicitly incorporating operating costs into the scheduling process.

The following (typical) input for these two scheduling tasks is assumed:

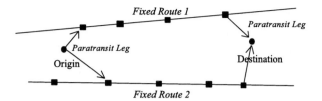

●   requested origin and destination
■   major time point (possible transfer point)

Figure 1. An integrated transit trip.

- the location of the passengers' pickup and dropoff points
- the passenger's requested times, and associated time windows, in which pickups and dropoffs must occur
- the location of fixed-route stops
- the schedules of all fixed-route vehicles
- the accessibility level of all fixed-route vehicles and transfer points
- the time windows in which paratransit vehicles are permitted to meet fixed-route vehicles at transfer points
- vehicle capacities
- passenger loading and unloading times
- the distance between stops
- minimum passenger level of service standards

The time windows for connecting between paratransit and fixed-route service may be based on local policy. In this case one must balance the need for flexibility and slack to accommodate variation in vehicle travel times with the need for short waiting periods at the transfer station. At the same time, the dwell time at the transfer point must be sufficiently long to load and unload passengers from the fixed-route service. Recent research has suggested that the elderly and disabled may require significantly longer time to board and alight, on the order of 1–3 minutes (Kittelson and Associates (1999)).

The tasks of passenger scheduling and vehicle scheduling are then performed sequentially. Typically, the passenger itinerary will be scheduled on-line, so that the itinerary can be relayed directly to the passenger when they are requesting a trip. The vehicle trip scheduling can be done off-line, once all passenger trips are scheduled. The following sections describe the passenger and vehicle scheduling methods, respectively.

2.2   Passenger Itinerary Development

In the first stage, the potential passenger trip from the origin to the fixed route, on the fixed route, and from the fixed route to the destination is scheduled. The itinerary development process is summarized in Figure 2. The method proposed below is a variant of more traditional public transit

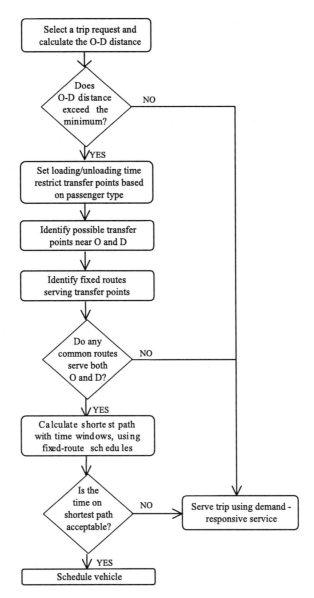

Figure 2. Passenger itinerary heuristic.

itinerary methods (e.g., Bovy and Stern (1990); Bander and White (1991); Han and Hwang (1992); Koncz et al. (1996)).

To develop an integrated itinerary, a passenger is selected and his/her requested times and locations are identified. Consistent with most existing paratransit scheduling software, the Euclidean distance between the origin and destination is calculated. The Euclidean distance allows a computationally fast estimate of the total travel time, although at a loss of precision when compared with the computationally burdensome but accurate shortest path techniques. The method here also uses a single vehicle speed, although one might have this value vary in peak periods versus off-peak periods in order to account for congestion.

The Euclidean distance must exceed some specified minimum distance; this screening is done to eliminate an inconvenient pair of transfers for very short trips, particularly when the paratransit legs of the integrated trip together form a very high percentage of the total origin-to-destination (O-D) distance. The distance between origin and destination can also be used to estimate the passenger's expected travel time for a direct paratransit trip. Also, the maximum allowable ride time for each passenger can be calculated as an incremental percentage above the expected travel time (e.g., 50% higher).

An additional screening is made based on the passenger's disability. The integrated service is intended to accommodate passengers traveling under provisions of the Americans with Disabilities Act (ADA) of 1990. This means that one could consider all types of passengers for integrated service. However, this requires that the stops, routes and vehicles on the fixed-route system are all able to accommodate ADA passengers (e.g., with wheelchair lifts, accessible shelters, appropriate curb treatments, etc.). As will be noted later, the ability of a transit agency to accommodate these ADA trips is an important determinant of the number of trips eligible for integrated service.

Based on the passenger's origin and destination, and any accessibility requirements, possible transfer points to the fixed route network must be identified. These should be less than some maximum distance from the origin or destination; in this way one may screen out trips where the fixed-route segment accounts for only a small percentage of the trip. The transfer points should also be farther than some minimum distance because, for the agency and for other passengers, it would be impractical to schedule a paratransit vehicle for a trip that is too short. Rather, such a request would be served directly by paratransit or by a single-transfer trip. To minimize passenger inconvenience, no more than two transfers are allowed, and so only two transfer points need to be identified.

The proposed method is a variation of that proposed by Liaw et al. (1996). One may construct circles geographically about both the rider's origin and destination and identify transfer points within these circles along a common fixed route. This technique can be used to identify any fixed routes that serve the origin or destination directly (i.e., within a very small walking distance), hence requiring only one or no paratransit legs.

In contrast to Liaw's method, however, it seems that integrated trips with two paratransit legs have a minimum, as well as a maximum, radius (i.e., a ring). The distance between a passenger's origin and destination, e.g., may be long enough that the passenger cannot make the trip without assistance but short enough that a single paratransit trip would be less expensive for the agency than a combination of paratransit and fixed-route trips. If served

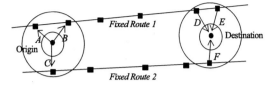

●   requested origin and destination
■   major time point (possible transfer point)

Figure 3. Possible integrated transit itineraries for a single request.

with a single paratransit vehicle and no fixed-route transfers, the passenger would not experience the onerousness of transfers or waiting at a fixed-route stop, and other passengers would not be unduly penalized by the need to schedule an additional integrated trip. Specifying a practical minimum distance between the origin and destination, therefore, improves passenger level of service for short trips and mitigates overall system scheduling inflexibility.

Figure 3 illustrates the proximity circles and the paratransit and fixed-route trips that might serve a single request. Possible paratransit connections are denoted with capital letters; these only connect the origin and destination to points within the proximity circle. One integrated transit trip might be *Origin → C → F → Destination*, via *Fixed Route 2*.

It is reasonable to expect that circles of different sizes will have different effects on the capabilities and cost-effectiveness of the integrated transit system. For example, large circles will include more fixed routes but may also require longer paratransit trips. At the same time, the more distant an origin and destination, the longer the paratransit trip legs can be without seeming an inefficient connection. Setting circle size at a percentage of the distance between a given origin and destination is one method for taking total travel distance into consideration. This percentage can be set at different levels for a given transit system and a sensitivity analysis performed to determine what radius provides a reasonable screening of itineraries.

After potential transfer points have been identified, common routes that connect the origin and destination must be found. This is accomplished through an explicit matching of fixed routes associated with major time points near both the origin and destination. If the time points near the origin and destination are not connected by a common route, then the trip request is served entirely with paratransit.

With the resulting sub-network of feasible paratransit legs and fixed-route services, feasible itineraries are constructed. Essentially, this involves solving a shortest path problem with time windows (Desrochers and Soumis (1988)) on this sub-network. For this, time windows at the origin, destination, and transfer points are used. Also, the published fixed-route schedule is used to estimate available time windows at transfer points; passengers must be picked up or dropped off within the time windows during which the transit vehicle is expected. Currently, the shortest path is generated by full enumeration (the size of the sub-networks are generally not too large). If the passenger has specified an appointment time at the destination, a backwards pass through the network is performed. If, instead, a departure time from the origin is specified, a forward pass through the network is performed.

For each such itinerary, a passenger's level of service measure must be evaluated; a "generalized time" calculates the sum of waiting, travel, and transfer time along each path. The waiting and transfer times can be estimated from the associated time windows; the fixed-route schedules give an estimate of travel times on the fixed-route service; and, a straight-line distance divided by an average vehicle speed is used to estimate in-vehicle travel times for the paratransit trips. It is also possible that different weights can be applied to these different components of travel time (e.g., if transfer time is more onerous than other types of time). In the example, a transfer "penalty" equivalent to 5 minutes of travel time is added. In total, a generalized time or disutility function Z can be described as follows:

$$\text{Minimize } Z = \beta_1 \cdot WT + \beta_2 \cdot IVT + \beta_3 \cdot XT + \beta_4 \cdot NX$$

where Z = generalized time or disutility, WT = total waiting time, IVT = in-vehicle travel time, XT = transfer time, NX = number of transfers, and β_1, β_2, β_3, β_4 = weights (coefficients) on each variable.

The path with the minimum generalized time is then compared to the generalized time of a direct paratransit trip (the baseline). The approach in the case study assumes that the *existing paratransit service* is the "default" or "baseline" service, should it prove infeasible or not cost-effective to serve the trip with the integrated service. In this case, the cost-effectiveness of full paratransit service is compared with the integrated transit trip. (One could just as easily define the default as fixed-route service, in order to examine the cost-effectiveness of paratransit "feeder" service. While not described in this paper, such a technique involves only minor modification of the proposed method.) The passenger trip is accepted if the generalized time (Z) is not more than the maximum allowable trip time. As one might expect, however, varying the maximum allowable trip time may have considerable impact on the likely number of passengers served with the integrated service.

An additional term could be added to the objective (passenger generalized time, Z) to include the disutility, or delay, to passengers on the fixed route service when waiting for, loading and (or) unloading integrated service passengers. Since many of these passengers may require considerable time to load and unload, the effect of this term may also be important. For the ease of analysis, this is not included here, but could easily be added.

2.3   Vehicle Trip Scheduling

In the Houston case study to be described later, it was not necessary to go to the detail of a full vehicle schedule. Rather, the results of the passenger trip scheduling technique were sufficient to evaluate the feasibility and potential advantages of the integrated service scheme. Below, an outline of the approach to completing the vehicle trip scheduling task is given; this is an area of ongoing research.

Once an integrated trip has been accepted, the vehicle trips from the origin to the fixed route and from the fixed route to the destination are added to a traditional paratransit vehicle routing and scheduling problem. While there are now a large number of heuristics that can be used for paratransit vehicle routing and scheduling, two were identified for use in this research. The first is a vehicle trip insertion heuristic (Jaw et al. (1986)); this method is an updated technique that naturally follows the original work by Wilson et al. (1976). This method essentially assigns an incremental "cost" to each vehicle itinerary to accept a new trip. The vehicle with the lowest cost receives the new passenger trip, as long as vehicle capacity constraints are not violated.

Also, given the large number of potential trips (over 3500 per day), and the existence of a reasonably good paratransit vehicle schedule, a mini-clustering and column generation technique (Ioachim et al. (1995)) is also possible. In this technique, the trips are grouped into clusters. Once so clustered, the algorithm uses a modified shortest path technique to re-optimize the allocation of clusters to individual vans.

For the case study, Houston METRO has provided an existing vehicle schedule that was created by their paratransit trip scheduling software. This provided a set of trip requests, locations, and time windows for these trips. It also provided "baseline" vehicle assignments of all the passenger trips.

Rather than re-scheduling all 3500 trips, the existing passenger trips were separated into two groups: those that could use the integrated service and those that could not. The potential integrated service trips may be removed from the existing vehicle schedules. This creates a subset of all passenger trips that are eligible for re-scheduling. Using the technique of Jaw et al. (1986), the re-scheduled trips may be re-inserted into the vehicle trip schedules based on a minimum cost insertion.

In some cases, however, this re-insertion is not desirable because of the ensuing geographic dispersion of vehicle trip segments. Rather, the remaining trips (those not eligible for integrated service) may be left as "clusters" of consecutive passenger trips served by a given vehicle, in the spirit of the "mini-clusters" described by Ioachim et al. (1995). These existing clusters can then be combined with the new integrated trip legs; i.e., there will be new trip "clusters" defined as the union of: *(1)* the individual integrated service trip "legs" (zero to two per integrated passenger trip); and, *(2)* the remaining "clusters" of consecutive trips served completely by door-to-door paratransit service. At this point, these mini-clusters can be optimized using the column generation technique of Ioachim et al. (1995). Greater detail on this technique is described in Hickman and Blume (2001).

3   Case Study

3.1   Background

The proposed scheduling heuristic is illustrated using the existing transit service in Houston, Texas. The transit agency in Houston (METRO) operates 94 fixed routes and a demand-responsive service for over 1750 passenger round trips per day, or about 3500 one-way trips per day. Much of the demand-responsive service is oriented to passengers qualifying under the Americans with Disabilities Act (ADA) of 1990, which specifies particular paratransit service requirements for these patrons. In this regard, 53% of METRO's demand-responsive passengers are ambulatory-impaired, and hence are really not eligible for integrated transit service as a result of METRO's own level of service requirements. This is because there are still fixed-route stops and vehicles that are not fully equipped for ADA service.

Because of a large service area (1400 sq km), trip lengths for the paratransit service average 13.3 km. Trip lengths over 40 km are not uncommon. With the requirements of the ADA, METRO is experiencing rapid growth in demand for the demand-responsive service, and is considering integrated service. Yet, METRO experiences greater costs for demand-responsive service ($10.28 per passenger trip, or $0.77 per passenger-km) than for fixed-route service ($2.24 per passenger trip, or $0.27 per passenger-km). As a result, there is reason to believe that the substitution of fixed-route service for part of the demand-responsive service may result in cost savings to the agency.

The primary questions to explore included:

- What number and percentage of trips could be served by integrated service?
- What impacts might be expected for passenger level of service, for eligible passengers?
- What potential cost savings might be realized?

For the purposes of this feasibility study, only the proposed passenger scheduling heuristic is applied. The passenger scheduling heuristic gives an initial estimate of the potential number of passengers served, the passenger level of service, and an upper bound on the potential reductions in paratransit vehicle kilometers and hours that might be possible under the integrated service strategy. The cost savings to the agency could be estimated based on the potential reduction in vehicle-km or vehicle-hours traveled. Assuming a constant utilization rate of vehicles, an upper bound on the cost savings is estimated as the total paratransit vehicle distance saved in the passenger itinerary, multiplied by the average paratransit cost per passenger-km. That is, the estimate of the cost savings is equal to the cost per passenger-km, multiplied by the difference in distance of the direct trip versus the sum of the new paratransit "legs."

3.2 Application

This case study explores the possible cost advantages and changes in passenger level of service with integrated service. Using the proposed scheduling method, the integrated service is compared with the existing fully demand-responsive service, using performance measures of the total number and percentage of trips served, the passenger level of service (travel time and transfers), and the potential agency cost savings.

Global parameters and assumptions for this case study included the following:

Eligible passengers: Only those passengers with no ambulatory impairments were considered eligible for an integrated trip. This corresponds to METRO's desire to serve these trips with the highest level of service, giving these passengers additional attention.

Minimum integrated trip length: The passenger's origin and destination must be at least 3 mi (4.8 km) apart in order to be considered for an integrated trip. Shorter trips are likely more easily served simply through a direct paratransit trip. Longer minima may also be considered; this is an area for further sensitivity analysis. The 3 mi restriction eliminates another 6% of the trips from consideration, with slightly under half of the trips (about 1700 of 3500) being eligible on the basis of having an ambulatory passenger with a sufficient trip length.

Average paratransit vehicle speed: This was set based on the distance between the passenger's origin and destination, and the value does not include intermediate stops. Distance-based values were provided by Houston METRO and ranged from 24 km/h for trips under 2 mi (3.2 km) to 66 km/h for trip lengths exceeding 20 mi (32 km).

Origin and destination time windows: 15-minute time windows were used for the pick-up at the origin and the drop-off at the passenger's destination.

Maximum waiting time at a fixed-route stop: Ideally, the paratransit vehicle would arrive at the transfer point at the same time as the fixed-route bus. However, to allow some flexibility in scheduling, a maximum waiting time for the paratransit passenger was set to five minutes. In other words, when dropping off a passenger, a paratransit vehicle could arrive to a fixed-route time point up to five minutes before the scheduled arrival of the fixed-route vehicle. Also, when picking up a passenger, a paratransit vehicle could arrive up to five minutes after the scheduled arrival time of the fixed-route vehicle.

Maximum ride time: METROLift, METRO's scheduling software, limits the amount of time that a rider spends on a vehicle to values that vary with the distance between the origin and destination. These values range from 30 minutes to 120 minutes for trips up to and exceeding 48 km.

Radius of proximity circles about origin and destination: A preliminary value of 30% of the distance between the origin and destination was selected. This was used to identify potential transfer points to the fixed-route system. Also, a minimum radius of 0.25 miles (0.4 km) was specified as the minimum distance eligible for a paratransit trip. Increasing this value would have the effect of reducing the number of integrated trips. Finally, for a direct connection to a fixed-route bus stop, a maximum walking distance of 0.1 mi (160 m) was used to restrict eligible fixed-route stops.

Penalty factors: It was also assumed that a penalty of 5 minutes of travel time would be applied for each transfer. With two transfers, a total of 10 minutes is added.

3.3   Example

Consider the following example trip for scheduling under the passenger itinerary heuristic. A fully ambulatory customer wishes to travel from their origin (home) to their destination (a doctor's office), a total (Euclidean) distance of 14.2 km. The passenger requests to leave home at 8:20 am, with 15-minute time windows on either side.

To begin, an initial screening of potential fixed-route stops indicated that none were within 0.1 miles (0.16 km) of the origin or destination. Second, "rings" around the origin and destination were generated from a radius of 0.4 km to 30% of the total O-D distance (4.25 km). From these rings, there were 6 timepoints near the origin and 365 timepoints near the destination (the destination is in a dense downtown area). These points have 5 routes in common.

With the current METRO operating parameters, the 14.2 km trip has a maximum allowable ride time of 59 minutes. Because a pick-up time is specified, the shortest path with time windows is determined using a forward pass in the network, from the origin to the destination. For this trip, the shortest travel time on the integrated service is 32 minutes (excluding the transfer penalties), comprised of two paratransit legs and a fixed-route leg:

- 3 min paratransit trip from the origin to a local transit center 0.6 miles away, traveling at an average of 15 mph (1.0 km at 24 km/h),
- 5 min total waiting time (one-half of the 5-minute time window at each fixed-route stop),
- 20 min on the fixed-route bus,
- 4 min paratransit trip from the second stop to the destination 0.8 miles away, averaging 15 mph (1.3 km at 24 km/h).

The pick-up at the origin is scheduled for 8:08 am with a fixed-route segment from 8:13 to 8:33. The final drop-off at the destination is scheduled for 8:40. Finally, for the level of service comparison, an additional 10 minutes is added as a transfer penalty (2 transfers at 5 min/transfer) to obtain a total time of 42 min. Note that this assumes that the demand-responsive legs of the integrated trip provide direct service from the origin to the fixed route, and from the fixed route to the destination. In this sense, the values from the passenger trip scheduling algorithm are lower bounds on the actual travel time once vehicle trips are scheduled.

As for the passenger level of service, the integrated trip described above can be compared with the "baseline" paratransit schedule. Interestingly, in this case, the scheduling software at METRO scheduled this passenger's trip for 43 minutes, which is longer than the direct trip on the integrated service. This occurs because the paratransit van also had an additional, intermediate stop between the passenger's origin and destination. As a result, even with the 10-min transfer penalty, this particular integrated trip (if served directly) provides the passenger with a slightly better level of service. From the operator's viewpoint, the trip is also beneficial, in that the total paratransit trip distance has been cut from 14.2 km (direct) to 2.3 km (for two legs), or a savings of 11.9 km.

3.4 Full Results

As a case study, the passenger scheduling heuristic was applied to a representative day of service at Houston METRO. The input to the heuristic was the existing schedule of trips, as output from the METROLift scheduling software. On the given day, a total of 3588 one-way passenger trips were taken on METROLift. Of those trips, 924, or about 26%, could be accommodated using the integrated service. This was a much higher percentage than originally anticipated. Of the trips that were not covered, 1925 (53%) were not covered due to passenger disability (e.g., a wheelchair prohibited the trip), 217 trips were too short for our heuristic (under 4.8 km or 3 mi total length), 312 trips could not be served by a single fixed route, and 211 could not meet the maximum travel time constraint. It is interesting that the trip length and total travel time constraints, while important, had a more modest effect in reducing the number of trips served.

As for the overall passenger level of service, 39% of the trips on the integrated service actually provided a shorter travel time than that produced by METRO's scheduling software. Yet, 61% will be slightly worse off. This comparison includes a 10-minute total transfer penalty (5 minutes per each transfer). Graphically, this result is illustrated in Figure 4, using a histogram of the time savings comparing the integrated trip versus the existing scheduled trip. The skew of this histogram to the left indicates that there are a number of passengers who would realize slight increases in travel time with the integrated service. However, the long tail to the right indicates that many passengers would realize substantial savings. The effect of this long tail is evidenced in the mean of the distribution, which is -3 minutes (i.e., an average 3-min disadvantage for the integrated trip versus the existing schedule).

Time savings appears to be more substantial for shorter trips where ride sharing occurs. In these cases, the integrated trip results in a less circuitous trip, and the passenger experiences a net time savings. Longer trips, on the

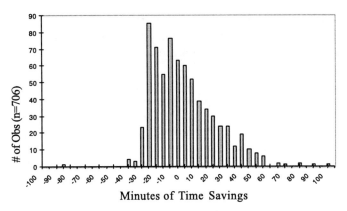

Figure 4. Histogram of travel time difference, existing – integrated service.

other hand, are less likely to have time advantages for the integrated trip. With an extensive freeway network, the assumed demand-responsive vehicle speeds are much higher than the fixed-route service for these long trips.

The potential cost advantages for METRO are stated in terms of the potential reduction in passenger-km of travel. When compared with the Euclidean distance, the integrated service reduces the total passenger-km of travel by 7380 km (4584 mi). This amounts to approximately 15% of the total passenger-km of travel at METROLift. At an average cost of $0.77 per passenger-km, the total cost savings has an upper bound of approximately $5682. This equates to about 15% of the daily operating cost of $36,000 at METROLift.

At the same time, one notes that this over-states the potential cost savings, since vehicle costs are likely to be highly non-linear with the costs per passenger-km. That is to say, the percentage reduction in passenger-km, particularly for trips where rides are shared, likely overstates the proportional reduction in vehicle-km. It is necessary to input the new integrated service trip legs through the vehicle scheduling heuristic to get a more accurate estimate of the vehicle operating cost savings.

3.5   Sensitivity Analysis

A sensitivity analysis to various assumed parameters in the passenger scheduling heuristic is appropriate. First, one might examine the effect of restricting the integrated service to persons that are ambulatory. As noted before, about 53% of the total passenger trips in the METRO case study involve ambulatory-impaired passengers. To examine this restriction, the passenger scheduling heuristic was also run with all 3588 trips. Fixed-route boarding and alighting times for the ambulatory-impaired passengers were set to 3 min each (Kittelson and Associates (1999)). The results are presented in Table 1.

Of the 3588 trips, 1805 (50.3%) could be accommodated using the integrated transit service. This is just slightly lower than for ambulatory passen-

Measure	All Trips	Ambulatory Trips
Trips accommodated with integrated service	1805	924
Total trips	3588	1664
Avg. paratransit (Euclidean) distance per trip (mi)	3.41	3.37
Avg. (Euclidean) O-D distance per trip (mi)	8.40	8.34
Avg. time difference, Original - integrated service (min)	-4.5	-2.7
Percent of trips, Original time > integrated time	35.5%	39.2%

Table 1. Sensitivity analysis on allowable passenger trips.

gers, with 924 (55.5%) of 1664 total trips accommodated. The average trip length, and length of the paratransit trip legs, are comparable. Between the two trip categories, there is a slight difference in the travel time when comparing the original demand-responsive service with the integrated service. This is reflected in the greater time savings with ambulatory trips (average -2.7 min savings versus -4.5 min for all trips) and a larger percentage of trips that are better off with integrated service with ambulatory trips (39.2% versus 35.5%). In summary, while the set of available passengers may change dramatically, the net passenger level of service is not noticeably different. The operator cost savings, while likely to be proportional to the number of potential trips served, is somewhat uncertain without a more detailed vehicle schedule.

Secondly, from Table 2, it appears that the number of trips with the integrated service is very sensitive to the assumed radius of the proximity circles for fixed-route bus stops about the origin and the destination. As the radius increases from 10% to 50% of the Euclidean distance from the origin to the destination, the number of trips accommodated grows markedly from only 63 (3.7%) with a radius of 10% to 1415 (85.0%) with a radius of 50%. As might be expected, the length of the paratransit trip legs increases with the radius, while the average O-D distance drops as the radius increases. Curiously, the percentage of trips that do better with the integrated service (versus the original service) increases with the value of the radius. This occurs because the operating speed of the demand-responsive service is much higher than that for the fixed-route service at longer distances. This implies that higher values of the radius will likely lead to more passengers being accommodated, and to greater average time savings for those passengers, assuming they receive a direct trip. However, the effect on the agency operating cost is uncertain; the large number of trips and the considerable length of paratransit trips raise questions about the net effect on operating costs.

Table 3 presents the results of a sensitivity analysis of the minimum allowable trip length, using a range from 1 mi (1.6 km) to 7 mi (11.3 km). Interestingly, the total trips accommodated do not vary substantially as the minimum trip length increases, particularly up to a minimum trip length of

Measure	Value of Radius				
	10%	20%	30%	40%	50%
Trips accommodated with integrated service	63	424	924	1284	1415
Total trips	1664	1664	1664	1664	1664
Avg. paratransit (Euclidean) distance per trip (mi)	1.43	2.23	3.37	4.58	5.83
Avg. (Euclidean) O-D distance per trip (mi)	9.73	8.49	8.34	8.22	8.13
Avg. time difference, Original - integrated service (min)	-0.6	-3.6	-2.7	-0.7	+3.0
Percent of trips, Original time > integrated time	38.2%	37.4%	39.2%	43.0%	49.4%

Table 2. Sensitivity analysis on radius of proximity circles.

4 mi (6.4 km). The number of trips then drops off more precipitously for minimum trip lengths of 5 to 7 mi. As might be expected, the total trip length and the average length of the paratransit trip legs also increase as the minimum trip length increases. Note also that the passenger level of service, as measured by travel time savings versus the original trip time, increases as the minimum trip length increases. This is caused by the combination of longer trips more generally, with corresponding higher speeds, as well as higher speeds for paratransit service versus fixed-route service. Overall, there appears to be slightly better passenger level of service, but for fewer passengers, as the minimum trip length increases. Again, the effect on agency costs is indeterminate from this analysis.

Measure	Minimum Trip Length (mi)						
	1	2	3	4	5	6	7
Trips accommodated with integrated service	940	933	924	851	781	681	602
Total trips	1664	1664	1664	1664	1664	1664	1664
Avg. paratransit (Euclidean) distance per trip (mi)	3.33	3.35	3.37	3.54	3.69	3.89	4.05
Avg. (Euclidean) O-D distance per trip (mi)	8.23	8.28	8.34	8.74	9.12	9.65	10.07
Avg. time difference, Original - integrated service (min)	-2.9	-2.9	-2.7	-1.4	-0.3	+2.2	+3.5
Percent of trips, Original time > integrated time	38.8%	38.8%	39.2%	41.8%	44.4%	49.5%	53.1%

Table 3. Sensitivity analysis on minimum trip length.

Lastly, from Table 4, the percentage of trips with improved service drops rather sharply as the assumed transfer penalty increases. With a higher penalty, the percentage of trips that are accommodated drops rapidly. A 10-min penalty per transfer (20 min total) reduces the number of possible trips accommodated to 597 of 1664 (35.8%), and a 15-min penalty reduces this to only 177 trips (10.6%). Most of the other results are clearly mixed, due to the dramatic change in the number of trips that are accommodated under the different transfer penalties. It is curious, nonetheless, that the percentage of trips that are better off with integrated service (versus the original baseline) does not drop off. Rather, the longer trips that remain with the 15-min transfer penalty (assumed to be served directly) still have advantages over the original trips with intermediate stops. Clearly, the potential number of trips and passenger level of service are very sensitive to the assumed transfer penalty.

Measure	Transfer Penalty (min per transfer)			
	0	5	10	15
Trips accommodated with integrated service	1045	924	597	177
Total trips	1664	1664	1664	1664
Avg. paratransit (Euclidean) distance per trip (mi)	3.48	3.37	3.35	3.64
Avg. (Euclidean) O-D distance per trip (mi)	8.42	8.34	8.79	10.88
Avg. time difference, Original - integrated service (min)	+4.7	-2.7	-5.6	-2.9
Percent of trips, Original time > integrated time	52.8%	39.2%	36.0%	40.9%

Table 4. Sensitivity analysis on transfer penalty.

4   Conclusions

This paper has described and illustrated a method for scheduling passenger and vehicle trips in an integrated transit service. It is suggested that the proposed two-stage heuristic for scheduling these trips allows more direct consideration of both passenger level of service characteristics and transit agency operating costs. Further sensitivity analysis is warranted on the proposed method. It appears that the potential cost savings and passenger level of service are sensitive to the parameters of (1) standards of passenger eligibility for the service; (2) the minimum and maximum passenger trip lengths for paratransit trip "legs"; and, (3) the assumed penalty for passenger transfers. Also, a full implementation with a vehicle scheduling heuristic is also warranted to obtain more detailed estimates of vehicle costs.

From the Houston case study, the number of eligible trips where fixed-route substitution is possible appears to be substantial. About 26% of the trips served by the existing demand-responsive service are eligible for the integrated service, upon consideration of the passenger disability, minimum trip lengths, maximum travel times, and the need for a single fixed route. Interestingly, a substantial minority (39%) of passengers will achieve travel time savings with the integrated service, when compared with the existing service. However, this result is heavily dependent on the assumed penalty to passengers for making transfers to and from the fixed-route service. Finally, preliminary indications are that the cost savings for integrated service can be bounded at about 15% of the total operating cost. However, the actual cost savings are likely to be lower.

Obviously, the next step in this evaluation is to compare these reductions in costs against the potential for degradation of the passenger level of service. Interestingly, some passengers will be made better off with the integrated service, because the existing baseline service has circuitous vehicle trips. However, the majority of passengers experience some degradation in the level of service, with longer total travel time (although still within stated maximum travel times). These increases in passenger travel times must then be balanced against the potential cost savings.

The analysis in this paper is clearly limited by the fact that it focuses on the passenger scheduling task. This task is useful and important in generating some preliminary figures about the cost and level of service of integrating service. Nonetheless, a full analysis using a vehicle scheduling heuristic is required to obtain more definitive cost implications for the operator.

Of course, cost and passenger travel time are not the only factors one might consider in deciding to implement such an integrated service, but the proposed method does allow evaluation of the cost and level of service implications. At the same time, potential increases in travel times, the effects of the requirement to transfer, and the resulting comfort and safety of passengers, must also be considered before such an integrated service is offered (Balog et al. (1996, 1997)). Also, the degradation of fixed-route service caused by waiting for, loading and unloading these transferring passengers also deserves further study.

Bibliography

Balog, J., J. Morrison, and M. Hood (1997). Integration of paratransit and transit services: Importance of vehicle transfer requirements to consumers. *Transportation Research Record 1571*, 97–105.

Balog, J., A. Schwarz, J. Morrison, M. Hood, J. Maslanka, and J. Rimmer (1996). *Guidebook for Attracting Paratransit Patrons to Fixed-Route Services*. Technical Report 24, Transit Cooperative Research Program (TCRP), Transportation Research Board, Washington, DC.

Bander, J. and C.C. White (1991). A new route optimization algorithm for rapid decision support. In *Proceedings of the Vehicle Navigation and Information Systems Conference, P-253*, , 2, 709–728.

Bovy, P.H.L. and E. Stern (1990). *Route Choice: Wayfinding in Transport Networks*. Kluwer, Dodrecht.

Crainic, T., F. Malucelli, M. Nonato, and S. Pallotino (1998). Heuristic approaches for flexible transit. In *Proceedings of the 6th Meeting of the EURO Working Group on Transportation*, Goteborg.

Desrochers, M. and F. Soumis (1988). A generalized permanent labelling algorithm for the shortest path problem with time windows. *INFOR 26*, 191–212.

Gerland, H. (1991). FOCCS: Flexible operation command and control system for public transport. *Proceedings of Seminar H held at the PTRC 19th Summer Annual Meeting*, P348, Sussex, 139–150.

Han, A. and C. Hwang (1992). Efficient search algorithms for route information services of direct and connecting transit trips. *Transportation Research Record 1358*, 1–5.

Hickman, M. and K. Blume (2001). *Methods for Scheduling Integrated Transit Service: Survey of Current Methods and a New Heuristic*. Technical report, Texas A&M University System, College Station. Draft Report for the Southwest Region University Transportation Center.

Horn, M. (1999). *Planning Multi-leg Urban Journeys with Fixed-schedule and Demand-responsive Public Transport Services*. Technical Report 99/62, CSIRO Mathematical and Information Sciences, Canberra.

Ioachim, I., J. Desrosiers, Y. Dumas, M. Solomon, and D. Villeneuve (1995). A request clustering algorithm for door-to-door handicapped transportation. *Transportation Science 29*, 63–78.

Jaw, J., A.R. Odoni, H.N. Psaraftis, and N.H.M. Wilson (1986). A heuristic algorithm for the multi-vehicle advance request dial-a-ride problem with time windows. *Transportation Research 20B*, 243–257.

Kittelson and Inc. Associates (1999). *Transit Capacity and Quality of Service Manual*. Technical Report Web Document 6, Transit Cooperative Research Program (TCRP), Transportation Research Board, Washington, DC.

Koncz, N., J. Greenfeld, and K. Mouskos (1996). A strategy for solving static multiple-optimal-path transit network problems. *Journal of Transportation Engineering 122*, 218–225.

Liaw, C., C.C. White, and J. Bander (1996). A decision support system for the bimodal dial-a-ride problem. *IEEE Transactions on Systems, Man, and Cybernetics – Part A: Systems and Humans 26*, 552–565.

Malucelli, F., M. Nonato, T. Crainic, and F. Guertin (2001). Adaptive memory programming for a class of demand responsive transit systems. This volume.

Wilson, N.H.M., R.W. Weissburg, and J. Hauser (1976). *Advanced Dial-a-ride Algorithms Research Project: Final Report*. Technical report, Massachusetts Institute of Technology, Cambridge.

Adaptive Memory Programming for a Class of Demand Responsive Transit Systems

Federico Malucelli[1], Maddalena Nonato[2], Teodor Gabriel Crainic[3], and Francois Guertin[4]

[1] DEI - Politecnico di Milano
   `malucell@elet.polimi.it`
[2] DIEI - Università diPerugia
   `nonato@diei.unipg.it`
[3] Dept. management et technologie, U.Q.A.M. and
   Centre de recherche sur les transports, Université de Montréal
   `theo@crt.umontreal.ca`
[4] Centre de recherche sur les transports, Université de Montréal
   `guertin@crt.umontreal.ca`

Abstract. In this paper we discuss Demand Adaptive Systems (DAS) which are intended as a hybrid public transportation system that integrates traditional bus transportation and on demand service. DAS lines regularly serve a given set of compulsory stops according to a predefined schedule and regardless of current demand. Between a compulsory stop and the next, optional stops can be activated on demand. Vehicles have to be rerouted and scheduled in order to satisfy as many requests as possible, complying with passage-time constraints at compulsory stops.

This paper provides a general description of DAS, and discusses potential applications and solution methods, emphasizing differences and analogies with classical Demand Responsive Systems. The particular mathematical structure of DAS requires innovative solution methods even when addressing its simplest version, the single vehicle, single line case. An efficient meta-heuristic algorithm based on adaptive memory ideas has been developed for this case. The method integrates sophisticated mathematical programming tools into a tabu search framework, taking advantage of the particular structure of the problem. The methodology is briefly discussed and experimental results are presented for the single line case. We show that the basic case can be efficiently solved, thus providing efficient algorithmic building blocks for more comprehensive approaches tackling the general case.

1 Introduction

Traditional public transport is evolving towards more flexible services and operation modes in order to capture additional demand and increase profitability. Several flexible services that target people with particular needs are regularly operated in several urban centers. *Dial-a-Ride (taxi-bus) services* and *on-demand door-to-door transportation* for handicapped and elderly people are examples of such *Demand Responsive Systems* (DRS) (Ioachim et al. (1995); Borndörfer et al. (1999)). Yet, the needs of several groups of customers are not addressed by these services. This is particularly the case when demand is relatively low, such as during evenings and night time, in suburban and rural areas.

To capture this additional demand, the service quality of regular transit lines should be significantly increased. However, this goal appears difficult to attain if the transit company must operate within reasonable economic limits. One approach increasingly explored by transit authorities is to introduce some flexibility into the regular line operations such that customers may be directly reached, while still controlling the operation costs. To provide personalized services on such a large scale with reasonable investments, new service models have to be designed, which exploit the integration of on-demand services with the existing transportation network, and improve the accessibility and efficiency of the whole system. Intelligent transportation systems offer the technological context for the deployment and operation of such systems through advanced location, communication, guidance, and information management systems. To render such user-oriented, flexible services efficient and profitable, one has also to develop new optimization models and effective algorithms that are the enabling factor of comprehensive management and control tools.

The contribution of this paper is to introduce *Demand Adaptive Systems* (DAS), a new type of public transportation service that integrates traditional bus transportation and personalized on-demand services, discuss analogies and differences between DAS and traditional DRS, propose potential application frameworks, and show how DAS can be efficiently managed by ad hoc solution methods despite the complexity of the associated mathematical representation.

The paper is organized as follows. The proposed service model is formally described in its most general case in Section 2. We then focus on the particular case of the efficient operation of one vehicle making a tour of one line. The mixed integer programming formulation of the related optimization problem is provided in Section 3. As shown in Crainic et al. (2001b), this problem constitutes the basic element that makes up the more general problem and, thus, efficient solution methods for this case are essential. The computation of upper bounds is discussed in Section 4, while Section 5 is dedicated to the description of the heuristic solution approach and its main ingredients. Computational results are discussed in Section 6. Finally, conclusions and further directions of work are discussed in Section 7.

2   Demand Adaptive Systems

The transportation system discussed in this paper, denoted *Demand Adaptive System*, is a hybrid system that combines the regularity of fixed-line transportation and the flexibility of on-demand service. A multiple line DAS

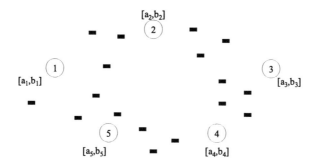

Figure 1. The compulsory and optional stops of a DAS line.

is designed as follows. Each trip of the line is described, as usually, in terms of a set of timetabled trips. The stops in the original timetable are called *compulsory stops*. Such points are locations where a demand for transportation may be naturally assumed to be present, such as, e.g., the stops of an express line, stops at rail or subway stations, or any other point that must be serviced according to the policies of the local transit authority. To introduce some flexibility into the vehicle routes, the vehicle is allowed to transit by each compulsory stop during a *time window*. Beside compulsory stops, a given set of stops to be activated on demand (*optional stops* hereafter) is available to the users. Each optional stop is assigned in a unique way to a pair of compulsory stops which are consecutive along the line and it can be serviced by the vehicle on its way from a compulsory stop to the next. Each line may be operated by several vehicles, starting from the line terminal according to a defined timetable. In Figure 1, optional and compulsory stops of a circular DAS line are depicted. Compulsory stops are progressively numbered according to the visiting order, and the associated time windows are shown. Black squares denote optional stops.

From the demand point of view, one can distinguish two classes of users. So called *passive users* make use of the transportation service in a traditional way, i.e., boarding and alighting at compulsory stops. No reservation is necessary since vehicles are guaranteed to serve each compulsory stop within a given time window. Users of the second class, so called *active users*, ask for a ride boarding or alighting at an optional stop. Active users must issue a *service request* and specify pick up and drop off stops, as well as earliest departure and latest arrival times.

In the absence of requests involving optional stops, a vehicle travels along the shortest path on the road network between each pair of consecutive compulsory stops. Accepting a request implies rerouting a vehicle to serve the pick up and the drop off stops within the required time windows. A vehicle detour may impact on the transit time at the following stops. In Figure 2, a possible itinerary of a vehicle on the line depicted in Figure 1 is shown; all optional stops visited along the itinerary correspond to boarding or alighting

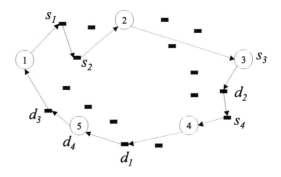

Figure 2. A vehicle tour on a DAS line and the set of served requests.

stops of requests. Requests may have either both stops or a single one located at optional stops. Requests r_1 and r_2 are examples of the first kind, as they board and alight at stops s_1, d_1 and s_2, d_2, respectively; conversely, r_4 alights at compulsory stop 5, while r_3 boards at compulsory stop 3.

Travel times on the arcs of the physical network are assumed known. Transfers between different lines, at line intersection points, are allowed at compulsory stops only, where intersection with fixed lines may also occur. If a request involves different lines, service time must allow for line transfer. Note that a request may be served by several possible itineraries, in particular when the flexible system is integrated with the fixed line system.

Several DAS models can be devised, depending on the policy used to deal with requests. In this paper, we focus on a specific case according to which requests are rejected if their acceptance causes the tour to become infeasible; a request rejection corresponds to the offer of a personalized service at a more expensive fare. On the other hand, once a request has been accepted, the user must be boarded and alighted exactly at the desired stops and within the requested time constraints.

In the basic model, namely DAS1 (Malucelli et al. (1999)), a single line circuit is traversed once by a single vehicle and the most profitable sub-set of requests that can be feasibly served must be selected. This model may be readily generalized to multiple intersecting lines with several vehicles traversing each line; here, users may travel from stops of a line to stops of another line on vehicles connecting at compulsory stops. The solution strategies developed for DAS1 provide the building blocks of more comprehensive formulations (Crainic et al. (2001b)).

2.1 Demand Adaptive Systems Versus Dial-a-Ride

Demand Adaptive Systems are intended to play a role in the family of Demand Responsive Systems that is quite different from the one of classical Dial-a-Ride systems (DAR). Differences arise not only relative to the service purpose, but they also involve the mathematical structure of the problem and the associated solution methods. We address these aspects in the following.

In DAS as in DAR, customers are allowed to request service between two given points and thus induce a detour in the vehicle route. However, in DAS the transportation service is always guaranteed, without any booking, at the compulsory stops of the line. Thus, a basic service is always assured to *passive users*.

Regarding intended users, DAR was originally introduced to serve people with special needs or, in general, as a highly subsidized service with precise targets. Once that broader classes of users have been entitled to the service, transit companies face the difficult challenge of how to cut down costs while not losing the acquired ridership. DAS, on the other hand, is not devoted to any special class of users in particular. Indeed, it represents a smooth way for transit agencies to migrate towards new transportation organizations, where users are expected to interact with the company call center. In fact, while *active* users personalize their itineraries by issuing a request, DAS still provides transportation without booking to *passive* users at compulsory stops. This may help users get progressively acquainted with the new kind of service.

Concerning the technological aspects and the required infrastructures, it is known how much flexible services rely on precise vehicle location systems as well as on information systems providing users with reliable information about times, itineraries, alternatives, etc. These issues are particularly central when the management of the requests is on-line and in real time. As far as the vehicle location system is concerned, which is usually a critical point especially in urban settings, note that in DAS, vehicle's position can be determined with a good approximation and without any expensive equipment, by exploiting the structure of the service. Since vehicles follow a basic itinerary, they can be located by knowing the last visited compulsory stop. The same information cannot be retrieved so easily in the DAR case because vehicles move with no restrictions in the service area, so that more sophisticated equipment must be utilized.

Regarding potential applications of DAS, one of the most suitable is given by a flexible version of the classical *feeder bus* (Crainic et al. (2000)). Feeder buses connect low demand areas to high capacity swift transport lines, either underground or light rail. In the flexible version, main line connections would occur at the compulsory stops of the flexible line and users could access the swift line by riding the feeder bus from an optional stop. Moreover, DAS lines could be operated to provide connection among multiple high capacity rapid lines. In this context, user itineraries would be composed of portions of trips operated by the flexible system and trips riding the regular, fixed line transit system. Furthermore, out of peak hours, several fixed lines operating in close by areas could be replaced by a single DAS line that would "virtually" cover the whole area, while guaranteeing regular connection at particular sites. In general, DAS is a system that may address very well the needs of relatively low demand areas (e.g., North American city suburbs) or periods (e.g., non-rush hours, evenings, weekends, etc.) or, broadly speaking, all those cases where a basic transportation service must be guaranteed even though a traditional type of transit system is not economically worthy and efficiency becomes a critical issue. In this context, it is clear how much the integration of the flexible and traditional transportation systems is important and how both systems can take advantage from being operated as a unique, seamless entity.

Significant differences between DAS and classical DAR also arise in the mathematical structure of the optimization problems. In the optimization problem associated with DAS (Section 3), time windows are present only at compulsory stops and the stop set is partitioned into subsets which must be visited in a predefined sequence. Therefore, the problem does not consist in partitioning and sequencing the nodes of a graph into disjoint routes that fulfill time windows and vehicle capacity constraints as in the DAR case. Rather, the problem requires to select which nodes have to be visited in each line trip, and the route definition can be decomposed into smaller subproblems, one for each portion of trip between two consecutive compulsory stops. In DAS, a single node is potentially associated to several requests, while in DAR each node is either a boarding or an alighting node of a single request. DAS is less constrained in the way nodes are sequenced; indeed, the feasible region encompasses any time feasible itinerary. On the other hand, DAS requires the solution of a selection problem of pairs of nodes. Therefore, the two problems have a rather different mathematical structure, which reflects on the search for taylored solution methods.

DAR problems are usually addressed with vehicle routing algorithms where the sequencing of nodes plays an important role. Due to the significant differences in the modelling and mathematical structure between the optimization problems related to DAS and DAR, a straightforward application of vehicle routing algorithms does not appear a viable approach for tackling DAS even in its simplest version. The challenge then is to devise new approaches that, based on innovative as well as classical methodologies, exploit the problem structure. In this paper, we concentrate on the basic problem, since efficient and effective algorithms for solving DAS1 constitute valid and essential tools to tackle the more general case. In fact, the same solution strategies developed for DAS1 may be used as building blocks for more comprehensive formulations. For example, the high parallelization potential of the general problem could be exploited by a decomposition approach where, at each step, the subproblems related to each single line trip are solved in parallel by way of the specific methods developed for DAS1.

3 Mathematical Formulations for DAS

Let us introduce some notation to formally state problem DAS1. Consider a line structured as a circuit, served by a single vehicle ending its tour at the starting terminal. Along the tour, the vehicle traverses a sequence $H = \{f_1, f_2, \ldots, f_{n+1}\}$ of $n + 1$ compulsory stops according to a given order, with

$f_{n+1} = f_1$. For each stop f_h a time window $[a_h, b_h]$ is defined, $h = 1, \ldots, n$; the vehicle must leave f_h not earlier than a_h and not later than b_h, but may arrive before a_h; b_{n+1} is the maximum tour completion time and $a_1 = b_1$ the starting time from the terminal.

A set F_h of optional stops is associated to each pair of consecutive compulsory stops $\langle f_h, f_{h+1} \rangle$. The vehicle passes by an optional stop only if a related boarding or alighting request at that stop has been issued. Sets F_h are mutually disjoint. Given any pair $\langle f_h, f_{h+1} \rangle$, we can define a directed graph $G_h = (N_h, A_h)$, such that $N_h = F_h \bigcup \{f_h, f_{h+1}\}$ is the stop set and $A_h \subseteq N_h \times N_h$ is the set of arcs connecting the stops. G_h will be referred to as *segment h*. Finally, $G = (N, A)$ is the whole graph, where $G = \bigcup_h G_h$. Travel times and costs of the physical network are given. Travel time τ_{ij} and travel cost c_{ij} for each $(i, j) \in A$ of consecutive stops are strictly positive, with τ_{ij} representing the duration of the shortest path from i to j including the stopping time at i. Then the subgraph induced by F_h is complete. We assume that triangular inequalities hold but time and cost matrices may be asymmetric.

In the following, we focus on a path formulation that offers a good insight in the problem nature and it is a natural support of efficient solution algorithms. An equivalent arc formulation is discussed in Malucelli and Nonato (2001). Let P_h be the set of paths in G_h from f_h to f_{h+1}. The vehicle itinerary in segment h is a path $p \in P_h$ with travel time $\tau(p)$ and cost $c(p)$ equal to the sum of travel times and costs of its arcs, respectively:

$$\tau(p) = \sum_{(i,j) \in p} \tau_{ij} \qquad \qquad c(p) = \sum_{(i,j) \in p} c_{ij} \qquad (1)$$

Let t_h be the starting time from f_h; w.l.o.g. we assume $t_1 = a_1$. A sequence of n paths, selected one from each segment, forms a *tour q* starting and ending at the terminal. Let $p_h(q) \in P_h$ be the path chosen in segment h. The arrival time at the end of the segment, i.e., at stop f_{h+1}, is then $t_h + \tau(p_h(q))$. Tour q is *time-feasible* if

(i) $t_h + \tau(p_h(q)) \leq t_{h+1}$ $h = 1, \ldots, n-1$
(ii) $t_n + \tau(p_n(q)) \leq b_{n+1}$
(iii) $a_h \leq t_h \leq b_h$ $h = 2, \ldots, n$

Then P_h is restricted to the set of all elementary paths from f_h to f_{h+1} with travel time less than or equal to $b_{h+1} - a_h$ for all segments $h = 1, \ldots, n$.

Let R be the *request set*, where request $r \in R$ is defined as a pair $\langle s(r), d(r) \rangle$ of boarding and alighting stops. We say that $s(r)$ is the *mate* of $d(r)$ with respect to request r, and vice versa. We assume that, for each request boarding and alighting at optional stops, the corresponding nodes do not belong to the same segment. Such an assumption is quite realistic, since any two optional stops within the same segment are relatively close to each other. Because of this assumption, no precedence constraints must be enforced between stops within the same segment, while precedence constraints regarding the boarding and alighting stops of each request are implicitly handled by the sequencing of compulsory stops. A *benefit* $u(r) > 0$ is associated with each request $r \in R$. It may represent the ticket fare or the value of a more complex utility function which can take into account also level of service aspects.

Let Q be the feasible tour set. Let $N(p_h(q))$, $h = 1, \ldots, n$, be the node set of $p_h(q)$, i.e., the *stops served* in segment h, and let $N(q) = \bigcup_h N(p_h(q))$ be the set of stops served in q. Then, $R(q) \subseteq R$, the subset of requests satisfied by tour $q \in Q$ is given by: $R(q) = \{r \in R : s(r), d(r) \in N(q)\}$. $u(q)$, the *benefit* of tour q, is given by $u(q) = \sum_{r \in R(q)} u(r)$ while the *global cost* $c(q)$ is the sum of the costs of the paths in q. DAS1 consists of finding a tour $q^* \in Q$ of maximum profit, i.e., maximizing the difference between benefit and cost: $u(q^*) - c(q^*) = \text{maximum}\{u(q) - c(q) : q \in Q\}$.

A path-based mathematical model can be given by introducing the following variables:

- z_p^h, $\forall\, p \in P_h$, $h = 1, \ldots, n$
- y_r, $\forall\, r \in R$
- t_h, $h = 1, \ldots, n$

where $z_p^h = 1$ if path $p \in P_h$ is chosen, $z_p^h = 0$, otherwise; $y_r = 1$ if request r is satisfied, $y_r = 0$, otherwise; and t_h is the starting time from f_h, and $t_1 = a_1$.

In the example depicted in Figure 3, boarding and alighting stops of four requests are shown; for each segment h, a variable z_0^h is associated to the direct path from f_h to f_{h+1} with no optional stop, which is depicted as a straight line. For segment $h = 1, 2, 3, 5$ an alternative path is provided, associated to variable z_1^h and depicted with a dotted line. If variables z_1^1, z_1^2, z_1^3, z_0^4, z_1^5 are equal to 1, then all given requests are satisfied; if $z_1^1 = 0$ or $z_1^3 = 0$, request r_1 is dropped, as at least one of its stops is not served; if $z_1^5 = 1$, request r_3 is satisfied, since its only optional stop is served.

$$(das1p): \quad Z(y, z) = \text{Maximize} \sum_{r \in R} u(r) y_r - \sum_{h=1}^{n} \sum_{p \in P_h} c(p) z_p^h$$

subject to

$$y_r \leq \sum_{p \in P_h} \delta_{s(r),p} z_p^h \qquad h = 1, \ldots, n \qquad \forall\, r : s(r) \in N_h \qquad (2)$$

$$y_r \leq \sum_{p \in P_h} \delta_{d(r),p} z_p^h \qquad h = 1, \ldots, n \qquad \forall\, r : d(r) \in N_h \qquad (3)$$

$$\sum_{p \in P_h} z_p^h = 1 \qquad h = 1, \ldots, n \qquad (4)$$

$$t_h + \sum_{p \in P_h} \tau(p) z_p^h \leq t_{h+1} \qquad h = 1, \ldots, n-1 \qquad (5)$$

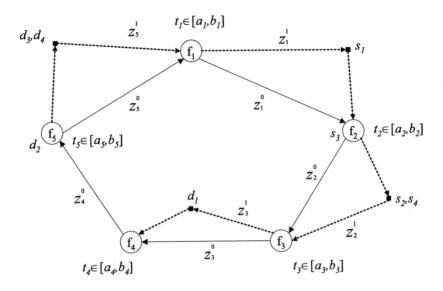

Figure 3. An example of paths and requests.

$$t_n + \sum_{p \in P_n} \tau(p) z_p^n \le b_{n+1} \tag{6}$$

$$a_h \le t_h \le b_h \qquad\qquad h = 1, \dots, n \tag{7}$$

$$y_r \in \{0, 1\} \qquad\qquad\qquad \forall\, r \in R$$

$$z_p^h \in \{0, 1\} \qquad\qquad h = 1, \dots, n \quad \forall\, p \in P_h$$

A request $r = \langle s(r), d(r) \rangle$ is satisfied only if both stops $s(r)$ and $d(r)$ are served. In the formulation, constraints (2) and (3) ensure this condition, as they link the choice of the path to served requests, coupling boarding and alighting stops for each request. $\delta_{s(r),p}$ is a constant equal to 1 if the boarding stop of the request is a node of path p, and 0 otherwise; similarly, $\delta_{d(r),p} = 1$ if $d(r) \in N(p)$, and 0 otherwise. Therefore, y_r can be 1 only if the selected path p in segment h, where $s(r) \in F_h$, is such that $\delta_{s(r),p} = 1$; the same applies for $d(r)$.

Constraints (4) impose the selection of one path for each segment, while constraints (5), (6), and (7) concern time-feasibility. Problem $(das1p)$ is NP-hard since the travelling salesman problem (TSP) reduces to a particular instance.

Note that in the present formulation the capacity issue is not addressed. In fact, in the DAS framework, passive users board the vehicle at compulsory stops without prior system knowledge; therefore, capacity can be dealt with for active users only. Even though the system is designed to operate in a low demand setting, where capacity should not be a tight constraint, the model and the proposed solution approach could be generalized to handle active users related capacity constraints.

Commercial mixed integer programming solvers fail to solve even reasonable size instances, as shown in Section 6. Even the computation of an upper bound is not a trivial issue: the linear relaxation of any arc-based integer linear programming model provides quite a loose bound. Heuristics are thus necessary to cope with practical problems. Yet, classical vehicle routing heuristics, constructive or improving (e.g., Shen et al. (1995); Laporte (1992); Cordeau et al. (2000); Savelsbergh and Sol (1995), etc.) much used for DAR problems, are not appropriate either. Indeed, such methods are based on arc and node manipulation, (principally insertion, deletion, and exchange of stops in and out of the current tour); the effect of these operations on the solution value cannot be uniquely determined in the DAS1 framework, since it depends on the solution at hand. Moreover, DAS1 requires the solution of a selection problem which involves pairs of nodes in different segments. This suggests the development of alternative methodologies and motivates the search for specific tools that exploit the particular structure of the problem. The structure of the formulation suggests to take advantage of path-based decomposition and work on paths, rather than on single nodes, and search good solutions by swapping paths in and out of the current tour. We present such a methodology in Section 5.

4   Upper Bounds

In this section, we briefly describe how to compute an upper bound to $(das1p)$. We also emphasize how the information provided as a by-product of the implemented bounding procedure may be used in the heuristic solution algorithms described in the next section. We consider three possible ways to limit from above the value of the optimal solution of $(das1p)$.

The first approach considers the LP-relaxation of $(das1p)$. This problem has potentially an exponential number of design variables z_p^h. Therefore, even the solution of the continuous relaxation is a difficult task. This kind of problems are usually tackled by column generation. Here, the formulation of the problem is not completely given, as an initial subset of columns is considered while more are added as needed. The generation of columns (corresponding to variables z_p^h) with positive reduced costs (i.e., potentially useful for improving the solution) requires the computation of an elementary path p for each segment h with maximum length and travel time no longer $b_{h+1} - a_h$ in a suitable graph. Lengths are given by reduced costs computed with respect to the dual variables of the LP-relaxation of $(das1p)$ defined on the subset of columns at hand.

A second approach considers alternative arc-based formulations characterized by a polynomial number of variables (Malucelli and Nonato (2001)). In such formulations it is necessary to introduce an additional time variable for each node in order to avoid cycles, as discussed in a more general setting in Desrosiers et al. (1995). As a side effect, the linear relaxation of this formulation is by far looser than the linear relaxation of model $(das1p)$. Nevertheless, using these models, it is easier to strengthen the formulation by introducing valid inequalities.

A third approach to compute an upper bound to DAS1 is the Lagrangean relaxation of model $(das1p)$, as summarized next. Apply a Lagrangean relaxation to constraints (5) with multipliers μ, and Lagrangean decomposition to variables y_r with multipliers λ. This yields a set of n separable subproblems, $(P(h)_{\lambda,\mu})$, $h = 1, \ldots, n$, which, although still NP-hard, are rather tractable. In fact, each $(P(h)_{\lambda,\mu})$ reduces to the search of a maximum weight path with a constraint on maximum travel time. Positive cycles may occur because of weight definitions, but, due to time window constraints, the problem is bounded and can be solved by using the techniques described in Desrosiers et al. (1995). Further details on the efficient computation of an upper bound to DAS1 are provided in Malucelli and Nonato (2001).

Since the solution of $(P(h)_{\lambda,\mu})$ for any set of multipliers λ and μ yields a path from h to $h+1$, each evaluation of the Lagrangean function yields a tour, generally not time-feasible, given by the set of the n paths associated with the solution of the n subproblems, one for each segment. The Lagrangean Dual, i.e., the problem of minimizing the Lagrangean function, is solved by way of a bundle algorithm (Carraresi et al. (1996)) which provides a vector of optimal multipliers, a collection of feasible paths for each segment, and an upper bound to the problem. The collection of these paths, called *promising paths* hereafter, provides a good starting base for heuristic methods that assemble feasible tours by selecting one path for each segment. Such *path-based meta-heuristics* are described in the next section.

5 A Meta-heuristic Solution Procedure

We present a meta-heuristic that integrates into a tabu search framework the Lagrangean relaxation procedure and a memory-enhanced multi-trial randomized greedy heuristic. Each individual procedure takes advantage of the particular structure of the problem, and has been extensively analyzed both as a stand-alone solution method and in a comparative framework (Crainic et al. (2001a)). In this paper we present the procedure that gave the overall best performance.

The meta-heuristic is built around a local search phase that iteratively attempts to improve a feasible tour. The core of the algorithm consists of swapping paths in and out of a tour. That is, starting from a feasible solution, the current tour is modified at each step by selecting one path not in the tour from a given set of promising paths and swapping it with the path currently in the tour, in the corresponding segment. The set of promising paths generated by the Lagrangean relaxation provides the initial pool of candidate paths. This set is dynamically updated by modifying, deleting, and adding new paths to the set in order to keep track of the evolution of the algorithm and, possibly, capture more profitable solutions. Two mechanisms are used to further enhance the search. On the one hand, the search is intensified around the most promising solutions found during the local search phase. On the other hand, a constructive greedy procedure which exploits the memory of the already generated solutions is used to diversify the search.

More formally, consider a pool of paths $P' = \bigcup_h P'_h$, $P'_h \subseteq P_h$, where P'_h corresponds to the set of available paths for segment h. Let q be the current time-feasible solution. At each step, the *best* path in P' with respect to q for a given segment h is selected according to an evaluation function – the *score* – which depends on the particular solution strategy and implementation. A new tour may then be built from q by *swapping* the selected path with $p_h(q)$, if the new tour is still time-feasible; the next best path is selected, otherwise. The path swapping operation thus represents the core move of the path-based meta-heuristics we introduce. Then, the neighborhood of the current solution q is the set of all time-feasible tours which can be obtained by way of a single path swap move, while the pool P' defines the restricted neighborhood on which swaps are actually evaluated and performed.

The proposed meta-heuristic (referred to as hybrid in the extensive study Crainic et al. (2001a)) proceeds according to three steps:

1. A *neighborhood search* (the "local" search phase) is performed until a given number of non-improving consecutive iterations are observed. At each iteration, a segment is first selected. The non-tabu path in the pool that ensures the best swap with the path currently in the segment is then identified, the move is implemented, and the memories and best solutions are updated. A local improvement of the current tour is performed by reordering the nodes of the selected paths and by inserting new nodes according to a specific priority function (mending step).

2. If the local best solution has been improved during the last neighborhood search phase, an *intensification* phase is performed to attempt to further improve it. The basic principle is to "lock" in the solution the elements that appeared most often in the best local solution. However, building this phase on paths may rapidly lead to situations where no paths may be moved out of or in the solution. Therefore, we work with requests: we forbid paths that do not contain requests that have appeared most often in the best local solution to enter the tour for a certain *tenure*. A new neighborhood search phase is started following intensification.

3. If the local best solution has not been improved during the last neighborhood search phase, we *diversify* the search. In this phase, a feasible tour is constructed from scratch, i.e., by performing n successful swap moves starting from the basic tour $\bar{q}$ as the initial solution. $\bar{q}$ is composed by the n basic paths $\bar{p}_h$, $h = 1, \ldots, n$, defined as the minimum travel time path from f_h to f_{h+1} without intermediate stops. Each successful swap is accomplished by a sequence of trials. First, the current *working* segment is randomly selected for swapping purpose among those not yet inspected. At each trial, the best candidate path not yet inspected, if any, is searched for among all paths of the working segment currently in the pool. The selected path is marked as inspected, and is tentatively

swapped in the current tour. Either the new tour is time-feasible, or the trial is repeated. If all paths of the working segment in the pool have already been inspected, the corresponding basic path will belong to the resulting solution (*null swap*), leading to the rejection of all requests involving an optional stop of that segment. If the new tour is time-feasible, it is improved by deleting profitless nodes, i.e., those nodes related to requests whose mate node has not been included in the tour; the pool is updated by inserting the new paths generated during the search process, and a new working segment is selected. A new neighborhood search phase is initiated from the resulting tour.

Several *memories* are defined to guide the search and to evaluate moves. All memories are shared by the tabu and greedy heuristics:

- A *short term tabu* memory. Implemented as a tag chosen randomly in a pre-specified interval, it indicates the iteration a path may be considered again for inclusion in the tour.
- A path *frequency* memory: for how many iterations a path has been in the best global tour.
- A request *frequency* memory in the current neighborhood search phase: for how many iterations a request has been served by the best local tour.
- *Promising* and *realized request* memories. The contribution of node i to the final solution depends on the selection of its mates. For each r such that $i = s(r)$ or $i = d(r)$ and such that the mate of i with respect to r has neither been chosen nor rejected in the current iteration, what matters is the probability of satisfying r once that i has been selected. Such a probability can be estimated by keeping a record of favorable cases (number of times the selection of i has been followed by the selection of its mate) over possible cases (number of times i has been selected before its mate was examined). To this purpose, for each request r, we record the number of times node $s(r)$ has been selected before $d(r)$ was processed, and the number of times the selection of $d(r)$ actually followed, thus satisfying r (the same applies conversely for the boarding node of each request). Let us call these *promising* and *realized (held)* *memories* $s_{pr}(r)$ and $s_{hd}(r)$, and $d_{pr}(r)$ and $d_{hd}(r)$, respectively. These counters are updated at the end of each iteration and are long term memories.

To compute the *score* of a potential move, the sum of the differences in the profits associated to the nodes of the paths and the associated transportation costs is calculated to obtain the *exact* move evaluation. However, this evaluation does not take into account any increasing benefit that may come from including extra nodes into the current tour. Moreover, in the constructive phase, we work on an incomplete solution, so that an exact evaluation is not possible. Therefore, we try to approximate the performance of each path on the basis of solutions visited previously during the search, using potential contribution of requests at nodes.

To evaluate the potential contribution of a node to the final solution we make use of the above mentioned promising and realized memories, as follows.

Call $\tilde{P}s(r)$ and $\tilde{P}d(r)$ the approximations of the probability of holding the boarding and alighting nodes, respectively, of request r. Then, $npr(i)$, the potential benefit of a node i, is computed as the sum of the revenue of r on all requests r such that i is either $s(r)$ or $d(r)$, times the approximation of the above mentioned probability.

$$npr(i) = \sum_{r:i=s(r)} u(r)\tilde{P}s(r) + \sum_{r:i=d(r)} u(r)\tilde{P}d(r) \qquad (8)$$

Such probabilities are estimated from the ratio of favorable cases over possible cases, whose record is kept in the promising and realized request memories.

In the local search phase, the node potential evaluation of a move is computed as the sum of the exact evaluation and the potential revenue of the not yet satisfied requests associated to the nodes of the incoming path. In the constructive phase, it is made of two components: for the segments for which a path has already been selected, the exact benefits are computed. For the other segments, the node potential benefit is used.

It is noteworthy that the *pool* of paths *evolves dynamically* during the search: new paths are added to the pool, others are modified, some may be discarded. Changes in the path set occur due to generation of paths by node deletion or insertion, extraction of subpaths from time-infeasible paths, path optimization by node reordering, path deletion due to pool overflow. Each of these issues is briefly addressed below.

Node deletion: Not all requests associated to a node of a given tour are necessarily served by that tour. In particular, after a swap move, some requests previously served may have lost their boarding or alighting node, or, similarly, some of the requests associated to the nodes of the entering path may not find their mate nodes in the paths of the other segments. Such requests are called *orphans* with respect to the current tour. A *profitless* node with respect to a given tour serves only orphan requests: as it yields no profit while increases the cost of the tour, it should be discarded. Clearing paths of profitless nodes in the current tour, possibly originates new paths thus enriching the pool.

Path reduction: New paths can also be generated during pool inspection, when searching for the best candidate path for swapping. If the selected path is longer than required by time-feasibility, any subpath extracted from it by deleting optional stops so as to meet time requirements, is a potential candidate for swapping.

Node reordering: In a post-optimization-like phase, paths may be modified regarding the order nodes are traversed, to improve the overall efficiency by decreasing path duration. Identifying the optimal node sequence in a segment belongs to the TSP family (Lawler et al. (1985); Laporte (1992)). Therefore, paths are checked only for local optimality with respect to the 2-opt neighborhood defined by the exchange of a single pair of arcs.

Node insertion: Tours are improved by various node deletion or reordering operations resulting in extra time for additional detours. The node insertion or *mending* step attempts to locally improve the solution by inserting additional nodes to increase the number of requests served and, therefore, the total benefit of the tour. Nodes are inserted sequentially, according to their potential benefit evaluation.

Pool management: As creation of new paths may cause pool overflow, a pool management strategy is needed, to identify the paths to be deleted. In our case, the total node potential benefit of a path is used as a measure.

6   Computational Results

The goal of the experiments was to verify whether problem DAS1 could be solved efficiently and effectively by the path based meta-heuristic, i.e., in terms of both computing time and solution quality, so that it can be utilized as a building block for more general cases (i.e., several vehicles on several lines).

6.1   Test Problem Data

Calibration and comparison efforts have been performed on a set of problems derived from actual transit lines. The lines have been selected from the transit networks of one Canadian and one European city, together with representations of the corresponding street networks. Each line has been the seed for several problem instances through variations in the number of segments, optional stops for each segment, the density of requests, the width of time windows, and the time available for each segment.

   The actual length of each street represented in the network (through a link) is known. These information have been used for computing distances between all the nodes of the network by way of all pairs shortest path algorithms; here, triangle inequalities hold, but distances are neither Euclidean nor symmetric. For each original transit line, a subset of stops (one stop for each segment) has been extracted to act as the compulsory stop set of the flexible line. Each street intersection laying within a maximum distance from any stop of the original transit line is eligible for being an optional stop of the flexible line. Optional stops are then assigned to the respective segment. The network topology of the corresponding instance is thus defined. Finally, costs and times on the arcs of the network have been defined proportional to distances. Beside reflecting a real feature, this allows the algorithm to discriminate among paths serving the same node set in a different order and to privilege the most efficient ones with respect to time. The benefits associated to requests have been defined such that the extra cost of any detour to serve a customer is always compensated by the additional benefit. With such settings, limited time availability is the only constraint which causes requests to be discarded. Time windows at compulsory stops have been set in such a way

Malucelli et al.

that extra time is provided for detours within each segment; in fact, for each
segment h the maximum duration of a path $(b_{h+1} - a_h)$ has been set equal to
one and a half the duration of the path from compulsory stop f_h to f_{h+1} that
was required by the original transit line. Time window widths at compulsory
stops vary from 60 to 120 seconds, which is considered a reasonable waiting
time for passive users who do not know in advance the arrival time of the
vehicle. The number of requests is proportional to the number of segments
(25 requests for each segment), and boarding and alighting stops have been
randomly chosen in the node set N according to a uniform distribution. Since
the problem involves a node selection, with this decision we intended to re-
produce the most difficult situation, where there is no substantial difference
concerning the number of requests associated to each optional node.

Some basic considerations hold. Instances with more segments are more
difficult not only because of the larger size: being time a shared resource
between adjacent segments, local decisions on path duration induce chain
reactions whose impact depends on the chain length. Other factors kept constant, the difficulty is also related to the width of the geographical area
encompassed by each segment. Such factor is controlled by the maximum
distance parameter used to select optional stops. Varying these factors several sets of problem instances have been generated. Three sets have been used
to calibrate the various parameters of the meta-heuristic and two others to
perform the experiments reported in the next section.

6.2   Testing the Hybrid Meta-heuristic Method

The experimental results reported in the following have been computed on
two classes of instances obtained from different lines of the same real transit
network of the two cities by the procedure given above; class A refers to
instances with 5 segments and maximum distance 0.4 of optional stops;[1] class
B is made by instances with 10 segments and maximum distance equal to
0.3; 20 different instances have been generated for each class. The number of
optional stops in each instance varies increasing with the number of segments
and the value of maximum distance.

In order to assess the quality of the heuristic solutions we exploited the
state of the art mathematical tools both to improve the bound as well as
to get an integer feasible solution: An arc based model, as mentioned in
Section 4, has been implemented and solved by a state of the art commercial
package (CPLEX (2000)) run on 10 processors of a 64-processor SUN E10000
computer for 10 hours each, for a total of 100 hours computing time, with
a limit of 500.000 nodes and a best bound strategy for the tree exploration.
The heuristic and the Lagrangean bound were computed on a Pentium II
machine with a 233 MHz processor and 128 MB memory.

---

[1] Instances A10 and A11 will be omitted from the results as they turned out to be
quite easy instances solved after 1 second.

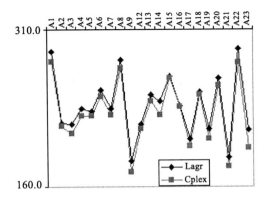

Figure 4. Lagrangean and CPLEX bound for class A.

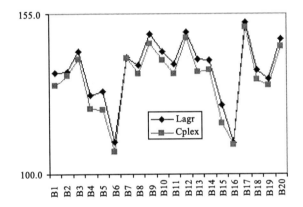

Figure 5. Lagrangean and CPLEX bound for class B.

Figures 4 and 5 compare the behavior of the upper bounds; for each instance the value of the Lagrangean bound and the value of the best bound obtained by CPLEX within the given time limit (*CPLEX bound*, hereafter) are reported. Further details are reported in Malucelli and Nonato (2001). CPLEX bound is always tighter than the Lagrangean bound, however, it is worth mentioning that, while the computation of the latter requires at most few minutes, CPLEX usually requires a considerable amount of running time before the former gets as low as the latter.

Tables 1 and 2 are devoted to the results of the heuristic approach; the solution values obtained after 1, 10, and 60 seconds of running time are reported. Since the procedure includes a number of random parameters, each experiment was repeated ten times for each instance and the average is reported; all solution values are expressed in terms of the percentage gap with respect to the best bound known for each instance. In order to evaluate the method robustness, the standard deviation σ is also given.

Instance	1 σ	1 gap	10 σ	10 gap	60 σ	60 gap	CPLEX gap	CPLEX time
A1	0.6	0.3	0.0	0.1	0.0	0.1	0.0	145
A2	4.2	13.2	2.0	10.9	1.3	10.4	35.9	1080
A3	4.8	6.0	4.6	3.0	0.5	1.2	2.3	680
A4	4.3	7.7	1.0	4.2	0.7	3.9	39.2	1950
A5	2.5	9.1	1.9	4.7	0.9	3.6	11.1	280
A6	2.6	10.5	2.0	7.9	1.3	7.4	30.4	980
A7	3.5	7.7	3.4	6.5	1.9	5.5	34.1	510
A8	3.4	11.6	2.8	10.0	2.3	8.8	31.9	250
A9	2.0	6.8	1.6	5.2	1.4	4.3	15.2	22640
A12	5.5	8.7	2.8	5.4	1.9	4.1	15.0	910
A13	1.6	7.0	1.5	4.8	1.1	4.4	23.7	520
A14	6.6	6.4	3.2	4.0	2.5	2.6	3.6	790
A15	5.9	6.2	1.2	3.9	1.8	3.5	22.6	4580
A16	3.8	15.0	3.1	13.2	2.1	12.5	42.3	370
A17	3.7	10.8	0.9	8.2	0.0	8.0	24.3	2500
A18	5.1	13.2	3.0	11.8	0.1	11.1	—	10 h
A19	4.6	12.9	3.3	10.4	1.0	8.8	29.9	1990
A20	3.4	6.5	0.5	5.1	0.3	5.0	35.6	440
A21	3.3	11.8	0.7	9.7	0.9	9.4	30.6	830
A22	1.9	1.9	2.7	1.5	3.0	1.1	0.0	611
A23	1.0	0.5	0.1	0.2	0.1	0.2	0.0	1414

Table 1. Average percentage gap between the heuristic solution and the best upper bound, and standard deviation over 10 runs: class A.

In order to verify the need for meta-heuristics, we also provide the value of the best integer solution obtained by CPLEX with the above mentioned procedure, and the time in seconds needed to compute it. When no value is reported, as for instance A18 in Table 1, CPLEX could not find a feasible solution within the given time limit (100 hours). When a 0 value is reported, as, e.g., in the case of instances A1, A22, A23 in Table 1, CPLEX found the optimal integer solution. From these results it can be seen that whenever CPLEX is able to solve the problem to optimality, the heuristic procedure is able to provide solutions within the 1.1% from the optimum quite faster; for more difficult instances, after 100 hours of running time the best bound provided by CPLEX is still rather far from the heuristic solution, but in turn,

Instance	1		10		60		CPLEX	
	σ	gap	σ	gap	σ	gap	gap	time
B1	0.3	1.9	0.1	1.8	0.1	1.7	1.6	2140
B2	1.0	5.3	0.6	4.4	0.3	4.3	31.0	2070
B3	1.5	1.1	1.4	1.0	1.3	0.8	0.0	98
B4	0.7	1.1	0.2	0.6	0.0	0.5	0.0	10720
B5	0.6	0.5	0.6	0.4	0.6	0.4	0.0	5469
B6	0.0	2.5	0.0	2.5	0.0	2.5	9.4	7730
B7	0.1	3.4	0.5	3.2	0.6	3.1	27.9	290
B8	1.2	1.4	1.0	1.1	0.7	0.6	0.0	925
B9	0.5	0.6	0.5	0.5	0.4	0.3	0.0	2842
B10	1.4	2.8	0.5	1.9	0.1	1.8	1.9	14590
B11	0.8	1.0	0.6	0.7	0.6	0.7	0.0	414
B12	1.2	2.2	0.6	1.5	0.5	1.2	0.0	855
B13	1.3	3.5	1.7	2.6	1.4	2.0	2.3	870
B14	1.0	1.4	1.1	1.1	1.0	1.1	0.0	200
B15	1.0	2.4	0.9	2.2	0.8	2.1	8.9	100
B16	1.1	5.1	0.8	4.6	0.7	4.6	27.9	150
B17	0.5	1.7	0.6	1.4	0.5	1.2	13.3	15320
B18	1.2	2.3	1.3	0.6	0.5	0.2	0.0	502
B19	1.1	1.2	0.9	0.5	0.9	0.4	0.0	288
B20	0.9	9.5	0.8	8.9	0.6	8.7	58.7	3490

Table 2. Average percentage gap between the heuristic solution and the best upper bound, and standard deviation over 10 runs: class B.

the best integer solution provided by CPLEX is considerably worse than the one provided by the heuristic.

7 Perspectives and Conclusions

In this paper a hybrid meta-heuristic method integrating several solution strategies has been proposed for the solution of the basic version of a new flexible transport system. Such a system shares some of the features of traditional fixed lines while offering at the same time a high degree of flexibility. The commercial mixed integer programming solvers are not efficient enough to solve the optimization problem associated to the service management even in the basic case, when requests must be selected for a single tour in a single line. The particular structure of the optimization problem makes even the basic case not suitable to be solved by classical dial-a-ride solution approaches. For these reasons we developed specific algorithmic tools for this problem.

The procedure has been tested on a set of problem instances derived from real life networks, and compared with a state of the art package for integer linear programming. Our method proved quite effective, although for same instances the quality of the solution has not yet been assessed.

Finally, we can state that the basic problem can be efficiently handled by specific methods, which can represent the building block of more general approaches to the multi-vehicle, multi-line case.

Acknowledgements

We are grateful to Professor Michael Florian who graciously gave us access to a number of data banks from which a set of realistic test problems could be derived. This project is supported by grants from NSERC Canada, the Fonds FCAR of the Province of Quebec, the Quebec-Italy cooperation program of the Ministry for Foreign Affairs of Italy and the Ministry for International Relations of the Province of Quebec.

Bibliography

Borndörfer, R., M. Grötschel, F. Klostermeier, and C. Küttner (1999). Telebus Berlin: Vehicle scheduling in a dial-a-ride system. In N.H.M. Wilson (Ed.), *Computer-Aided Transit Scheduling, Lecture Notes in Economics and Mathematical Systems*, 471, Springer, Berlin, 391–422.

Carraresi, P., A. Frangioni, and M. Nonato (1996). Applying bundle methods to the optimization of polyhedral functions: An applications oriented development. *Ricerca Operativa 25*, 5–49.

Cordeau, J.-F., G. Laporte, and A. Mercier (2000). *A unified tabu search heuristic for vehicle routing problems with time windows*. Technical Report CRT-2000-03, Centre de Recherche sur les Transports, University of Montreal, Montréal, Canada.

CPLEX (2000). *ILOG CPLEX 7.0 Reference Manual*. ILOG Inc., CPLEX Division.

Crainic, T.G., F. Guertin, F. Malucelli, and M. Nonato (2001a). *Metaheuristics for a class of demand responsive transit systems*. Technical report, Centre de Recherche sur les Transports, University of Montreal, Montréal, Canada.

Crainic, T.G., F. Malucelli, and M. Nonato (2000). A demand responsive feeder bus system. In *Proceedings of the 7th World Congress on Intelligent Transport Systems, 6–9 November 2000, Turin, Italy*.

Crainic, T.G., F. Malucelli, and M. Nonato (2001b). *Many-to-few + few-to-many = an almost personalized transit system*. TRISTAN IV, São Miguel, Portugal, June 2001.

Desrosiers, J., Y. Dumas, M.M. Solomon, and F. Soumis (1995). Time constrained routing and scheduling. In M.O. Ball, T.L. Magnanti, C.L. Monma, and G.L. Nemhauser (Eds.), *Network Routing, Handbooks in Operations Research and Management Science*, 8, Elsevier, Amsterdam, 35–139.

Ioachim, I., J. Desrosiers, Y. Dumas, M. Solomon, and D. Villeneuve (1995). A request clustering algorithm for door-to-door handicapped transportation. *Transportation Science 29*, 63–78.

Laporte, G. (1992). The travelling salesman problem: An overview of exact and approximate algorithms. *European Journal of Operational Research 59*, 231–247.

Lawler, E.L., J.K. Lenstra, A.H.G. Rinnooy Kan, and D.B. Shmoys (1985). *The Travelling Salesman Problem: A Guided Tour of Combinatorial Optimization*. Wiley, New York.

Malucelli, F. and M. Nonato (2001). *Formulations and bounding procedures for optimization problems in demand adaptive transit systems*. Technical report, Politecnico di Milano.

Malucelli, F., M. Nonato, and S. Pallottino (1999). Some proposals on flexible transit. In T. Ciriani, S. Gliozzi, E.L. Johnson, and R. Tadei (Eds.), *Operations Research in Industry*, Macmillan Press, London, 157–182.

Savelsbergh, M.W.P. and M. Sol (1995). The general pickup and delivery problem. *Transportation Science 29*, 17–29.

Shen, Y., J.-Y. Potvin, J.-M. Rousseau, and S. Roy (1995). A computer assistant for vehicle dispatching with learning capabilities. *Annals of Operations Research 61*, 189–211.

A Cycle Based Optimization Model for the Cyclic Railway Timetabling Problem

Leon Peeters and Leo Kroon

Erasmus University Rotterdam, Rotterdam School of Management,
PO Box 1738, 3000 DR Rotterdam, The Netherlands
{l.peeters,l.kroon}@fbk.eur.nl.

Abstract. The paper presents an optimization model for cyclic railway timetabling that extends the feasibility model by Schrijver and Steenbeek (1994). We use a mixed integer non-linear programming formulation for the problem, where the integer variables correspond to cycles in the graph induced by the constraints. Objective functions are proposed for minimizing passengers' travel time, maximizing the robustness of the timetable, and minimizing the number of trains needed to operate the timetable. We show how to approximate the non-linear part of the formulation, thereby transforming it into a mixed integer linear programming problem. Furthermore, we describe preprocessing procedures that considerably reduce the size of the problem instances. The usefulness and practical applicability of the formulation and the objective functions is illustrated by several variants of an instance representing the Dutch intercity train network.[1]

1 Introduction

The demand for railway transportation in Europe has been increasing rapidly during the past few years, partly influenced by the ongoing efforts of European governments to persuade people to use public transportation. Because of the resulting increase in railway traffic, the process of constructing railway timetables is becoming more and more complex. At the same time, the introduction of competition on the European railroads has pushed the management of railway companies to consider various methods for improving their business planning processes, one important planning process being the construction of timetables. Therefore, railway companies are interested in methods for generating several timetables, preferably of high quality, in a reasonable amount of time, so as to be able to evaluate several alternatives for their future timetables.

These developments have led to an increasing interest in the problem of constructing cyclic railway timetables, as operated in several European

---

[1] This work was partially supported by the Human Potential Programme of the European Union under contract no. HPRN-CT-1999-00104 (AMORE), and by the DONS R&D project, being jointly carried out by Railned and NS Travelers.

countries. In such a cyclic timetable, train connections are operated on a regular basis, e.g., a train for a certain destination will leave every hour at 5 minutes past the hour. Recently, several authors have published work on this problem, with a first successful attempt by Schrijver and Steenbeek (1994), who studied the problem of constructing a feasible timetable, and more recent publications by Nachtigall (1999); Goverde (1999); Lindner (2000) on the optimization of cyclic railway timetables. These authors all use models that are based on the Periodic Event Scheduling Problem (PESP) introduced by Serafini and Ukovich (1989).

This paper proposes some new objective functions that can be used within the above described models. These objective functions aim at minimizing the passengers' travel time, constructing a robust timetable, and minimizing the number of train compositions to operate a timetable. Also, any combination of these objectives is possible, with a possible weight factor.

A cyclic railway timetabling problem can be represented by a so-called constraint graph. At the core of the mixed integer linear program describing the cyclic railway timetabling problem lies a basis of the cycle space of this constraint graph. In this paper, we also show the considerable impact that the choice for a certain type of cycle basis has on the computation times.

1.1 Practical Background

During the past decade, the Netherlands' largest passenger railway operator NS Travelers and the Dutch railway capacity manager Railned have been putting a lot of effort into the development of the automatic timetabling system DONS (Designer Of Network Schedules). One of the intelligent modules of DONS, called CADANS, was developed by Schrijver and Steenbeek (1994) to assist timetable planners in generating a tentative timetable based on constraints deduced from the rough layout of the railway network, the train line system, safety regulations, and quality requirements. The timetable constructed by CADANS is cyclic with a cycle length of one hour.

The DONS system has been in use at NS Travelers and Railned for some years now, and CADANS generally performs well, both from a mathematical point of view and in the view of the planners that use it on a daily basis. However, the operational use has also indicated some problems. Firstly, CADANS is an algorithm based on constraint satisfaction, that searches for *some* timetable that satisfies all requirements. After finding a feasible timetable, provided that one exists, the algorithm terminates without giving information on the quality of the returned timetable, or on what other timetables may exist that also satisfy the requirements. Although it is possible to locally optimize the obtained timetable, such a local optimization ignores many other, possibly better, timetables. Secondly, when no feasible timetable exists, a minimal set of mutually conflicting constraints is returned, but practical experiences have shown that this information does not completely answer the planners' requirements. In this case, one would ideally like to receive information on how the specified requirements should be adjusted.

A solution to these two problems is offered by using an optimization approach, rather than just searching for a feasible timetable. Such an optimization approach enables one (i) to search for a timetable that is optimal with respect to the defined objective function, and (ii) to gain insight into the necessary changes to an infeasible instance, by allowing a penalized violation of the constraints or some subset of them.

1.2 Outline of the Paper

The remainder of the paper is outlined as follows. The next section gives an overview of the literature on related railway timetabling problems. Section 3 presents the optimization extension to the existing railway timetabling model, resulting in a mixed integer program. In Section 4 we propose some preprocessing procedures to reduce the size of instances, and thus the number of integer variables. The presented methods were tested on several variants of the Dutch intercity train network, the results of which are discussed in Section 5. The final section contains the conclusions and some ideas for further research.

2 Literature Review

Several researchers have studied problems in the field of railway timetabling or railway scheduling. Here, we only give an overview of the literature that is related to our research: *network based cyclic railway timetabling* for the types of railway networks that one finds in several European countries, i.e., generally one track for each direction of traffic, and high frequency traffic that is inter-related by many connections.

The model that lies at the basis of the cyclic railway timetabling model was developed by Serafini and Ukovich (1989). Their paper describes the PESP, that considers the scheduling of a set of periodically recurring events under periodic time window restrictions on pairs of these events. An objective function was not taken into account; their main interest was in finding feasible periodic schedules. They proved that the general PESP is NP-complete, and proposed a Branch&Bound procedure for finding feasible solutions. They also presented some applications in job shop scheduling, transportation scheduling and traffic light scheduling. Hassin (1996) describes an optimization model for the Network Synchronization Problem (NSP), a mathematical model that provides an optimization formulation for PESP, but does not take into account constraints as hard rules. Instead, forbidden time window values are highly penalized.

Schrijver and Steenbeek (1994) solve a PESP formulation for the Dutch railway timetabling problem with a constraint propagation algorithm. The basis of this algorithm is enumeration: whenever an arrival or departure time has been fixed, the consequences for all other times are propagated. Moreover, some clever branching and backtracking tricks were invented to speed up the solution process. Hurkens (1996) uses the same formulation to test a polyhedral solution method. He concludes that the method works, but computation times increase rapidly with the instance sizes. Odijk (1996, 1997) uses a cutting plane algorithm to generate timetables. The objective of his research is to quickly generate a family of timetables, in order to evaluate possible infrastructure lay-outs around and within stations. His results are promising, but only for rather small problems, i.e., a station and a small network immediately surrounding it. Nachtigall (1999) studied optimization for cyclic railway timetabling, using as objective minimal waiting times, minimal travel times, and a multi-criteria optimization. He developed two classes of facet defining inequalities for the problem, and uses these in a cutting plane method, which is able to solve medium sized real life problems. We refer to Nachtigall (1999) for a more detailed overview of cyclic railway timetabling problems. The construction of minimum cost timetables was studied by Lindner (2000). His model combines the assignment of train types and train units to train lines with the timetabling problem.

3   The Cyclic Railway Timetabling Problem

Very briefly stated, the Cyclic Railway Timetabling Problem (CRTP) consists of finding cyclic arrival and departure times for all trains at their corresponding relevant stations in a railway system, taking into account aspects such as safety, service levels, connecting trains, operating efficiency, etc. In this paper we will mainly focus on passenger train timetabling, although it is possible to handle cargo trains more or less in the same way.

This section gives a description of a model for the CRTP. We start with an overview of the assumptions, and continue with a description of the constraints in our model. Next, a cycle based formulation for these constraints is presented. § 3.4 describes the objective function, that may be non-linear. A method for linearizing the objective function is proposed in § 3.5. The final subsection describes a heuristic for finding a good cycle basis.

3.1   Assumptions

Since the 1930's, a cyclic timetable for passenger trains has been in use in the Netherlands. This means that trains leave (almost) *every hour* of the day *at the same time* from a certain station for a certain direction (with a possible deviation of a few minutes). Because of the cyclic character of the passenger service timetable, one usually also reserves cyclic time paths for cargo trains; these are assigned to specific trains at some later point in time. In order to construct a cyclic timetable, one only needs to construct a cyclic timetable for one hour, called an hourly pattern. This hourly pattern forms the basis for the actual timetable, where adjustments are made for rush hours, weekends, holidays, special trains, etc. For constructing such an hourly pattern, we assume the following information to be given:

- **Infrastructure:** We assume the infrastructure to consist of *nodes* (stations, junctions, bridges, crossings etc.) and *tracks* that connect these nodes. It is possible that there are multiple parallel tracks between two nodes; in that case, however, we assume that each train has been assigned to one of these tracks a priori. We do not consider the infrastructure within stations in detail.
- **Train lines:** The information on trains is given in the form of *train lines*. A train line is a direct train connection between an origin and a destination station that is operated with a certain frequency (number of trains per cycle time). Furthermore, a train line has a certain type (intercity, interregional, local, cargo) that determines what intermediate stations the trains of that line will serve. Finally, we are also given the fixed trip times between every consecutive pair of nodes on the train line.
- **Constraints:** The timetable should satisfy three types of constraints: *safety* constraints, *marketing* constraints and *trip time* constraints. The safety constraints are used to ensure that trains will not collide, requiring that there should be a certain time buffer between possibly conflicting train movements. Marketing constraints are used to model any other requirements that the timetable should have in order to either make it as attractive as possible for passengers, or to be cost effective. With respect to the passenger-friendliness one should imagine transfer possibilities to connecting trains, fixed departure times for certain trains, and synchronization of the departure times of trains heading for the same direction. Regarding the cost-effectiveness there are matters like short turning times for trains at destination stations, so as to use the rolling stock as efficiently as possible. The trip time constraints, finally, relate the departure and arrival times of a train at subsequent stations.

3.2   The Basic Model

As mentioned, we want to construct a cyclic pattern which consists of arrival and departure times for every train at each traversed node in the railway network. To model this, we introduce the concept of an *event*. An event is a combination of (i) either an arrival or a departure, (ii) a train, and (iii) a node in the railway network. The set of events, denoted by N, can be derived from the infrastructure and from the train line data. In order to construct a timetable, we then need to determine the time instant at which each event takes place. These time instants should respect the constraints that were described in the previous section, which all turn out to be constraints on *pairs of events*, as will be explained in more detail below. The set of pairs of events for which we have a constraint is denoted by A. The model uses the following sets, decision variables, and parameters:

N the set of events that need to be scheduled

$A \subset (N \times N)$ the set of event-pairs for which a constraint exists

$v_i \in \{0, \ldots, T-1\}$ the integer time instant within period T at which event $i \in N$ takes place

$p_{ij} \in \mathbb{Z}$ the integer variable modeling the cyclic nature of the constraint relating events i and j

T the cycle time of the timetable, which for our purposes will usually equal 60 minutes

$[l_{ij}, u_{ij}]$ the time window relating the events i and j, meaning that event j should take place between l_{ij} and u_{ij} minutes after event j takes place

All constraints that were mentioned in the previous section, safety constraints, marketing constraints, and trip time constraints, can be represented by constraints on pairs of events. Such a constraint on the event-pair $(i, j) \in A$ is written in the following way:

$$v_j - v_i \in [l_{ij}, u_{ij}]_T. \tag{1}$$

This constraint states that event j should take place between l_{ij} and u_{ij} minutes after event i takes place. The notation $[.]_T$ means that the constraint is cyclic with cycle time T, i.e., each constraint should be taken modulo T. We assume that $0 \leq u_{ij} - l_{ij} < T$, since otherwise the constraint would have no meaning in a cyclic setting with cycle time T. Multiple constraints may exist for the event pair (i, j), these will be ignored to keep the notation simple. A representative set of example constraints for the CRTP can be found in the Appendix A.

We model the cyclic nature of the constraints by the integer decision variable p_{ij}. The p-variables will be used to subtract, when necessary, an integer multiple of the cycle time T from the time window $[l_{ij}, u_{ij}]$. The feasible region for the railway timetabling problem can be described as

$$\mathcal{V} = \left\{ v \in \mathbb{Z}^n \; \middle| \; \begin{array}{ll} v_j - v_i + Tp_{ij} \in [l_{ij}, u_{ij}] & \forall (i,j) \in A \\ 0 \leq v_i \leq T - 1 & \forall i \in N \\ p \in \mathbb{Z}^m & \end{array} \right\} \tag{2}$$

where n is the number of events and m the number of constraints.

The constraints of the timetabling problem can also be represented by a directed constraint graph $G = (N, A)$, with a node i for every event that needs to be scheduled, and an arc (i, j) whenever there exists a constraint $v_j - v_i \in [l_{ij}, u_{ij}]_T$. The arc (i, j) will be associated with the time window $[l_{ij}, u_{ij}]$. Figure 1 shows an arc in the constraint graph G.

Theorem 1. *If a convex objective function is used, the variables v_i in (2) can be relaxed to real values.*

Proof. If the integer variables p_{ij} are fixed, we are left with an integer program with time window constraints on the v-variables only. The constraint matrix for this integer program is the node-arc incidence matrix for the constraint graph G. This means that all vertices of the polyhedron $\mathcal{V}$ are integral, and we can drop the requirement that v should be integer. ∎

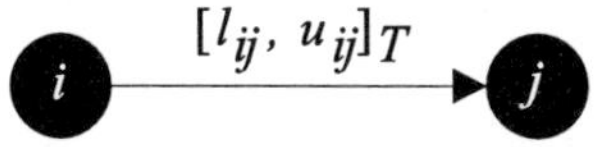

Figure 1. Arc (i, j) in the constraint graph G.

Serafini and Ukovich (1989), Odijk (1997) and Nachtigall (1999) showed that the PESP, which considers the problem of finding a feasible solution to $\mathcal{V}$, is NP-complete.

3.3 A Cycle Based Formulation

The previous section introduced cyclic constraints involving an integer variable to mathematically model the CRTP. These constraints prove to be very powerful for clearly expressing timetable requirements by relations between pairs of event times. For solving the problem, we shall, however, use a different model. Instead of studying the feasible region (2), we shall use a formulation based on cycles in the constraint graph G, as proposed by Nachtigall (1999) and Schrijver (1986).

The variables v_i, corresponding to the nodes in the graph G, are also known as potentials, see, e.g., Rockafellar (1984). Correspondingly, an arc variable $x_{ij} = v_j - v_i$ is a so-called *tension*. It is well-known that a vector x is a feasible tension if and only if the oriented sum of tensions x_{ij} along any cycle in the constraint graph G equals zero. It is, however, sufficient to consider a basis $\mathcal{C}$ of the cycle space. For a cycle c, let γ_{ij}^c be 1 for forward arcs, -1 for backward arcs, and 0 for arcs that are not in the cycle. Then x is a feasible tension if and only if

$$\sum_{(i,j)\in A} \gamma_{ij}^c x_{ij} = 0 \qquad \forall c \in \mathcal{C}.$$

Rewriting $\mathcal{V}$ in terms of tensions x_{ij} gives

$$\mathcal{X} = \left\{ x \in \mathrm{R}^m \left|
\begin{array}{l}
\sum_{(i,j)\in A} \gamma_{ij}^c x_{ij} = 0 \quad \forall c \in \mathcal{C} \\
l_{ij} \leq x_{ij} + Tp_{ij} \leq u_{ij} \ \forall (i,j) \in A \\
p \in \mathrm{Z}^m
\end{array}
\right. \right\} \qquad (3)$$

Note the strong relation between potentials v_i and tensions x_{ij}. Given potentials v_i, tensions are straightforwardly calculated. Conversely, given tensions x_{ij}, a matching set of potentials can be constructed by fixing one of them, say $v_0 = 0$, calculating all other potentials by the formula $v_j = v_i + x_{ij}$, and afterwards taking all potentials modulo T.

Since we are considering a cyclic problem with cycle time T, there is no difference between studying values x or values $x + Tp$. Therefore, the above set can be formulated as

$$
\mathcal{X} = \left\{ x \in \mathbb{R}^m \left|
\begin{array}{ll}
\sum_{(i,j)\in A} \gamma_{ij}^c x_{ij} = T \sum_{(i,j)\in A} \gamma_{ij}^c p_{ij} \ \forall c \in \mathcal{C} \\
l_{ij} \le x_{ij} \le u_{ij} & \forall (i,j) \in A \\
p \in \mathbb{Z}^m
\end{array}
\right. \right\} \tag{4}
$$

The first condition in this expression states that the oriented sum of tensions along each cycle should equal a multiple of the cycle time T. This can also be expressed using a new integer variable q_c for each $c \in \mathcal{C}$, giving

$$
\mathcal{X} = \left\{ x \in \mathbb{R}^m \left|
\begin{array}{ll}
\sum_{(i,j)\in A} \gamma_{ij}^c x_{ij} = T q_c \ \forall c \in \mathcal{C} \\
l_{ij} \le x_{ij} \le u_{ij} & \forall (i,j) \in A \\
q \in \mathbb{Z}^{m-n+1}
\end{array}
\right. \right\} \tag{5}
$$

where $m-n+1$ is the number of cycles in a cycle basis. Note that a cycle basis may be constructed by starting with a spanning tree of G, and iteratively creating a cycle by adding one of the $m - (n - 1)$ non-tree arcs.

Finally, we can obtain bounds on the q-variables by substituting the bounds on the x-variables. This gives

$$
\sum_{(i,j):\gamma_{ij}^c=1} l_{ij} - \sum_{(i,j):\gamma_{ij}^c=-1} u_{ij} \le T q_c \le \sum_{(i,j):\gamma_{ij}^c=1} u_{ij} - \sum_{(i,j):\gamma_{ij}^c=-1} l_{ij}. \tag{6}
$$

Because of the integrality of q, we can divide by T and round, which yields the following bounds $l_c \le q_c \le u_c$

$$
l_c = \left\lceil \frac{1}{T} \left(\sum_{(i,j):\gamma_{ij}^c=1} l_{ij} - \sum_{(i,j):\gamma_{ij}^c=-1} u_{ij} \right) \right\rceil \tag{7a}
$$

$$
u_c = \left\lfloor \frac{1}{T} \left(\sum_{(i,j):\gamma_{ij}^c=1} u_{ij} - \sum_{(i,j):\gamma_{ij}^c=-1} l_{ij} \right) \right\rfloor \tag{7b}
$$

Putting all the above together, the feasible region of the CRTP is described by

$$
\mathcal{X} = \left\{ x \in \mathbb{R}^m \left|
\begin{array}{ll}
\sum_{(i,j)\in A} \gamma_{ij}^c x_{ij} = T q_c \ \forall c \in \mathcal{C} \\
l_{ij} \le x_{ij} \le u_{ij} & \forall (i,j) \in A \\
l_c \le q_c \le u_c & \forall c \in \mathcal{C} \\
q \in \mathbb{Z}^{m-n+1}
\end{array}
\right. \right\} \tag{8}
$$

The model based on the feasible region $\mathcal{X}$ turns out to solve much faster than the model based on $\mathcal{V}$. The remainder of the paper focuses on the model described by (8). For experiences with the original model, we refer to Peeters (1999).

3.4   Objective Functions

Clearly, railway companies are interested in a high quality timetable. Several aspects can influence the quality of a timetable. One could think of passenger satisfaction (e.g., the available time to transfer between trains), operating efficiency (e.g., running trains at the highest possible speeds so that traveling times are minimal and the rolling stock usage is maximal), and robustness (e.g., there should be sufficient buffer time in the timetable to absorb delays). All of these desired characteristics can be included in one specific timetable only to a certain extent; some of them are conflicting, in which case a trade-off between them has to be made.

The first part of the objective function is aimed at directing certain tensions towards certain values. Let A_o denote the subset of tensions that are incorporated into the objective function. The objective function will be the sum over functions of the tensions in A_o. We define a function $f_{ij}(x_{ij})$ for every $(i,j) \in A_o$ to represent the preferred values of the tension x_{ij}. The functions f_{ij} will be linear or quadratic. In Figure 2 these two functions are illustrated. In Figure 2.I the value in the middle of the window $[l_{ij}, u_{ij}]$ is preferred, in Figure 2.II the value l_{ij} is preferred. Both preferred values are assigned a function value of zero. Practically, one can think of functions that

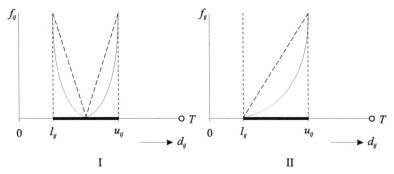

Figure 2. Linear and quadratic functions $f_{ij}(x_{ij})$.

reflect a preference for the middle of a frequency window or the middle of a safety window (see example constraints 4 and 5 in Appendix A), which would result in a function resembling Figure 2.I. Another possibility is a preference for minimal dwell and connection times (see example constraints 2 and 3 in Appendix A), which gives Figure 2.II. Linear functions f_{ij} can be used for situations in which each minute of deviation from the preferred value is equally important. Quadratic functions f_{ij} are convenient for expressing passengers' perceptions of waiting: it is well known that each additional minute of waiting is valued heavier than the previous one. Moreover, quadratic functions can be used to create a difference between solutions that would yield the same solution value when using linear functions. With the functions f_{ij}, the first part of the objective function is expressed as "Minimize $\sum_{(i,j)\in A_o} f_{ij}(x_{ij})$."

Minimizing the number of trains needed to operate a timetable will be the second objective. Consider Figure 3, which illustrates the outward and return trip for a train, and the so-called turn-around constraints that tie these two trips together. These last constraints attach a time-window to the time that a train spends at a terminal station before executing its return trip. Such a time-window is needed to clean the train, to shunt it, or to replace the crew. The task-cycle [outward trip→turnaround waiting→return trip→turnaround waiting] is called a rolling stock circulation cycle, or just circulation cycle. The q_c-variable for a circulation cycle corresponds exactly to the number of hours it takes a single train to execute one circulation, assuming a planning cycle time of one hour. Therefore, q_c equals the number of trains that will be needed to execute the part of the timetable corresponding to the circulation cycle, since every hour one train has to leave the origin node. Let C_t be the set of all such circulation cycles in the constraint graph G. Minimizing the number of trains needed to operate a timetable can then be expressed as "Minimize $\sum_{c\in C_t} q_c$." Note that more complicated circulation structures can be dealt with similarly, e.g., circulation structures in which a train executes several other trips before returning to the origin station of its initial outward trip.

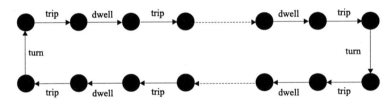

Figure 3. Rolling stock circulation cycle in G.

The following mixed integer non-linear program then defines the CRTP:

$$\text{Minimize } \alpha \sum_{(i,j)\in A_o} f_{ij}(x_{ij}) + \beta \sum_{c\in C_t} q_c$$

subject to

$$\sum_{(i,j)\in c} \gamma_{ij}^c x_{ij} = T q_c \qquad \forall c \in C$$

$$l_c \le q_c \le u_c \qquad \forall c \in C \qquad (9)$$

$$l_{ij} \le x_{ij} \le u_{ij} \qquad \forall (i,j) \in A$$

$$x \in R^m, q \in Z^{m-n+1}$$

Here, α and β are weights for the two objectives. Of course, one can also associate weights with certain train cycles, or incorporate weights in the functions f_{ij}. Note that in this formulation, the set of train cycles C_t must be part of the cycle basis C. One can also construct a cycle basis that does not contain the train cycles explicitly, and write the q-variables for the train cycles as linear combinations of q-variables for cycles in the cycle basis.

Theorem 2. *The mixed integer non-linear program (9) can be transformed to a mixed integer linear program by replacing all quadratic functions $f_{ij}(x_{ij})$ by linear approximations f'_{ij} that are exact for integer values of x_{ij}. The resulting mixed integer linear program will have exactly the same optimal solution and solution value as (9).*

Proof. Once the integer vector q is fixed, the constraint matrix for $\mathcal{X}$ becomes a cycle-arc incidence matrix, and is, therefore, totally unimodular. This means that the polyhedron represented by $\mathcal{X}$ has integral vertices only. The only non-linear part of (9) are quadratic functions f_{ij}. If we approximate these by linear functions f'_{ij} that are exact for integer values of x_{ij}, the solution and solution value of the approximated mixed integer linear program will be the same as the solution and solution value of (9). ∎

3.5 Linearizing Quadratic Objective Functions

We can use Theorem 2 to linearize quadratic objective functions f_{ij}, and thereby formulate the CRTP as a pure mixed integer program. Consider the quadratic function in Figure 2.II. For all integer values $x_{ij} \in [l_{ij}, u_{ij}]$, the quadratic function $f_{ij}(x_{ij}) = x_{ij}^2$ passes through the following points $(x_{ij}, f_{ij}(x_{ij}))$

$$(l_{ij} + \delta, \delta^2) \qquad \delta = 0, \dots, u_{ij} - l_{ij}.$$

The line through the consecutive points

$$(l_{ij} + \delta, \delta^2) \text{ and } (l_{ij} + \delta + 1, (\delta + 1)^2) \qquad \delta = 0, \dots, u_{ij} - l_{ij} - 1.$$

is described by

$$y = a(\delta)x_{ij} + b(\delta) \qquad \delta = 0, \dots, u_{ij} - l_{ij} - 1$$

with

$$\begin{cases} a(\delta) = 2\delta + 1, \\ b(\delta) = -\delta(\delta + 1) - (2\delta + 1)l_{ij}. \end{cases}$$

We will linearize the quadratic function f_{ij} by introducing an auxiliary variable f'_{ij}. These variables are bounded from below as follows

$$f'_{ij} \geq a(\delta)x_{ij} + b(\delta) \qquad \delta \in \Delta_{ij}, \tag{10}$$

with $\Delta_{ij} = [0, \dots, u_{ij} - l_{ij} - 1]$. For the quadratic function in Figure 2.I both the right and left half can be approximated similarly by linear equations.

The above approximation yields the following mixed integer programming formulation for the CRTP

$$\text{Minimize } \alpha \sum_{(i,j)\in A_o} f'_{ij} + \beta \sum_{c\in \mathcal{C}_t} q_c$$

subject to

$$\sum_{(i,j)\in c} \gamma^c_{ij} x_{ij} = T q_c \qquad \forall c \in \mathcal{C}$$

$$l_c \leq q_c \leq u_c \qquad \forall c \in \mathcal{C} \tag{11}$$

$$l_{ij} \leq x_{ij} \leq u_{ij} \qquad \forall (i,j) \in A$$

$$f'_{ij} \geq a(\delta) x_{ij} + b(\delta) \qquad \forall (i,j) \in A_o, \delta \in \Delta_{ij}$$

$$x \in \mathrm{R}^m, q \in \mathrm{Z}^{m-n+1}$$

Note that, in order to apply Theorem 2, the quadratic functions f_{ij} must be approximated such that the approximation is exact for every integer value in the time window $[l_{ij}, u_{ij}]$. For wide time windows, this will result in many linear inequalities that bound the variables f'_{ij} from below. This in turn may lead to very large linear programming relaxations in a Branch&Bound procedure. A compromise might then be to approximate the quadratic functions such that the approximation is exact only for $\delta = 0, \ldots, \delta_{\max}$, and to approximate to remainder of f_{ij} by the last linear inequality

$$f'_{ij} \geq (2\delta_{\max} + 1) x_{ij} - \delta_{\max}(\delta_{\max} + 1) - (2\delta_{\max} + 1) l_{ij}.$$

3.6   Calculating a Cycle Basis

Consider the following quantity, the so-called width of a cycle basis $\mathcal{C}$

$$W(\mathcal{C}) = \prod_{c\in\mathcal{C}} (u_c - l_c + 1). \tag{12}$$

The width of a cycle basis is an indication of the size of the solution space of the CRTP, since it counts the possible values of the vector q. Therefore, it is clear that a good cycle basis would be one with small $W(\mathcal{C})$. However, constructing a cycle basis that minimizes $W(\mathcal{C})$ is hard because of the rounding in calculating u_c and l_c. We use a heuristic proposed by Nachtigall (1999) to construct a cycle basis with small width. First, compute a spanning tree that is minimal with respect to the quantities $w_{ij} = u_{ij} - l_{ij} + 1$, the width of the time windows. Then, iteratively add a non-tree arc, which generates a cycle. Each of the wide non-tree arcs will, therefore, only appear in one cycle. Moreover, the part of a cycle that is contained in the tree will have a relatively small width. Therefore, the total width of the cycle basis will hopefully be small. A similar procedure was used by Serafini and Ukovich (1989) for their algorithm to solve PESPs.

If we want to include rolling stock circulation cycles into the objective function (see § 3.4), we proceed as follows in constructing a so-called circulation cycle basis. For all arcs corresponding to trip time constraints, dwell time constraints, and turn-around constraints, we set $w_{ij} = 0$. For all other arcs, $w_{ij} = u_{ij} - l_{ij} + 1$ as before, and we iteratively add a non-tree arc as before. This way, all rolling stock circulation cycles have width zero, and will, therefore, be included in the cycle basis. After all circulation cycles have been added to the cycle basis, the procedure complements the cycle basis with small width cycles as in the previous procedure. Clearly, the circulation cycle basis may be much wider than the cycle basis constructed by the heuristic, especially for trains that stop frequently, resulting in many dwell time constraints, and wide turn-around windows.

4   Preprocessing

The instances that we use are obtained from the decision support system DONS. The size of instances can be reduced drastically by applying preprocessing. By deleting arcs from G, we may also delete some cycles, thereby reducing the number of integer variables, which can considerably improve the solution process.

Removing parallel constraints: The DONS instances typically contain many parallel constraints, e.g., safety constraints for trains that belong to the same train series, and that are, therefore, already separated by their frequency constraints.

Nodes with degree one or two: Nodes with degree one in the constraint graph can be deleted: they are no part of any cycle, and the tension that corresponds to the deleted arc can be calculated in a post-processing phase. Nodes with degree two can be contracted, i.e., their two adjacent arcs are merged into a single arc, the window of which is the sum of the previous two windows. Consider the following example

$$v_j - v_i \in [l_{ij}, u_{ij}]_T, v_k - v_j \in [l_{jk}, u_{jk}]_T.$$

Here, node j is assumed to have degree two, so that the arcs (i, j) and (j, k) can be merged. This contracted arc then represents the constraint

$$v_k - v_i \in [l_{ij} + l_{jk}, u_{ij} + u_{jk}]_T.$$

A contracted arc can be expanded during post-processing to calculate the tensions for the two original arcs. A contraction can, however, only be executed if $(u_{ij} + u_{jk}) \mod T \geq (l_{ij} + l_{jk})$. Otherwise, the contracted arc represents two disjunct time windows. Consider the example

$$v_j - v_i \in [20, 40]_{60}, v_k - v_j \in [20, 40]_{60},$$

resulting in the merged constraint

$$v_k - v_i \in [40, 80]_{60} = [0, 20]_{60} \cup [40, 59]_{60}.$$

The two disjunct windows may be modeled by two new parallel arcs, but we may then just as well not contract the two original arcs.

Removing subsequent safety constraints: Consider the situation of two trains, 1 and 2, running from station A to station B, and using the same track. Let the trip times along track AB be r_1 and r_2 for train 1 and 2, respectively, and denote the departure times from station A by v_1 and v_2. The arrival times at station B will then be $(v_1 + r_1)$ and $(v_2 + r_2)$. Upon leaving and entering a station, there should be a buffer time between the two trains (see also Example 5 in Appendix A). Practically, this buffer time is often the same for both entering and leaving a station. Supposing the identical buffer times equal h, we get the following two constraints

$$v_2 - v_1 \in [h, T - h]_T \qquad \text{for leaving station A,}$$
$$(v_2 + r_2) - (v_1 + r_1) \in [h, T - h]_T \text{ for entering station B,}$$

which can be rewritten as

$$v_2 - v_1 \in [h, T - h]_T \qquad \text{for leaving station A,}$$
$$v_2 - v_1 \in [h, T - h]_T + (r_1 - r_2) \text{ for entering station B.}$$

If $r_1 = r_2$, clearly one of the constraints can be removed. If $r_1 \neq r_2$, then both constraints can be replaced by the following constraint

$$v_2 - v_1 \in [h, T - h]_T \cap [h + (r_1 - r_2), T - h + (r_1 - r_2)]_T.$$

Shrinking systems of trip time and safety constraints: Consider the case where several trains, not necessarily belonging to the same series, use a certain segment of railway track. Suppose that these trains do not stop at each station they pass. This means that we can contract their trip times over the segment, under the condition that the safety constraints are also adjusted. Such a situation is illustrated in Figure 4, where the nodes (1,2,3)

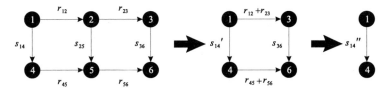

Figure 4. Shrinking a system of trip time and safety constraints.

correspond to one train, and the nodes (4,5,6) to the other. The windows for all safety constraints are assumed to be $[h, T - h]$, and are denoted by s_{14}, s_{25}, s_{36}. The contraction of the equality constraints, respecting all safety buffers, can be done by moving the safety constraint between the heads of two arcs to the tails of the arcs. For the trip arcs (1,2) and (4,5) this means that s_{14} should be intersected with $s_{25} + r_{12} - r_{45}$, giving

$$s'_{14} = s_{14} \cap (s_{25} + r_{12} - r_{45}).$$

Note that this can also be seen as strengthening the window s_{14} by considering the path 1-2-5-4. Repeating this procedure gives

$$s''_{14} = s'_{14} \cap (s_{36} + r_{12} + r_{23} - r_{45} - r_{56}).$$

Especially for large subgraphs of this structure, involving several trains and segments, the shrinking procedure considerably reduces the number of arcs and, more importantly, the number of cycles in the constraint graph.

5 Computational Results

The model was tested on the Dutch intercity network for 1997/1998, as obtained from NS Travelers. We will first describe the instance in more detail, then discuss the used objective functions, and conclude with an overview of the computational results.

5.1 The IC97 Instance

The IC97 instance contains all the intercity trains in the Netherlands in the hourly pattern for the 1997/1998 timetable. The instance consists of 50 stations and 25 train lines. The cycle time is one hour, i.e., $T = 60$. Although all trains have frequency one, many routes are visited by multiple train lines, and frequency constraints between these different lines are defined so as to have a half hour service on the majority of the routes. Connections are defined such that it is possible to travel with a good connection between any two intercity stations. The instance also contains many connecting and disconnecting events between train lines, to model the situation where two trains are connected to travel a common part of their trips as one train. The initial constraint graph, as obtained from the DONS system, consists of 1475 nodes and 3394 arcs. After applying the preprocessing procedures of Section 4, the size of the constraint graph is reduced to 217 nodes and 586 arcs. Consequently, a cycle basis for the constraint graph will contain 370 cycles.

We consider four variants of the instance, see Table 1, that differ in the width of the time windows for the three types of market constraints: connection constraints, dwell time constraints, and turn-around constraints. Instance init is the initial instance, that turned out to be infeasible. In the

	connection	dwell	turn-around
init	$[2,5]_{60}$	$[1,3]_{60}$	$[10,40]_{60}$
1	$[2,5]_{60}$	$[1,5]_{60}$	$[10,40]_{60}$
2	$[2,8]_{60}$	$[1,8]_{60}$	$[10,40]_{60}$
3	$[2,12]_{60}$	$[1,11]_{60}$	$[10,40]_{60}$
4	$[2,5]_{60}$	$[1,5]_{60}$	$[10,59]_{60}$

Table 1. Variants of the IC97 instance.

first three variants, the widths of the connection and dwell time windows increase. The purpose of increasing the width of these time windows is two-fold. Firstly, relaxing the market constraints' time windows is what we propose as a possible solution for dealing with infeasible instances. It is, however, not clear beforehand how large such a relaxation should be. Studying these three variants shows the results of small, middle and large relaxations. Secondly, it allows us to study what happens when the instances become harder, since the solution space increases. This issue is of course also related to the first purpose. The fourth variant considers wide turn-around time windows. Note that the lower bounds for the time windows remain the same for all variants. We assume that the lower bounds represent hard operational constraints that may not be relaxed, i.e., they represent the absolute minimum time needed for the associated activity. For all these variants the frequency constraints are perfect; their time windows consist of the singleton 30. We will also study variants, indicated by the letter F, for which the frequency time window is relaxed to $[25,35]_{60}$. For all variants, the headway time equals three minutes.

5.2 Objective Functions

The following objective functions were considered in our computational experiments:

L   minimizing the sum of connection and dwell times by a linear objective function

Q   minimizing the sum of connection and dwell times by a quadratic objective function

R_i maximizing robustness with $\delta_{\max} = i$

T   minimizing the number of trains needed to operate the timetable

TL   combination of T and L

TQ   combination of T and Q

For the objective functions L, Q, and R, the weight parameters are $\alpha = 1, \beta = 0$. For L and Q, the set A_o consists of all connection and dwell time constraints, that are penalized linearly and quadratically, respectively. These objective functions aim at minimizing the travel time of passengers, since each minute above the absolute lower bound of a connection or dwell time window is time that a trains spends waiting at a station, and, therefore, adds to the total travel time of the passengers in it. For R, the set A_o consists of the safety constraints, and the deviation from the middle of the safety time windows is penalized quadratically. The idea here is to push trains apart as much as possible, so that it becomes less likely that they conflict with each other in case of delays. These constraints typically have very wide time windows, e.g., for the IC97 instance $[3, 57]_{60}$. We will, therefore, consider three variants, that differ in $\delta_{\max}$, the number of inequalities that are used to approximate the quadratic objective function. For the objective function T the weight parameters are set to $\alpha = 0, \beta = 1$, and a rolling stock circulation cycle basis is used. Finally, for TL and TQ, the weight parameters are set to $\alpha = 0, \beta = 10,000$.

5.3 Computational Results

We used the mixed integer programming solver of CPLEX 6.6, ILOG (2001), to solve problem (11). The tables below contain the computation times in seconds (time), the number of Branch&Bound nodes (nodes), the number of iterations required to solve the linear programming relaxations (iterations), and the optimal objective value (z_{opt}) for each of the variants. All computations were done on a Pentium III 667 MHz with 128 Mb of memory.

Each line in the tables represents a combination of an objective function and a variant of the instance. For example, L1 means objective function L for variant 1, Q3$_3$ means objective Q with $\delta_{\max} = 3$ for variant 3, and TL4 objective TL for variant 4. Whenever an F appears, the frequency time windows have been set to $[25, 35]_{60}$ instead of the default $[30]_{60}$.

Table 2 compares linear and quadratic objective functions. It is clear that linear objective functions result in faster computation times, due to the smaller size of the linear programming relaxations. However, a closer inspection of the optimal solutions shows that quadratic objective functions yield solutions in which the deviations from the lower bound of the connection and dwell time windows are smaller. A nice compromise is to approximate the quadratic function only partially exactly, as is shown by comparing Q3, Q3$_3$, and Q3$_7$. Computation times decrease, while the approximation error is small. Finally, it can be concluded that the computation times are acceptable for practical use.

The results for maximizing the robustness of a timetable are placed in Table 3. From these results it is clear that restricting the approximation of the quadratic functions f_{ij} to low values of $\delta_{\max}$ is essential for computation times to stay within reasonable limits. For R1$_8$, i.e., $\delta_{\max} = 8$, the IC97 instance could not be solved within four hours. However, from a practical point of view this restricted approximation is no problem. Indeed, with a headway time of three minutes, approximating exactly up to $\delta_{\max} = 3$ results in a timetable for which the approximation is exact up to twice the minimal required safety buffer time, which is more than adequate for practical purposes.

	time (sec)	nodes	iterations	z_{opt}
L1	11.99	4564	17589	104
L2	15.16	6502	21799	87
L3	38.62	12442	29513	87
LF1	15.91	2686	17866	93
LF2	55.59	7844	95899	83
LF3	229.34	37127	292591	83
Q1	57.04	6146	62488	233
Q2	71.34	6131	80531	225
Q3	291.11	16968	189701	225
$Q3_3$	105.43	13557	126376	199
$Q3_7$	180.29	13360	145608	225
QF1	83.21	5020	82013	199
QF2	140.01	6484	135220	196
QF3	438.05	18228	228780	196

Table 2. Results for the IC97 instance: minimizing passenger travel time.

	time (sec)	nodes	iterations	z_{opt}
$R1_3$	44.95	4530	56844	31
$R1_5$	599.12	42187	618784	159
$R1_8$	(11% gap after 4 hrs of computing)			

Table 3. Results for the IC97 instance: maximizing robustness.

Table 4 shows the results for minimizing the number of trains needed to operate a timetable, possibly in combination with minimizing passengers' travel time. In all solutions, 65 trains are needed to operate the timetable. Note that the actual number of trains needed is larger, since we did not include international trains and some other special cases into the objective function. The computation times are again very acceptable, although a lot of extra time is needed for TL1 and TQ1 in comparison to L1 and Q1. This is because the cycle bases are much wider when we are minimizing the number of trains, since they must contain the quite wide circulation cycles. Also, computation times increase when relaxing the turn-around time windows from variant 1 to variant 4, but this also results in a decrease in the total passenger travel time.

Finally, Table 5 illustrates the importance of choosing a good cycle basis. The first three lines of the table are the same as the corresponding lines in Table 2, the last three lines represent the same objective function and

	time (sec)	nodes	iterations	z_{opt}
T1	30.33	5593	26413	65000
TL1	83.12	13165	51711	65127
TQ1	81.9	6578	83438	65295
T4	61.75	12641	81644	65000
TL4	52.02	10440	35892	65091
TQ4	53.18	3994	61014	65219

Table 4. Computational results for the IC97 instance: minimizing train usage combined with minimizing passenger waiting time.

variant as the first three, only based on the much wider train circulation cycle bases (the subscript c stands for circulation). From the table it is clear that computation times increase drastically when using the train cycle bases. Quite remarkable is the factor four difference in computation time between $Q1_c$ in Table 5, and TQ1 in Table 4. Both consider variant 1 and objective Q, but the latter one also includes objective T. Apparently, the inclusion of minimization of the number of trains in the objective leads to a much faster solution process.

	time (sec)	nodes	iterations	z_{opt}
Q1	57.04	6146	62488	233
Q2	71.34	6131	80531	225
Q3	291.11	16968	189701	225
$Q1_c$	342.38	28655	281537	233
$Q2_c$	741.43	52153	589643	225
$Q3_c$	968.18	45306	532134	225

Table 5. Computational results for the IC97 instance: comparing cycle bases.

The infeasibility of the initial variant init is found within a fraction of a second in CPLEX's presolve phase, regardless of the used objective function. The tables show that the penalized relaxation of the constraints offers a quite adequate tool for creating a timetable that minimally violates the initial constraints. With respect to the aspect of computing an optimal timetable, it is interesting to remark that for most variants of the IC97 instance, up to four or five integer feasible solutions are found during the Branch&Bound process, where the value of the first found integer solution can be about 3 to 4 times worse than the optimal integer solution. Clearly, a lot might be missed when one just searches for a feasible solution.

6   Conclusions and Further Research

The cycle based model for the CRTP seems to work quite well on this sample of test instances. Solving the model yields timetables that have a considerably better objective value than other, merely feasible, solutions that might be found by CADANS. However, CADANS finds a feasible timetable (for the relaxed variants) within a computation time varying from 5 to 15 seconds, which is faster than our optimization model needs. Moreover, the presented model provides an easy and effective concept for dealing with infeasible instances. An important factor to make the method work well is the application of the preprocessing procedures described in Section 4. Without these procedures, computation times can easily increase up to several hours.

Obviously, the IC97 instance and its variants presented here do not give a full view of the performance of the model on general CRTP's. A first direction for further research is, therefore, to test the model on other, more complicated instances. A second direction of research is to derive more general cases for which shrinking of the constraint graph is possible. One idea is to also contract dwell time constraints, viewing the result as a train with a *variable*, instead of a fixed, trip time. Under certain conditions, which are described in Kroon and Peeters (1999), safety constraints involving only the departure and arrival surrounding such a variable trip time are necessary and sufficient to meet all safety restrictions along the route. Other directions of research include the investigation of methods to construct a good, maybe even optimal cycle basis, and the incorporation of the classes of valid inequalities by Nachtigall (1999) into the model. Finally, it is not our intention to start from scratch in constructing a timetable, without using the very well performing CADANS algorithm. We, therefore, want to investigate how to use the knowledge presented in this paper to improve the practical use of CADANS and DONS.

Acknowledgements

The authors would like to thank both referees for their useful comments.

A   Example Constraints

This appendix presents some representative examples of practical railway timetable constraints. To improve the clarity of these examples, we will use the variable d for departure times and a for arrival times. The examples below are illustrated in Figure 5.

Example 1. Trip time constraint.
Intercity train A runs from Rotterdam to Utrecht, which takes 30 minutes. We have

$$a_{A,Utrecht} - d_{A,Rotterdam} = [30]_T.$$

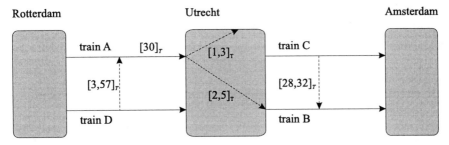

Figure 5. Illustration of example constraints.

Example 2. Dwell time constraint.
Suppose that upon arrival in Utrecht, train A has to stop between 1 and 3 minutes for the boarding and alighting of passengers. This gives

$$d_{A,Utrecht} - a_{A,Utrecht} \in [1,3]_T.$$

Example 3. Connection constraint.
Next, suppose that train B leaves from Utrecht for Amsterdam, and that some passengers of train A want to transfer to train B in Utrecht. Train B should leave between 2 and 5 minutes after train A arrives, so that the passengers have enough time to transfer. We then have

$$d_{B,Utrecht} - a_{A,Utrecht} \in [2,5]_T.$$

Example 4. Frequency constraint.
Now let train B belong to some train series with frequency 2 (twice per hour), and suppose we want to have a flexibility of (plus or minus) 2 minutes in this frequency. Let train C be the other train in this series. Since twice per hour means that a train should leave every 30 minutes, and taking into account the 2 minutes flexibility, we obtain

$$d_{C,Utrecht} - d_{B,Utrecht} \in [28,32]_T.$$

Example 5. Safety constraint.
Finally, suppose that train D is a Stoptrain, running from Rotterdam to Utrecht, and using the same track as train A. If train A leaves Rotterdam, then train D should *not* leave Rotterdam in the following 3 minutes, and vice versa. This yields the following two constraints

$$d_{D,Rotterdam} - d_{A,Rotterdam} \notin [0,3)_T \text{ AND } d_{A,Rotterdam} - d_{D,Rotterdam} \notin [0,3)_T,$$

which together give

$$d_{D,Rotterdam} - d_{A,Rotterdam} \notin (-3,3)_T.$$

Because of the cyclic nature of the timetable, this is the same as

$$d_{D,Rotterdam} - d_{A,Rotterdam} \in [3,57]_T$$

since the complement of the interval $(-3,3)_T$ is $[3,-3]_T$, and $[3,-3]_T = [3,57]_T$, assuming that the cycle time equals 60 minutes.

Bibliography

Goverde, R.M.P. (1999). Improving punctuality and transfer reliability by railway timetable optimization. In P.H.L. Bovy (Ed.), *Proceedings TRAIL 5th Annual Congress, 2*, Delft.

Hassin, R. (1996). A flow algorithm for network synchronization. *Operations Research 44*, 570–579.

Hurkens, C. (1996). Een polyhedrale aanpak van treinrooster problemen (A polyhedral approach to railway timetabling problems). Unpublished. In Dutch.

ILOG (2001). *ILOG CPLEX.* http://www.Ilog.Com/products/cplex/, 07.04.2001.

Kroon, L.G. and L.W.P. Peeters (1999). *A variable running time model for cyclic railway timetabling.* Erasm management report series 28-1999, Erasm, Rotterdam, The Netherlands.

Lindner, T. (2000). *Train Schedule Optimization in Public Rail Transport.* Ph.D. thesis, Technical University Braunschweig, Braunschweig, Germany.

Nachtigall, K. (1999). *Periodic Network Optimization and Fixed Interval Timetables.* Habilitation thesis, University Hildesheim.

Odijk, M.A. (1996). A constraint generation algorithm for the construction of periodic railway timetables. *Transportation Research B 30*, 455–464.

Odijk, M.A. (1997). *Railway Timetable Generation.* Ph.D. thesis, Delft University of Technology, Delft, The Netherlands.

Peeters, L.W.P. (1999). An optimization approch to railway timetabling. In P.H.L. Bovy (Ed.), *Proceedings TRAIL 5th Annual Congress, 1*, Delft.

Rockafellar, R.T. (1984). *Network Flows and Monotropic Optimization.* Wiley, New York.

Schrijver, A. (1986). *Theory of Linear and Integer Programming.* Wiley, New York.

Schrijver, A. and A. Steenbeek (1994). *Dienstregelingontwikkeling voor Railned (Timetable construction for Railned).* Technical report, CWI, Center for Mathematics and Computer Science, Amsterdam. In Dutch.

Serafini, P. and W. Ukovich (1989). A mathematical model for periodic event scheduling problems. *SIAM Journal of Discrete Mathematics 2*, 550–581.

Minmax Vehicle Routing Problems: Application to School Transport in the Province of Burgos

Cristina R. Delgado Serna and Joaquín Pacheco Bonrostro

Departamento de Economía Aplicada, Universidad de Burgos Fac C EE y EE,
C/Parralillos s/n. BURGOS 09001, Spain
cdelgado@ubu.es, jpacheco@teleline.es

Abstract. School transport is a delicate problem from an economic, social and political viewpoint. In previous works (Pacheco et al. (1999, 2000)) the authors approach the problem from an economic objective, i.e., to minimize the costs of the routes which should not exceed a fixed duration (60 min). In this work, we take advantage of the structure of algorithms used previously, approaching the problem from a "social" objective: minimize the duration of the longest route, with a fixed maximum number of vehicles. Results are provided for data from secondary education in the province of Burgos.

1 The Problem

Let us consider the problem of transporting a set of pupils who must be picked up at a series of geographically distributed locations and taken to a school. Let n be the number of locations (1 denotes the school, and $2, 3, \ldots, n$ the rest of the locations). Let $q(i)$ be the number of pupils to be picked up at each point $i, i = 2, \ldots, n$. Pupils from one location must be transported in the same vehicle and the duration of each route must not exceed a specified maximum time t_{max}. To fulfill these requirements there is a specific number of vehicles m with capacity Q. The distances d_{ij} and travel times t_{ij} between each pair of points $i, j \in \{1, 2, \ldots, n\}$ are known.

1.1 Antecedents: The Economic Question

In a recent work, Pacheco et al. (2000) posed the problem from an "economic" objective function: to design a set of routes with minimum cost, respecting maximum duration for each route and the capacity of each vehicle. The transport cost for each route is basically given by the number of kilometers covered, although the number of pupils and the number of stops are also important. Specifically, at 13 December 1997 the Ministry of Education and Culture, through the Secretary General for Education and Professional Training, suggested a formula that could be used as a reference for the calculation of the

costs necessary for the contracting of school transport. These costs (in Pesetas) are broken down into three sections *Kilometers (km)*, *Pupils*, and *Stops*
as follows:

$$\text{Cost for km} = \begin{cases} Pr \cdot m & \text{if } k \leq 35 \\ Pr \cdot 35 + (k - 35) \cdot (1.33) \cdot Pr & \text{if } k > 35 \end{cases}$$

where Pr is the reference price per km, which ranges between 125 and 163
(depending on the quality of the vehicle), and k the number of km. The total
for this concept cannot exceed 14,250.

$$\text{Cost for pupils} = \begin{cases} 0 & \text{if } M \leq 33 \\ 100 \cdot (M - 33) & \text{if } M > 33 \end{cases}$$

where M is the number of pupils.

$$\text{Cost for stops} = \begin{cases} 75 \cdot (L1 - 6) & \text{if } L1 > 6 \\ 0 & \text{if } L1 \leq 6 \end{cases}$$

where $L1 = min\{L, M/3\}$ and L is the number of stops.

However, in other cases, the people responsible for school transport in
each province have to negotiate the amount by different tariff systems, i.e.,
payment per km. The price depends on the number of pupils on the route. In
any case, the above formula is a good approximation and will be used in this
work. Finally, the distance covered and the time taken are counted from the
first pick-up point, and not from the point from which the bus leaves.[1] Thus
the problem is a specific case of the well-known Vehicle Routing Problem
(VRP), or to be more precise it is a VRP with load and time constraints (see
Laporte et al. (1984, 1985)).

There are many solution algorithms for the VRP (and/or variants, principally of the VRP with time windows) in the literature. Collections of the main
ones can be found in the works of Bodin and Golden (1981); Desrochers et al.
(1988); Haouari et al. (1990); Laporte (1992); Laporte and Osman (1995);
Laporte and Semet (1998); Laporte et al. (1999); Desaulniers et al. (1999). In
the last few years the development of algorithms based on processes known
as metaheuristics have become important, such as Genetic Algorithms, Simulated Annealing, Tabu Search, GRASP, Guided Local Search (Voudouris
and Tsang (2001)), Ant Colonies etc. especially since the works of Gendreau
et al. (1994) and Osman (1993), and more recently in the works of Potvin
et al. (1993); Potvin and Bengio (1994); Thangiah et al. (1993, 1994); Campos and Mota (2000); Kontoravdis and Bard (1995); Rochat and Taillard
(1995); Kilby et al. (1999); de Backer et al. (2000); Bullnheimer et al. (1999);
Rego (1998).

Pacheco et al. (2000) proposed a two-level heuristic strategy: the first level
mainly consists of a series of local search processes, giving rapid solutions; the
second level consists of a tabu search process which is applied to the solution
obtained in the first.

---

[1] Obviously from here on $d_{1i} = t_{1i} = 0$ for $i = 2, \ldots, n$.

1.2 The Social Question: The Minmax Problem

The people responsible for school transport in the province of Burgos stressed the need to design routes which, apart from saving money, would shorten the current route times as much as possible. This need was echoed by the Parents' Associations (APAS) who, especially in the recent past (March 2000), have protagonized various protests about the excessive journey time of many of the present routes.

As a result of these worries and needs, an attempt was made to approach this problem from a more "social" objective: minimize the duration of the longest route while complying with the requirements established earlier; in other words, to determine just how much one can reduce journey times. So the idea is to reduce the duration of the routes, while making them as balanced as possible.

We use the following notation to give a formal description of the problem. For $i = 1, \ldots, m$ let $R_i = \{r_i(1), \ldots, r_i(n_i)\}$ denote the route for bus i, where $r_i(j)$ is the index of the jth location visited and n_i is the number of pick-up locations in the route. Obviously $r_i(n_i + 1) = 1$. By $length(i)$ we denote the length of route i (which is also the maximum travel time corresponding to the students picked up at the first location). Note that according to our definitions, the length of route i is calculated as $length(i) = \sum_{j=1}^{n_i} t_{r_i(j)r_i(j+1)}$.

The "social" problem is to find a set of routes in order to:

$$\text{Minimize} \quad t_{max} = \max_{i=1,\ldots,m} \{length(i)\} \qquad (1)$$

subject to

$$\sum_{j=1}^{n_i} q(r_i(j)) \leq Q \;\; \forall\, i = 1, \ldots, m \qquad (2)$$

$$\sum_{i=1}^{m} n_i = n - 1 \qquad (3)$$

$$r_i(j) \neq r_k(j') \qquad \forall\, i, k = 1, \ldots, m; \quad i \neq k \qquad (4)$$
$$\forall\, j = 1, \ldots, n_1, \forall\, j' = 1, \ldots, n_k$$

The objective function (1) minimizes the maximum duration of any tour. Constraints (2) enforce the physical capacity of each bus. Constraints (3) and (4) indicate that all the students must be picked up and that a given location cannot be assigned to more than one route.

A heuristic is presented which consists of two parts: a constructive algorithm to find a feasible solution to which a local search procedure is applied and an algorithm based on a tabu search process which is applied to the solution obtained in the first part. Both parts are adaptations for this problem of that proposed in Pacheco et al. (2000).

In the next section we describe the different neighbourhoods used. Then we describe the algorithms used in the first part and in Section 4 the tabu

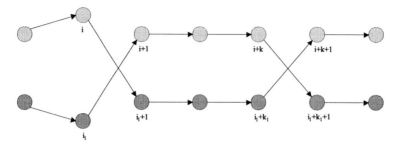

Figure 1. Elimination of arcs $(i, i+1)$, $(i+k, i+k+1)$, (i_1, i_1+1), (i_1+k_1, i_1+k_1+1) and incorporation of arcs (i, i_1+1), $(i+k, i_1+k_1+1)$, $(i_1, i+1)$, $(i_1+k_1, i+k+1)$.

search. Finally, in Section 5, we describe results obtained with the different algorithms, together with the calculation times used for a series of experiments with real problems using data obtained in the province of Burgos.

These computational results combine the two strategies and the two types of problem: social and economic. In a first phase, PHASE A, the "social" problem is solved, which establishes a new maximum journey time for the routes, which is used as a constraint (new value of t_{max}) for solving the economic problem (PHASE B).

Henceforth, S will be used to denote the set of feasible solutions to the problem, and f the function to be minimized defined in S (in this case the duration of the longest route).

2   Neighbourhoods Used

2.1   CROSS Exchanges

This type of neighbourhood, proposed by Taillard et al. (1997), considers exchanges of chains of points (of sizes k and $k1$) between two different routes, (CROSS exchanges) as shown in Figure 1. For each $s \in S$, $N_2^k(s)$ will be used to denote the set of feasible solutions generated by these CROSS exchanges of chains with at most k elements; similarly $N_2^\infty(s)$ will denote the set of feasible solutions generated by all these CROSS exchanges.

Note that each exchange of the subsection is determined by the two chains belonging to different routes r and r'; at the same time each chain is determined by its initial and final elements. Therefore, in a solution s one must consider $O(m^2)$ pairs of routes and in each route approximately n/m elements, i.e., $O(n^2/m^2)$ chains. In this way the number of exchanges one must consider to obtain $N_2^\infty(s)$ is of $O(m^2 \cdot (n^2/m^2) \cdot (n^2/m^2)) = O(n^4/m^2)$.

Checking for feasibility and the evaluation of each exchange, demands an excessive calculation time. In the work of Pacheco and Delgado (1997) the use of global variables is proposed, with which the number of operations necessary to check and evaluate each exchange is constant, in other words, independent of the problem size.

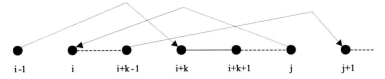

Figure 2. Relocation of a chain of k elements.

2.2 Or Type Neighbourhood

When a CROSS exchange is performed the two routes modified can be improved by means of interchange of elements. In this work we use the method of interchange proposed by Or (1976) which are feasible. Or's interchange method is a variant of the so-called r optimal interchanges developed by Lin (1965) and Lin and Kernighan (1973) for the symmetrical TSP. It will clearly be seen that Or's method can be used in asymmetrical problems. The efficiency of this method for the TSP has been verified in works such as that of Nurmi (1991).

Or proposes to limit the search for exchanges to 3-exchanges in which chains of one, two or three consecutive points are relocated between another two. Note that with these exchanges the direction of the different sections is not altered. In our case, we use the same idea, but only consider forward relocations and place no limitation on the size of the chain to be relocated (except limits determined by the size of the problem itself). Furthermore, the feasibility of each possible relocation must be checked with regard to the constraints of the problem. Figure 2 shows the relocation of a chain of k element starting at i between j and $j + 1$.

For each route $r, N_1^k(r)$ will denote the set of feasible routes obtained by forward relocations of chains of at the most k elements in r. $N_1^\infty(r)$ will denote the set of feasible routes obtained by all the relocations.

Let nr be the number of points which make up a route r, as each exchange is determined by the i and j points position and the size k of the chain to be relocated, then for a finite k the number of exchanges to be considered to obtain $N_1^k(r)$ is $O(nr^2)$, and for $N_1^\infty(r)$ is $O(nr^3)$.

3 Algorithm Design

As mentioned earlier, the algorithm proposed in this section consists of a local search which is applied to a solution obtained by a constructive method. The local search procedure proposed is based on examining the CROSS exchanges and choosing the best neighbour in each iteration.

3.1 Description of Local Search Procedures

Let $s \in S$ be a solution, s' of $N_2^\infty(s)$ an exchange, $r1$ and $r2$ the routes of s involved in this exchange and $r1'$ and $r2'$ the new routes, then we define

$\Delta(s') = \max\{tpo(r1), tpo(r2)\} - \max\{tpo(r1'), tpo(r2')\}$; where $tpo(r)$ is defined as the duration of route r. Δ will be the measure of "quality" of each exchange s' of $N_2^\infty(s)$. One interesting aspect is that we are going to look at exchanges between all the pairs of routes and not only those in which the longest route intervenes. In other words, exchanges are allowed, apparently unnecessary, since they do not reduce the objective function (duration of the longest route). However, we understand that this measure of quality, which is applied to all the exchanges, is interesting for the following reasons: it allows one to explore greater areas, avoids the rapid "stagnation" of the algorithm in local minima which are close to the initial solution, and favours the global equilibrium of the set of routes. These suppositions will be empirically analysed in § 3.2.

Let rf be any route and k a pre-set non-negative integer number, and let $sf \in S$ be a solution, then we define:

$\rightarrow$ Procedure <u>Or Local Search</u> (k, rf)
    Repeat
        Determine $r' \in N_1^k(rf)$ checking $tpo(r') = min\{tpo(r)/r \in N_1^k(rf)$
        If $tpo(r') < tpo(rf)$ then make $rf = r'$
    Until $tpo(r) \geq tpo(rf)$, $\forall r \in N_1^k(rf)$

$\rightarrow$ Procedure <u>Ta Local Search</u> (sf)
    Repeat
        Determine $s' \in N_2(sf)$ checking $\Delta f(s') = max\{\Delta f(s)/s \in N_2(sf)\}$
        If $\Delta f(s') > 0$: make $sf = s'$
            execute Or_Local_Search(3,r') and Or_Local_Search(3,r'')
            where r' and r'' are the two routes of sf modified to give s'
    until $\Delta f(s) \leq 0$, $\forall s \in N_2(sf)$

These local search procedures are analogous, although it should be pointed out that in the second case when one performs a CROSS exchange, the first is executed to improve each of the two routes involved in this exchange.

3.2 An Empirical Analysis

In this subsection we show the results of a series of tests to compare the *Ta_Local_Search* procedure when all the exchanges are considered (therefore admitting changes which do not affect the longest route), to the basic approach when only those longest routes are considered (only admitting changes which improve the objective function value). Simultaneously we compare the use of both *Best-Improvement* and *First-Improvement* strategies. Therefore, we consider four variants in these tests:

Variant_1: Only exchanges involving the longest route are considered with
    Best-Improvement strategy
Variant_2: All exchanges are considered with *Best-Improvement* strategy
Variant_3: Only exchanges involving the longest route are initially considered with *First-Improvement* strategy

Variant_4: All exchanges are initially considered with *First-Improvement* strategy

All variants use the same initial solution obtained by an adaptation of an *insertion* algorithm.[2] A series of instances were generated with 10 different sizes: 11 points, 21, until 101 points (destination included). For each size 20 instances were generated. The data for each instance is defined as follows:

- Each point of the problem is assigned two co-ordinates x and y, whose values are generated at random with uniform distribution between 0 and 100. (The locations of the real problems could be considered uniformly distributed). The distance between each pair of points is defined as the corresponding Euclidean distance. The journey times (in minutes) are taken equal to the distance (in km), i.e., we consider a speed of 60 km/h.
- To each $q(i), i = 2, \ldots, n$, we assign an integer value uniformly generated at random between 0 and 12.
- Vehicle capacity is assumed to be $Q = 20$.

Table 1 shows the average results of the function f (duration of the longest route **LR**, shown boldfaced), the average duration of all routes(AR), and the calculation times in seconds (*CT*, shown in italics), according to size. The number of vehicles available is that obtained by the insertion algorithm. All algorithms described have been programmed using BORLAND PASCAL (version 7.0) and BORLAND DELPHI (3.0) compilers. The equipment used is a PENTIUM II 300 MHz Personal Computer.

The results confirm the claims of § 3.1: Variants considering all exchanges (*Variant_2* and *Variant_4*) obtain better results in duration of longest route and, especially in the average duration of all routes. This last effect is very important and interesting for the problem considered in this work.

---

[2] This is a simple constructive algorithm which at each step inserts an element of the current solution. Define at each step:

S Set of points not yet inserting in the current solution
R_j Route j, $j = 2, \ldots, n$; (at first as many routes as points)
$C(j, i)$ Increase in distance resulting from inserting i in the jth route (at the best position)
$A_i = C(j^{**}(i), i) - C(j^*(i), i)$
where $j^*(i) = \text{argmin} \{C(j, i)/j = 2, \ldots, n\}$
                  $j^{**}(i) = \text{argmin} \{C(j, i)/j = 2, \ldots, n; j \neq j^*(i)\}$.

→ Initial Algorithm
    Initially set $S = \{2, \ldots, n\}$; $R_j = 1\ 1$ (only the origin), for $j = 2, \ldots, n$.
    Repeat
        Determine $A_{i^*} = min \{A_i\ /\ i \in S\}$
        Insert i^* in $R_{j^*(i^*)}$
        Execute Or_Local_Search (3, $R_{j^*(i^*)}$) to improve $R_{j^*(i^*)}$
        Set $S = S - \{i^*\}$
    until $S = \emptyset$.

Size		Insertion	Variant_1	Variant_2	Variant_3	Variant_4
11	LR	229.450	178.350	176.700	175.550	175.850
	AR	173.108	161.979	155.888	158.508	157.696
	CT	0.002	0.004	0.004	0.01	0.005
21		248.300	178.100	176.700	177.900	177.100
		173.387	158.460	150.951	157.654	150.103
		0.011	0.004	0.023	0.01	0.023
31		253.700	191.550	189.550	191.150	188.400
		175.832	165.876	152.163	165.622	151.136
		0.017	0.012	0.055	0.01	0.046
41		271.600	193.150	190.900	193.050	191.800
		178.821	167.021	150.766	166.210	154.785
		0.037	0.022	0.097	0.02	0.077
51		261.300	189.050	186.500	189.300	187.650
		181.447	164.112	150.583	161.330	150.749
		0.067	0.034	0.117	0.03	0.084
61		268.550	188.350	186.250	189.850	186.350
		179.557	162.896	145.709	165.256	147.158
		0.080	0.043	0.194	0.03	0.135
71		271.050	187.650	186.350	187.200	186.350
		179.077	160.537	144.348	162.329	143.400
		0.123	0.055	0.276	0.050	0.187
81		276.400	199.000	198.300	200.950	198.150
		185.059	170.833	151.984	174.350	153.952
		0.180	0.063	0.378	0.04	0.239
91		268.400	189.750	188.100	190.450	188.100
		182.438	164.383	145.309	164.315	146.665
		0.257	0.088	0.505	0.060	0.290
101		272.300	197.300	195.800	197.050	196.100
		183.836	166.876	146.312	169.140	146.494
		0.347	0.093	0.640	0.050	0.399

Table 1. Numerical results.

Variant_2 obtains slightly better results than Variant_4 in both aspects: duration of longest route and average duration of all routes; on the other hand, the calculation time is greater, although it is rather short. Therefore, in the next sections we consider all exchanges and use the *Best-Improvement* strategy.

It should be pointed out that in order to accelerate the execution of *Ta_Local_Search* a strategy was used, known as *Fast Local Search,* proposed by Bentley (1992). The adaptation of his idea to this model and neighbourhood is explained in detail in Pacheco and Delgado (2000b).

3.3   Description of the Algorithm

Let $s^* \in S$ be the best solution found at each moment, then the algorithm proposed for this first part would be as follows:

$\rightarrow$ Procedure $\underline{\text{Part\_1\_Algorithm}}$ (Output $s^* \in S$)
   Step 1: Obtain an Initial Solution S by the constructive method
   Step 2: *Execute Ta_Local_Search* (s^*)
   Step 3: To each route r of s^* apply Or_Local_Search(∞, r)

In order to obtain an initial solution we use an adaptation of Fisher and Jaikumar (1981) for this method. Although in this case acceptable solutions are also obtained with an adaptation of the *nearest insertion* algorithm.

The routes obtained in step 2 after the execution of the *Ta_Local_Search* are local optima with respect to a N_1^3. In other words, they cannot be improved on with forward relocations of chains of three elements. In step 3 local optima are achieved with respect to N_1^∞, i.e., relocations of any chain. This step, as will be seen in the results, supposes only very slight improvements in the results obtained. However, the calculation time is very short.

4   Improvement with a Tabu Search Procedure

The algorithm described in the previous part gives us satisfactory results markedly improving the solutions used nowadays in the schools analyzed, with a reasonable calculation time (see the following section). However, the solutions obtained can still be slightly improved, in some cases, adding some steps in which a metaheuristic developed in the last years is applied. Specifically, a tabu search procedure is proposed which is described as follows.

Tabu Search is a procedure or strategy which became known with the works of Glover (1989, 1990), and has been used widely and successfully in recent years. According to its creator it is a procedure which *"explores the space of solutions beyond the local optimum,"* Glover and Laguna (1993). It allows for *upward exchanges,* or exchanges which worsen the solution once a local optimum has been reached. Simultaneously, the last moves are qualified as tabu during the next iterations to avoid a return to previous solutions and prevent the algorithm from cycling. The term *tabu* makes reference to *"an inhibition to something for cultural or historic reasons and which can be overcome under certain conditions...,"* Glover (1996). Recent extensive tutorials on Tabu Search including all types of applications can be found in Glover and Laguna (1997, 2001).

4.1   The Basic Algorithm

The algorithm proposed works in the following way:

$\rightarrow$ Basic Tabu Search Procedure

    Read initial solution s_f (obtained from the algorithm described in Section 3)

    Set $s^* = s_f$, $T = \emptyset$, *niter* $= 0$, *kiter* $= 0$

    Repeat

        niter:=niter+1;

        Choose $s \in N_2(s_0)/s \notin T$ *or* s *verify the 'aspiration' criterion with*

        $\Delta(s)$ *maximum*

        Set $s_0 = s$

        Execute Or_Local_Search (3,r') and Or_Local_Search (3,r")

        where r' and r" are the two modified routes

        If $f(s_0) < f(s^*)$ *or* $(f(s_0) = f(s^*))$ *and the total duration of* s *routes*

        is less than that of s^* *then set* $s^* = s_0$ *and kiter* $=$ *niter*

        Update T

    until niter $-$ *kiter* $\geq$ *maxiter*

The best solution found is denoted by s^* and the current solution at each moment by sf. T is the set of tabu moves and is obtained by determining which set of solutions have certain *active tabu attributes*. Therefore, these tabu attributes must be defined and for how many iterations they are to remain active (and hence the solutions they contain).

The purpose of applying the *aspiration criterion* is to determine under what conditions a tabu move may be accepted. Usually, a solution s_0 is considered to fulfill the *aspiration criterion* if $f(s_0) < f(s^*)$. This move facilitates a new search direction and guarantees that cycles are not produced.

Furthermore, any local move supposes the incorporation of a set of arcs and the elimination of others (as illustrated in Section 3). In order to not to return to previously visited solutions, moves which imply the incorporation of some of these arcs will be prevented. In other words the *tabu attributes* will be the arcs which make up each solution and a move is tabu if it implies the incorporation of an arc eliminated in recent iterations (*active tabu attribute*). It must be pointed out that in this case the definition of active tabu attributes is only applied to arcs eliminated in the exchange of elements between routes, but it is not applied to arcs eliminated in the improvements in each of the routes involved (i.e., execution of *Or_Local_Search* for r' and r'').

In order to identify which tabu attributes (arcs) will be active, define *arco_tabú A*, a $n \times n$ matrix with entries *arco_tabú(r,l)* equal to the number of the last iteration in which the arc (r, l) was eliminated. A specific arc(r, l) will be an active tabu attribute if *niter* $-$ *arco_tabú*(r, s) $<$ *maxiter_tabu*, with *maxiter_tabu* being the number of iterations which remain active as tabu attributes from the moment when it is eliminated from the current solution.

Initially define: *arco_tabú*$(r, l) = 0$, for each arc(r, l) in the initial solution and *arco_tabú*$(r, l) =$ *maxiter_tabu* (or a more negative value) for the rest. This prevents arcs which do not form part of the initial solution being declared *active tabus* in the first iterations. So initially it is ensured that $T = \emptyset$. In addition set *maxiter_tabu* to $2 \cdot n^{1/2}$. As a stopping criterion we define *maxiter* $= 10 \cdot n$.

This section has described a basic tabu search algorithm. It could be complemented and enlarged with procedures based on what is usually known as long term and medium term memory such as *diversification and intensification;* it could also be enriched with the possibility of controlled visits to infeasible solutions by means of penalty functions *(strategic oscillation)*.

4.2   Intensification Phase

As its name indicates, this phase intensifies the exploration of regions and neighbourhoods where the best solutions in the initial phase were found (basic algorithm) with the hope of finding even better solutions. This phase can be designed in different ways. In our case it is inspired by the works of Rosing (1997); Rosing and Revelle (1997); Rosing et al. (1998) on *Heuristic Concentration,* a strategy for finding solutions to combinatorial problems in two phases. In the first a local search procedure is executed several times, recording the best local optima obtained; in the second, a *Concentration Set, CS,* is constructed with elements of the best solutions obtained in the first, and an exact or heuristic algorithm is executed, but only considering the elements of *CS*.

This work uses the idea of constructing a Concentration Set with the elements of the best solutions obtained in the basic algorithm. More precisely, this model considers as elements of the solutions the arcs which form it. For example, the solution made up of the two routes

  route 1:   1 – 3 – 5 – 1   *and*   *route 2:   1 – 4 – 2 – 1*

which can be expressed as the sequence 1–3–5–1–4–2–1, and will be made up of the arcs (1,3), (3,5), (5,1), (1,4), (4,2) and (2,1).

To form the concentration set we use the following procedure:

→ Procedure for <u>Construction of Concentration Set</u>
  Order the solutions obtained up to now in a list according to the f value
  (beginning with the best)
  Set i = 0 and CS = ∅
  Repeat
    set i = i+1
    Add the elements of the ith solution from the list to CS
  until Cardinal(CS) ≥ num_arcs

Note that unlike Rosing's proposal, a predetermined number of solutions whose elements are introduced is not set, but rather a minimum number of elements to be introduced, *num_arcs*, is set.

Next a procedure is designed which concentrates the search for elements in CS. Initially the following auxiliary times matrix $t1$ is defined as follows:

$$t1(i,j) = \begin{cases} d(i,j) & \text{if } (i,j) \in CS \\ d(i,j) + 10 \cdot max\_t & \text{if } (i,j) \notin CS \end{cases}$$

where $max\_t = max\{t(i,j)/i,j = 1,\ldots,n\}$. Analogously, define $tpo1(r)$ as the duration of each route r corresponding to $t1$. Finally, given a solution s, the quality of each exchange s' of $N_2^\infty(s)$, denoted by $\Delta1(s')$, defined the same as $\Delta(s')$ but changing tpo for $tpo1$. Let $sf \in S$ be a solution, the following search procedure is proposed taking into account both matrices t and $t1$:

$\rightarrow$ Procedure <u>Local Searchtowards CS</u>(sf)

 Set $s_0 = sf$

 Repeat

  previous cost $= f(sf)$

  Repeat

   Search $s' \in N_2(s_0)/\Delta1(s') = max\{\Delta1(s)/s \in N_2(s_0)\}$

   If $\Delta1(s') > 0$: Set $s_0 = s$

    Execute Or_Local_Search$(3,r')$ *and* Or_Local_Search$(3,r'')$

    (using t1 instead of t where r', r'' are the two modified routes of s_0)

    Search $s'' \in N_2(s_0)/\Delta(s'') = max\{\Delta(s)/s \in N_2(s_0)\}$

    In s'' Execute Or_Local_Search$(3,r')$ *and* Or_Local_Search$(3,r'')$

     where r' and r'' are the two modified routes to get s''

    If $\Delta(s'') > 0$ then set $sf = s''$

   until $\Delta1(s') \leq 0$

  set $s_0 = sf$

 until $f(sf) = previous\ cost$

This is a local search procedure within a loop: at each step the current solution $s1$ is substituted by a better one according to $\Delta1$, i.e., according to $t1$, therefore, searching for solutions that contain elements of CS; when there is no improvement in $\Delta1$, s_0 is substituted by sf, the best solution according to f observed in the neighbourhoods explored, and the local search is restarted. The procedure comes to an end when there is no improvement in $f(sf)$.

It is really a search procedure "directed" by $\Delta1$, in other words by $t1$, towards solutions that contain the greatest number of elements of CS possible. Note that $t1$ "penalizes" arcs not belonging to CS, but does not "prevent" their selection at each step, since this could excessively reduce the number of solutions to be considered and "box in" the procedure. $\Delta1$ is taken instead of Δ and the *Local_Search_towards_CS* procedure is executed directed by $\Delta1$, instead of the usual local search. In fact, what happens is that the search for solutions in the regions with elements of CS is "concentrated" or intensified. It could in some ways be considered a *path relinking*. This procedure is inserted in the intensification phase which takes the following form:

$\rightarrow$ Procedure <u>Intensification</u>

 Execute Construction *of_* Concentration_Set;

From i:=1 until num_solutions set
    Take s_i the ith solution of the ordered list obtained in the last
       execution of Basic_Tabu_Search
    Execute Ta_Local_Search(s_i)
    For each route r of s_i execute Or_Local_Search, $(k = \infty)$
    Execute Local_Search_towards_CS(s_i);
    If $f(s_i) < f(s^*)$ set $s^* = s_i$

We choose $num\_solutions = 50$. As for the value of the parameter num_arcs, the work of Pacheco and Delgado (2000a) describes the results of different experiments which advise fixing num_arcs to 10% of the total number of arcs (i.e., $num\_arcs = 0.1 \cdot n \cdot (n-1)$). When the intensification phase is added, the *Tabu_Search_Algorithm* takes the following form

→ Procedure <u>Main Tabu Search</u>
    Execute Basic_Tabu_Search
    Execute Intensification

The following considerations must be kept in mind. During the execution of the basic phase it is necessary to build and update a list containing the best solutions to be used in the intensification phase. This list will include, in each iteration, the best of all solutions explored in *Basic_Tabu_Search*, independently of whether they contain tabu items or not. On the other hand, a *Diversification* phase can be added to the *Main_Tabu_Search* procedure and this can be executed again once or several times. However, in this work we will not do this to avoid a long computational times.

5 Computational Results

In this section we analyse the different problem instances of secondary school transport in the Province of Burgos. Each problem is defined by a school, by the locations where the pupils who go to this school are to be picked up and by the number of pupils to be picked up at these locations. In all cases the maximum traveling time $t_{max} = 60$. Each of these instances has been solved using the algorithms described earlier.

To calculate the distance matrix and the route between each pair of points of the problem the distance of each section has been weighted according to the type of road such that main roads are chosen before secondary ones, secondary ones before unpaved roads, etc. In this way, on many occasions the routes obtained are not the shortest but they are the most comfortable and fastest.

As mentioned in Section 1, in a first phase, PHASE A, the "social" problem or *minmax* was solved using the strategy described in this paper; in a second phase, PHASE B, the value of the duration of the longest route of the solution obtained in PHASE A is used as a new value of t_{max} (constraint of maximum duration of the routes) to solve the economic problem. PHASE B

Pr.	School	Loca-tion	Chil-dren	Vehi-cles	min	mean	max	σ
1	Aranda de Duero	57	429	14	1	7.526	31	7.696
2	Belorado	24	175	8	1	7.292	43	10.799
3	Briviesca	24	101	7	1	4.208	44	8.485
4	L. de Mendoza - B.	19	127	5	1	6.684	42	11.168
5	Diego Marín - B.	22	58	4	1	2.636	10	2.326
6	Diego Siloé - B.	32	141	6	1	4.406	32	6.219
7	S. de Colonia - B.	36	158	7	1	4.389	24	5.198
8	Lerma	53	302	11	1	5.698	30	6.511
9	Medina de Pomar	39	182	7	1	4.667	56	9.183
10	Melgar	22	71	6	1	3.227	16	3.246
11	Miranda	13	41	5	1	3.154	20	5.051
12	Quintanar	4	105	4	21	26.250	30	3.562
13	Roa	28	207	7	2	7.393	40	7.423
14	Salas	23	81	5	1	3.522	30	6.013
15	Villadiego	31	128	8	1	4.129	34	5.934
16	Villarcayo	9	23	3	1	2.556	10	2.793
Total		436	2329	107	1	5.342	56	7.600

Table 2. Description of 16 secondary (middle) school problems (B.: Burgos).

uses the strategy described in an earlier work, Pacheco and Delgado (2000a). Therefore, priority is given to the social aspect over the economic aspect although the latter is not forgotten.

In Table 2, for the problems corresponding to secondary school transport (16 in all), we show where each school is located, the number of places where pupils are to be picked up, the number of children and the number of vehicles available. Likewise, some statistics corresponding to the distribution of students with respect to pick-up stops are provided.

Table 3 shows, for posterior comparisons, the present solutions, (i.e., the manual solutions generated by the transit authority) and the solutions obtained by minimizing the total costs (also reported in Pacheco et al. (2000)). Specifically, it shows the duration of the longest route and the cost.[3]

Table 4 shows the results obtained by the two-phase strategy proposed in this work. It indicates the value of the solutions obtained in each step (duration of the longest route in PHASES A and B, and costs in PHASE B), and the calculation times (considering the cost function described in Section 1 and the information supplied by the Dirección Provincial de Educación).

---

[3] LR: Duration of the longest route; BTB: Basic Tabu Search; BTI: Intensification (1) Evaluated according to the objective function described in §1.1; (2) Information supplied by the Dirección Provincial de Educación de Burgos.

Prob.	Present Solution		Minimizing the total costs (1)			
	Cost (1)	LR (2)	Part1: Cost	BTB: Cost	BTI: Cost	Final LR
1	61177.5	70	59033.38	59033.38	57680.75	58
2	21045	45	16962.50	16962.50	16962.50	32
3	25051.3	60	23404.00	23404.00	23404.00	59
4	18608.8	70	17542.38	17299.25	16863.63	49
5	15902.5	60	14614.88	14614.88	14614.88	55
6	30720	80	29742.00	29742.00	28484.00	58
7	34835	60	22561.25	22561.25	22539.13	60
8	51112.5	75	43441.50	43441.50	40754.50	58
9	31075	90	32855.00	31824.38	31267.38	56
10	19402.5	60	18887.00	18847.88	18847.88	56
11	16622.5	60	18847.88	18847.88	18847.88	50
12	8025	25	3312.50	3312.50	3312.50	9
13	22975	45	19550.00	19550.00	19437.50	36
14	21488.8	60	18586.75	18550.00	18550.00	53
15	29088.8	50	25231.25	25231.25	25226.75	49
16	8252.5	60	10847.00	10847.00	10847.00	53
Total	415382.7	970	375419.27	374069.65	367640.28	791

Table 3. Numerical results.

Table 5 shows the behaviour of the cost in PHASE B, to measure the impact of PHASE B on solution quality.

As for the data of the problems (see Table 2), the low number of students picked up at each location is striking (average 5.34). This is because the rural areas of the province are sparsely populated. Secondly, the present solutions are noteworthy (Table 3). In many cases the route duration norms are not complied with ($t_{max} = 60$), which has led to the protests by parents' associations commented on in § 1.2. The reason for the present durations is because the solutions are designed and/or brought up to date manually with no technical support.

As for the results obtained, the first point is that both sets of solutions, those obtained by minimizing costs (Table 3) and those obtained by the two-phase strategy described in this paper (Table 4) substantially improve both the cost and the duration of the longest routes of the present solutions. In the first case the total cost is reduced from 415,382.7 Pesetas to 367,640.8 (an 11.49% reduction), while in the second case the cost is only reduced to 409,290.4 (barely 1.5%). However, the total duration of the longest route in the first case is reduced from 970 minutes to 791 (i.e., 18.45%), while for the second case it is reduced to 634 (a 34.64% reduction). Both solutions

Pr.	PHASE A/Part 1, Step			Tabu Search		PHASE B	
	1	2	3	BTB	BTI		
1	LR: 60	51	51	48	48	C: 66792.13	LR: 48
	TC: 6.42	0.39	0.05	44.82	29.28	TC: 123.86	
2	46	29	29	29	29	17962.50	29
	0.65	0.06	0	5.27	5.71	12.52	
3	60	47	47	45	45	25160.00	45
	0.77	0.11	0	5.22	4.06	12.09	
4	53	36	36	36	36	21212.50	36
	0.22	0.05	0	2.69	2.64	5.11	
5	55	45	45	44	44	16291,13	43
	0.27	0.11	0	6.43	4.06	9.45	
6	60	51	51	47	47	31743.38	45
	0.99	0.22	0	11.97	9.28	25.71	
7	60	38	38	37	37	26845.25	37
	1.04	0.17	0.05	17.3	12.97	40.59	
8	60	51	51	51	51	41753.63	51
	3.57	0.388	0	43.5	27.9	101.29	
9	60	53	53	53	52	31373.88	52
	1.97	0.28	0	23.07	16.42	72.39	
10	58	41	41	41	41	19975.00	41
	0.5	0.11	0	4.06	3.68	8.02	
11	55	45	45	45	45	21338.88	45
	0.11	0	0	1.04	2.26	2.96	
12	15	9	9	9	9	3312.50	9
	0	0	0	0.22	0.72	0.71	
13	53	32	32	29	29	22000.00	29
	0.66	0.16	0	11.87	8.57	11.80	
14	60	53	53	48	48	19191.00	46
	0.39	0.05	0	5.39	3.9	11.97	
15	59	42	42	40	40	30812.50	40
	1.48	0.06	0	12.91	6.7	15.32	
16	55	40	40	38	38	13526.13	38
	0.06	0	0	0.65	1.32	1.27	
Tot.	869	663	663	640	639	C: 409290.41	634
	19.1	2.15	0.1	196.41	139.47	455.05	

Table 4. LR Duration of the longest route; TC Calculation Time; C Cost.

Prob.	Initial Solution (Obtained in Ph. A)	Part 1	Basic Tabu Search	Intensification in T.S.
1	73751.63	68663.5	68001	66792.13
2	18562.5	17962.5	17962.5	17962.5
3	27450	25655.75	25547.63	25160
4	21212.5	21212.5	21212.5	21212.5
5	16978.25	16291.13	16291.13	16291.13
6	34448.38	31902.13	31902.13	31743.38
7	27778.25	27657.25	27045.25	26845.25
8	50328.63	44419.88	44383.38	41753.63
9	33525.88	32160.63	31390.5	31373.88
10	20100	20062.5	19975	19975
11	21338.88	21338.88	21338.88	21338.88
12	3312.5	3312.5	3312.5	3312.5
13	22025	22000	22000	22000
14	19265.25	19265.25	19216	19191
15	30812.5	30812.5	30812.5	30812.5
16	13526.13	13526.13	13526.13	13526.13
Total	434416.28	416243.03	413917.03	409290.41

Table 5. Evolution of the costs in PHASE B.

obviously fulfill the maximum duration requirements for the routes. It is clear that the first strategy is more appropriate for the economic objective and the second for the social one. The strategy proposed in this paper achieves great improvements in the maximum duration of routes, and thus for the conditions for students.

As can be seen in the results in Table 4, in Part 1 solutions are obtained quickly (in about 21 seconds for all the problems of secondary schools) which significantly reduces the maximum duration of the average current solutions: from more than 60' to just over 41'. If more calculation time is available, the tabu search can reduce these times to less than 40'. In PHASE B the cost of current solutions is only reduced by about 1.5%, but with much better conditions regarding the duration of the routes.

As can be seen in Table 5, the impact of PHASE B is highly positive, as the total cost obtained in PHASE A is reduced from 434,416.2 to 409,290.4, i.e., an improvement of 5.78% which we consider noteworthy if one takes into account that the new values of t_{max}, given in PHASE A, reduces the solution space.

Finally, we point out some aspects that could be improved in future investigations:

- The initial solutions used, Step 1 in PHASE A, are notably improved by local search, Step 2 and Step 3. Rapid initial algorithms need to be designed which provide better solutions.
- In Table 3 it can be seen how the intensification (in tabu search) of the strategy proposed for the economic objective function reduces costs by more than 6,300 Pesetas; while the basic tabu search only reduces costs by little more than 1,000. Equally, in Table 4, this intensification reduces the cost by more than 4,600, while the basic tabu search only reduces costs by about 2,300. In other words, the intensification phase designed is more efficient than basic tabu search when minimizing the cost function.
- On the other hand, in Table 4, the intensification phase proposed for the social objective only achieves a reduction of 1 minute in total, while basic tabu search achieves a reduction of 23. In other words, when attempting to minimize the social objective this intensification phase is not efficient. Perhaps other more classic intensification and/or diversification strategies should be tried in this case.
- Finally, it can be seen in Table 4 that PHASE B, which attempts to reduce costs, also reduces the duration of the longest routes by 4 minutes. This suggests modifying or widening the guide function Δ proposed in § 3.1. Regarding this, it should be pointed out that some initial tests are carried out which consist in adding a second guide function $\Delta'(s') = tpo(r1) + tpo(r2) - tpo(r1') - tpo(r2')$, in Ta_Local_Search, such that a move s' can be acceptable if $\Delta(s') > 0 \vee \{\Delta(s') = 0 \wedge \Delta'(s') > 0\}$.

6 Reflections

With regard to the above results, one must take into account that the real duration of current routes, which are those which have been used in this work, is in many cases greater than that initially foreseen by the authorities. This is clear from the recent complaints and protests of the Parents' Associations.

However, for the execution of the algorithms proposed, we have assumed fairly moderate speeds for each type of road; 80 km/h for motorways and dual carriageways, 70 for main roads, 60 for first class local roads, 55 for second class local roads, 50 for third class local roads, 40 for unpaved roads, 30 for link roads and 20 for crossings. Obviously these speeds can in reality be reached easily and even exceeded. The important thing is that the solutions planned in theory can in fact be put into practice with some margin (the opposite of what happens in some solutions used at present).

Acknowledgement

We are extremely grateful to the people responsible for transport in the Delegación Provincial del Ministerio de Educación y Ciencia de Burgos, for the information supplied.

Bibliography

Bentley, J.L. (1992). Fast algorithms for geometric traveling salesman problems. *ORSA Journal on Computing 4*, 387–411.

Bodin, L.D. and B.L. Golden (1981). Classification in vehicle routing and scheduling. *Networks 11*, 97–108.

Bullnheimer, B., R.F. Hartl, and C. Strauss (1999). Applying the ant system to the vehicle routing problem. In S. Voß, S. Martello, I.H. Osman, and C. Roucairol (Eds.), *Meta-Heuristics: Advances and Trends in Local Search Paradigms for Optimization*, Kluwer, Boston, 285–296.

Campos, V. and E. Mota (2000). Heuristic procedures for the capacitated vehicle routing problem. *Computational Optimization and Applications 16*, 265–277.

de Backer, B., V. Furnon, P. Shaw, P. Kilby, and P. Prosser (2000). Solving vehicle routing problems using constraint programming and metaheuristics. *Journal of Heuristics 6*, 501–523.

Desaulniers, G., J. Desrosiers, M.M. Solomon, and F. Soumis (1999). *The VRP with time windows*. Les Cahiers du GERAD G-99-13, GERAD, Montréal, Canada.

Desrochers, M., J.K. Lenstra, M.W.P. Savelsbergh, and F. Soumis (1988). Vehicle routing with time windows: Optimization and approximation. In B.L. Golden and A.A. Assad (Eds.), *Vehicle Routing: Methods and Studies, Studies in Management Sciences and Systems*, 16, North-Holland, Amsterdam, 65–84.

Fisher, M.L. and R. Jaikumar (1981). A generalized assignment heuristic for vehicle routing. *Networks 11*, 109–124.

Gendreau, M., A. Hertz, and G. Laporte (1994). A tabu search heuristic for vehicle routing problem. *Management Science 101*, 1276–1290.

Glover, F. (1989). Tabu search: Part I. *ORSA Journal on Computing 1*, 190–206.

Glover, F. (1990). Tabu search: Part II. *ORSA Journal on Computing 2*, 4–32.

Glover, F. (1996). Búsqueda tabú. In A. Diaz (Ed.), *Optimización Heurística y Redes Neuronales*, Paraninfo, Madrid, 105–142.

Glover, F. and M. Laguna (1993). Tabu search. In C. Reeves (Ed.), *Modern Heuristic Techniques for Combinatorial Problems*, Blackwell, Oxford, 70–150.

Glover, F. and M. Laguna (1997). *Tabu Search*. Kluwer, Boston.

Glover, F. and M. Laguna (2001). Tabu search. In P.M. Paradalos and M.G.S. Resende (Eds.), *Handbook of Applied Optimization*, Oxford Academic Press. To appear.

Haouari, M., P. Dejax, and M. Desrochers (1990). Les problems de tournées avec contraintes des fenêtres de temps: L'etat de l'art. *Recherche Operationnelle/Operations Research 24*, 217–244.

Kilby, P., P. Prosser, and P. Shaw (1999). Guided local search for the vehicle routing problem with time windows. In S. Voß, S. Martello, I.H. Osman, and C. Roucairol (Eds.), *Meta-Heuristics: Advances and Trends in Local Search Paradigms for Optimization*, Kluwer, Boston, 473–486.

Kontoravdis, G. and J.F. Bard (1995). A grasp for the vehicle routing problem with time windows. *ORSA Journal on Computing 7*, 10–23.

Laporte, G. (1992). The vehicle routing problem: An overview of exact and approximate algorithms. *European Journal of Operational Research 59*, 345–358.

Laporte, G., M. Desrochers, and Y. Nobert (1984). Two exact algorithms for the distance-constrained vehicle routing problem. *Networks 14*, 161–172.

Laporte, G., M. Gendreau, J.-Y. Potvin, and F. Semet (1999). *Classical and moderns heuristics for the vehicle routing problem*. Les Cahiers du GERAD G-99-21, GERAD, Montréal, Canada.

Laporte, G., Y. Nobert, and M. Desrochers (1985). Optimal routing under capacity and distance restrictions. *Operations Research 33*, 1050–1073.

Laporte, G. and I.H. Osman (1995). Routing problems: A bibliography. *Annals of Operations Research 61*, 227–262.

Laporte, G. and F. Semet (1998). *Classical heuristics for the vehicle routing problem*. Les Cahiers du GERAD G-98-52, GERAD, Montréal, Canada.

Lin, S. (1965). Computer solutions to the traveling salesman problem. *Bell Syst. Tech. Journal 44*, 2245–2269.

Lin, S. and B.W. Kernighan (1973). An effective heuristic algorithm for the traveling salesman problem. *Operations Research 20*, 498–516.

Nurmi, K. (1991). Traveling salesman problem tools for microcomputers. *Computers & Operations Research 18*, 741–749.

Or, I. (1976). *Traveling Salesman Type Combinatorial Problems and their Relations to the Logistics of Blood Banking*. Ph.D. thesis, Department of Industrial Engineering and Management Sciences, Northwestern University.

Osman, I.H. (1993). Metastrategy simulated annealing and tabu search algorithms for the vehicle routing problem. *Annals of Operations Research 41*, 421–451.

Pacheco, J., A. Aragón, and C. Delgado (1999). *Diseño de un sistema que facilite soluciones racionales al problema del Transporte Escolar en la provincia de Burgos*. XIII Reunión ASEPELT España, Burgos.

Pacheco, J., A. Aragón, and C. Delgado (2000). Diseño de algoritmos para el problema del transporte escolar. Aplicacin en la provincia de Burgos. *Qüestiio 24*, 55–82.

Pacheco, J. and C. Delgado (1997). Problemas de rutas con ventanas de tiempo y carga y descarga simultnea: Diseño de filtros para algoritmos de intercambio (caso de un sólo vehí culo). *Estudios de Economía Aplicada 7*, 79–100.

Pacheco, J. and C. Delgado (2000a). Diseño de metaheuristicos hibridos para problemas de rutas con flota heterogénea:. *Estudios de Economia Aplicada 14*, 137–151.

Pacheco, J. and C. Delgado (2000b). Resultados de diferentes experiencias con bsqueda local aplicados a problemas de rutas. *Rect@ Revista Electrónica de Comunicaciones y Trabajos de ASEPUMA 2*(1), 53–82.

Potvin, J.Y. and S. Bengio (1994). *A genetic approach to the vehicle routing problem with time windows.* Technical Report CRT953, Centre de Recherche sur les Transports, University of Montreal, Montréal, Canada.

Potvin, J.Y., T. Kervahut, B.L. Garcia, and J.M. Rousseau (1993). A tabu search heuristic for vehicle routing problem with time windows. *Management Science 40*, 1276–1290.

Rego, C. (1998). A subpath ejection method for the vehicle routing problem. *Management Science 44*, 1447–1459.

Rochat, Y. and E.D. Taillard (1995). Probabilistic diversification and intensification in local search for vehicle routing. *Journal of Heuristics 1*, 147–167.

Rosing, K.E. (1997). Heuristic concentration: An introduction with examples. In *The 10th Meeting of the European Chapter on Combinatorial Optimization, Tenerife, Spain.*

Rosing, K.E. and C.S. Revelle (1997). Heuristic concentration: Two stage solution construction. *European Journal of Operational Research 97*, 75–86.

Rosing, K.E., C.S. Revelle, E. Rolland, D.A. Schilling, and J.R. Current (1998). Heuristic concentration and tabu search: A head to head comparison. *European Journal of Operational Research 104*, 93–99.

Taillard, E., P. Badeau, M. Gendreu, F. Guertain, and J.Y. Potvin (1997). A tabu search heuristic for the vehicle routing problem with soft time windows. *Transportation Science 31*, 170–186.

Thangiah, S.R., I.H. Osman, and T. Sun (1994). *Hybrid genetic algorithm, simulated annealing, and tabu search methods for the vehicle routing problem with time windows.* Working paper UKC/OR94/4, Institute of Mathematics and Statistics, University of Kent, Canterbury, UK.

Thangiah, S.R., R. Vinayagamoorty, and A. Gubbi (1993). Vehicle routing with time deadlines using genetic and local algorithms. In S. Forrest (Ed.), *5th International Conference on Genetic Algorithms*, Morgan Kaufmann, San Mateo, CA, 506–513.

Voudouris, C. and E. Tsang (2001). *Guided Local Search (GLS) Project.* http://cswww.essex.ac.uk/Research/CSP/gls.html, 07.04.2001.

An Approach Towards the Integration of Bus Priority, Traffic Adaptive Signal Control, and Bus Information/Scheduling Systems

Pitu Mirchandani[1], Anna Knyazyan[1], Larry Head[2], and Wenji Wu[1]

[1] Systems and Industrial Engineering Department,
  The University of Arizona, Tucson, AZ 85721
  pitu@sie.arizona.edu, wenji@u.arizona.edu
[2] Siemens-Gardner Transportation Systems, Inc.,
  6375 E. Tanque Verde #170, Tucson, AZ 85750
  Larry.Head@gts.sea.siemens.com

Abstract. This paper addresses the integration of adaptive traffic signal control and bus priority. In bus priority, several possible approaches are used for giving more "weight" to the buses: *(1)* passive priority, when signal timings are set, ahead of time, so that buses incur less delays, *(2)* active priority, where buses are detected at approaches to the intersection and phase splits are adjusted to accommodate the bus, and *(3)* "optimization-based" priority where the current state of the system is estimated and the signals are changed as per active priority schemes. Our work is related to the last approach where the signals are set based on real-time optimization of the phasing that considers all the vehicles on the network, the passenger counts in the buses, and the schedule status of the buses.

The architecture for phase optimization is based on the *RHODES* *TM* traffic adaptive signal control system developed at the University of Arizona. *RHODES* *TM* (Real-time, Hierarchical, Optimized, Distributed, and Effective System) takes second-by-second data from loop-detectors at the intersections as input, and outputs the durations of the phases. Objectives for optimizing phase durations include, among others, "minimize average delay per vehicle."

When bus priority, referred to as *"BUSBAND,"* is introduced into *RHODES* *TM*, it is assumed that exact locations of buses are available in the network, as well as passenger counts through an advanced communication/information system. In this way, the bus is given a weight that depends on the number of passengers, and whether the bus is behind schedule. The *RHODES* *TM*/*BUSBAND* scheme was analyzed using a micro-simulation modeling package known as CORSIM.

RHODES *TM*, with and without bus priority, significantly increases average travel speeds and decreases total traffic delays as well as average and variance of bus delays. With *BUSBAND* there is an additional decrease in bus delays and passenger travel times with little effect on the rest of the traffic.

On-time performance of the buses will depend on how well the route travel times and ridership are estimated when the bus schedules are developed. Future efforts are planned on developing the bus schedules with consideration of ridership and traffic adaptive control.

1   Introduction

Traffic congestion and traffic signals cause significant delay and increase operating costs for bus service. *Signal priority* has been a promising method to improve bus operations and service quality, but it has not seen widespread deployment. On the other hand, *signal preemption*, which effectively gives a green phase to the "preempting vehicle" while ignoring other vehicles, has been used when the preempting vehicles are emergency response units at intersections or trains at grade crossings. Signal preemption for buses has been deployed in few places in the USA, but has more widespread applications in Europe and elsewhere. The resistance to implementation in the USA for signal priority/preemption for transit is based on a concern that overall traffic performance may be unduly compromised when signal timing intended to optimize traffic flow is traded off with the desire to provide a travel advantage to buses. A major objective of the research reported here was to investigate if in fact well-designed signal priority is detrimental to overall traffic performance.

Traditional traffic signal systems have had limited capabilities, resulting in simplistic bus priority strategies, such as extending the green phase. Recent advancements in the field of Intelligent Transportation Systems (ITS) have created new capabilities to support transit priority in traffic signal systems. These advancements cover a wide range of features including smart buses, detection, communications, control hardware, optimization algorithms, and simulation modeling.

Traffic signal priority for transit must contribute to the objectives set for operation of the transportation system. Some of these objectives (e.g., reducing emissions) will be attractive to both the transit agency and the agency responsible for the signal system. Other objectives (e.g., reducing bus-operating costs) will be principally attractive to the transit agency. Some objectives may be partially conflicting (e.g., reducing average delay of all vehicles, transit and passenger vehicles, and reducing total person delay). In designing transit priority for a traffic signal system, operating objectives must be determined and, if needed, a balance found between conflicting objectives.

Strategies for signal priority can be classified into two general categories (Sunkari et al. (1995)): *(1)* Passive Priority and *(2)* Active Priority. Real-time, or traffic adaptive priority is a third type strategy that has received some attention (Yagar (1993); Chang et al. (1995)). These strategies may be summarized as follows:

- **Passive Priority Strategies**
  - Adjustment of Cycle Length
  - Phase Splitting
  - Area-wide Timing Plans
  - Metering Priority
- **Active Priority Strategies**
  - Phase Extension
  - Early Phase Activation

  – Special Transit Phase (e.g., queue jump)
  – Phase Suppression (lift strategy)
  – Unconditional Priority
  – Conditional Priority
- **Real-time Priority Strategies**
  – Delay Optimizing
  – Intersection Control
  – Network Control

Passive priority strategies attempt to accommodate bus operations by considering factors such as bus link travel times in headway computations, reducing the cycle length to reduce delay, providing phase sequences designed to more frequently serve a phase that has high bus demand, or by providing bus by-pass at metering locations.

Active priority strategies require the ability to detect or identify buses at the signalized intersections. The most basic active strategies provide bus priority by either extending the current phase or activating a phase early (from a vehicle based control point of view). Other active strategies include the inclusion of special phases, such as a short bus passage phase, that are actuated by the detection of a bus, or phase suppression of non-main street phases with little or no demand.

Depending on the location and capability of bus sensors, active priority may be either conditional or unconditional. Unconditional strategies provide priority regardless of the status of a transit vehicle. Generally, the conditional decisions are made based on the schedule or headway adherence of an arriving vehicle. This assumes that the system, either a "smart vehicle" or a "smart controller," knows the operating status of an arriving transit vehicle.

Real-time strategies attempt to provide transit priority based on optimizing some performance criterion, primarily delay. Delay measures may include passenger delay, vehicle delay, weighted vehicle delay or some combination of these measures. Real-time priority strategies use actual observed vehicle (both passenger car and bus) arrivals as input to a traffic model that either evaluates several alternative timing plans to select a most favorable option, or optimizes the actual timing in terms of phase durations and phase sequences.

Other than a few applications of *SCOOT* (Hunt et al. (1981); Bretherton (1996)) and *SCATS* (Luk (1984); Cornwell et al. (1986)), no real-time traffic-adaptive signal control systems with bus priority are reported in the literature and only a few cities have bus priority capability. The *SCOOT* system is essentially an on-line version of the *TRANSYT* (Robertson (1969)) signal optimization algorithm. *SCOOT* considers either active strategies (phase extension and early phase activation) or, implicitly, passive strategies. Reported results indicate a 22% reduction in bus delay per intersection with as much as 70% reduction in light volumes. *SCATS* accomplishes traffic adaptive control by first dividing a large network into small zones, where definition of each zone is based on contiguous intersections having similar "degrees of saturation," and then choosing a common cycle time for each zone. Bus priority is accomplished through application of active priority strategies. Reported results state a 6-10% improvement in bus travel times with little significant effect on travel times of other vehicles.

The focus of our work is to consider bus priority within $RHODES^{TM}$, a traffic adaptive signal control system developed at the University of Arizona. $RHODES^{TM}$ (Real-time, Hierarchical, Optimized, Distributed, Effective System) takes as input second-by-second data from loop-detectors at the intersections, and outputs the durations of the phases (including the duration of zero if phase skipping is allowed).

When bus priority is introduced into $RHODES^{TM}$, it is assumed that exact location of buses is available in the network, (e.g., via detectors at bus stops, or through GPS), as well as passenger counts through an advanced communication/information system. In this way, the bus is given a weight that increases *(1)* with the number of passengers, and/or *(2)* when the bus is behind schedule. Likewise, when the bus is early and/or has very few passengers then its weight is decreased. We refer to this as the *"weighted bus"* priority approach and the corresponding $RHODES^{TM}$ component as *"BUSBAND"* since it attempts to provide an effective green band through the network for buses that are on time and utilized as expected.[1]

Another approach for providing bus priority is not to consider passenger counts in adjusting phase durations because one does not know if there are passengers waiting to board downstream. In this case, we need to minimize the likelihood that it will stop at the signals and the simplest approach is to provide a constraint from the *"network flow control"* logic of $RHODES^{TM}$ to try and get the appropriate phase for the given bus movement when it approaches an intersection; we refer to this as the *"phase constrained"* approach.

In the next section we provide a brief background on $RHODES^{TM}$. In Section 3 we discuss the integration of bus priority within $RHODES^{TM}$. Evaluation of $RHODES^{TM}/BUSBAND$ using simulation modeling is discussed in Section 4. Finally we summarize our findings and discuss areas of future research.

2 $RHODES^{TM}$

The current approaches to control traffic signals on arterials are *(1) fixed time*, perhaps based on time-of-day traffic conditions, and *(2) actuated* (or *semi-actuated*) where sensors on the road (e.g., loop detectors) detect traffic on specific lanes and/or movements and based on some programmed logic provide pre-specified phases, phase skips, phase extensions, force-offs and gap-outs to allow for the movement of the detected traffic. The major deficiency for such types of strategies is that there is no way for the control system to

---

[1] If the passenger counts are not available then $RHODES^{TM}$ can include an estimation algorithm that estimates passenger counts although this is currently not implemented in *BUSBAND*.

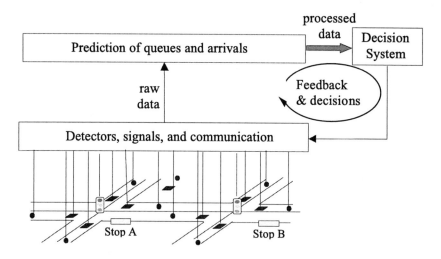

Figure 1. A simplified diagram of the *RHODES* ^{TM} architecture.

tradeoff or optimize signal settings to respond to anticipated arrival volumes – by varying phase durations and/or using more appropriate cycle times and phase sequencing – even though detectors may have identified unusual traffic conditions (either unusually large volumes or very small volumes, due to, e.g., events and incidents). *RHODES* ^{TM} traffic-adaptive signal control system (Head et al. (1992)) attempts to address this deficiency by processing the detector data in real-time and then setting phase durations to optimize a given measure of performance.

A simplified architectural diagram for *RHODES* ^{TM} is depicted in Figure 1. Basically, there are two main processes within *RHODES* ^{TM}: *(1) estimation and prediction* which takes the detector data and estimates the actual flow profiles in the network and the subsequent propagation of these flows, and *(2) decision system* where the phase durations are selected to optimize a given objective function, the optimization being based on dynamic programming and decision trees. Objectives that can be used are "minimize average delay per vehicle," "minimize average queues at the intersections," "minimize number of stops" and so on. In the computation of the objective function, each vehicle is given a weight, which increases when the vehicle is too long in queue if delays and queue lengths are considered in the objective function.

There are three aspects of the *RHODES* ^{TM} approach that make it an effective system to adaptively control traffic signals. First, it recognizes that recent technological advances in communication, control, and computation *(a)* make it possible to move data *quickly* from the street to the computing processors (even now most current systems have communication capabilities that are not utilized to their potential), *(b)* make processing of this data to algorithmically select optimal signal timings *fast*, and *(c)* allow the *flexibility* to implement through modern controllers a wide-variety of control strategies.

Second, *RHODES*[TM] recognizes that there are natural stochastic variations in the traffic flow and, therefore, one must *expect the data to stochastically vary* (by simply smoothing the data and working with mean values does not make the actual traffic that the system sees smooth and average, as assumed by some real-time traffic control schemes). And third, *RHODES*[TM] proactively responds to these variations by explicitly *predicting* individual vehicle arrivals, platoon arrivals and traffic flow rates.

Basically, the developed *RHODES*[TM]/*BUSBAND* system predicts arrivals and queues of individual vehicles at the arterial approaches, and tracks and predicts the movement of buses; and based on these predictions and a given criterion of performance determines the optimal phasing of the signals at the intersections.

2.1 Prediction of Vehicle Arrivals in *RHODES*[TM]

For proactive traffic control, it is important to predict vehicle arrivals, turning probabilities and queues at intersections, in order to compute phase timings that optimize a given measure of effectiveness (e.g., average delay). Consider an intersection with several approaches. Associated with each approach are several possible traffic movements: left turn, right turn and a through movement. Any non-conflicting combination of movements that can share the intersection at any one time can be assigned a signal phase that allows those movements protected use of the intersection. The *PREDICT* algorithm of Head (1995) uses the output of the detectors on the approach of each upstream intersection, together with information on the traffic state and planned phase timings for the upstream signals, to predict future arrivals at the intersection under *RHODES*[TM] control. Note that the *PREDICT* model is based on processing arrival data as it becomes available. At any point in time the predicted arrival flow pattern at a downstream detector accounts for vehicles that have already passed upstream detectors. The benefit of this vehicle-additive process of the predictor is that it constantly provides, for a given prediction horizon, *(1)* nearly complete information of anticipated vehicle arrivals in the very near future (of those vehicles that have already passed the upstream intersections) and *(2)* partial information of anticipated vehicles in remaining part of the prediction time horizon (of those vehicles that have not passed the upstream intersections, since some new vehicles may still arrive that will effect the delays in the prediction time horizon). Results of an evaluation study of the *PREDICT* algorithm for arrivals at an intersection have been reported by Head (1995).

2.2 Estimation of Parameters

To use the *PREDICT* model, several parameters (given in italics) need to be provided: *(1)* travel times on links (detector to detector) which depends on the *link free-flow speed* and current traffic volumes, *(2) queue discharge rates* which also depends on volumes (as well as on queue spillbacks and opposing- and cross-traffic volumes), and *(3) turning percentages.* In addition, to estimate arrivals and demand for various phases we also need to have *estimates of queues* at the intersections. Included in the *RHODES* ^{TM} system are algorithms to estimate these parameters (Mirchandani and Head (2001)).

2.3   Control Algorithms in *RHODES* ^{TM}

Fixed control strategies are based on a signal-timing plan defined in terms of operating parameters for traditional signal control, namely *cycle time, splits,* and *offsets.* These parameters are generally developed based on traffic studies and standard procedures, such as the Highway Capacity Manual, or signal timing software such as *TRANSYT* and *PASSER.* The traffic studies result in estimates of traffic conditions, link volumes and turning percentages, for specified time periods. Signal timing *parameters* are developed for each of these time periods and, typically, implemented on a time-of-day basis with no consideration of current actual traffic conditions. In many cases, even the use of standard procedures for the development of signal timing plans is abandoned and traffic engineers operate in a judgment-based fashion with moderate levels of success. None of these approaches is truly traffic-adaptive or even attempt to actually minimize some measure of traffic performance such as average vehicle-delay.

Most currently available traffic responsive systems attempt to address the problem of responding to actual traffic conditions by switching these *parametric* signal timing plans based on current wide-area traffic conditions rather than time of day. This requires that signal-timing parameters be developed for a variety of possible traffic conditions. Nevertheless, implicit in the usage of *parametric* timing plans is the assumption that for the next several minutes, or even hours, the traffic in the network can be well characterized by the measured *average* flows and parameters. No account is taken of the fact that the second-by-second and minute-by-minute variabilities of traffic are significant and plans based on averages produce unnecessary delays for some traffic movements when the traffic on conflicting movements is absent, or very small, during some periods.

The *RHODES* ^{TM} approach is to predict both the short-term and the medium term fluctuations of the traffic (in terms of individual vehicle arrivals and platoon movements, respectively), and explicitly set phases that maximize a given traffic performance measure. Note that we do not set timing plans in terms of cycle times, splits and offsets, but rather in terms of phase durations for any given phase sequence. (*RHODES* ^{TM} does not necessarily require a pre-specified phase sequence, but since many traffic engineers prefer a pre-specified sequence, *RHODES* ^{TM} has been developed to allow the traffic engineer to specify a desired sequence.) In other words, in the *RHODES* ^{TM} control strategy, the emphasis shifts from changing timing parameters in reacting to traffic conditions just observed to *proactively* setting phase durations for *predicted* traffic conditions.

In the simulation experiments of this project, only the intersection control level of $RHODES\ ^{TM}$ was used which utilizes a dynamic-programming algorithm formulation of Sen and Head (1997); for description of traffic control at other levels see Mirchandani and Head (2001). There are other signal timing schemes which have been experimented that do not provide parametric timing plans but instead provide phase durations, notably $OPAC$ (Gartner (1983); Gartner et al. (1991)) and $PRODYN$ (Khoudour et al. (1991)) and $UTOPIA$ (Mauro and Di Taranto (1990)). In some ways, these also use dynamic programming or related optimization schemes, but, in their current implementations, the underlying models are more approximate and the systems are not truly real-time traffic responsive. Also, there are no results available in the literature on their application to transit priority.

3   Implementation of Bus Priority in $RHODES\ ^{TM}$

As we mentioned earlier, in the computation of the objective function value for the $RHODES\ ^{TM}$ dynamic program, each vehicle is treated alike. That is, they all have a "weight" of unity. Hence, $RHODES\ ^{TM}$ gives green phase to the movement which has more "delay" associated with it, where this delay could depend on the number of vehicles needing this movement and the time in queue for these vehicles. In the standard $RHODES\ ^{TM}$ algorithms, a bus is also given a unity weight regardless of the number of passengers in it and whether or not it is late.

On the other hand, since the $RHODES\ ^{TM}$ algorithms give individual weights to vehicles, it is not hard to imagine modifying $RHODES\ ^{TM}$ that explicitly provides additional consideration or weight to detected buses. The simplest approach is to provide a constraint from the network flow control logic of $RHODES\ ^{TM}$ to try and get the appropriate phase for the given bus movement when it approaches the intersection – we refer to this as the "*phase constrained*" approach. The other approach is to give each bus a variable weight that depends on the number of passengers it has and on how late it is, if it is behind schedule – we refer to this as the "*weighted bus*" approach. Here we let n_i be the number of passengers on bus i, and its "lateness" be denoted by d_i, which is negative when the bus is early, and positive when it is late. Then weight w_i for bus i given to $RHODES\ ^{TM}$ is defined by the function $w_i = n_i(1 + f_i)$, with delay factor,

$$f_i = \begin{cases} 0 & \text{if lateness } d_i \leq 0 \\ Kd_i & \text{if lateness } d_i > 0 \end{cases}$$

where K is some constant.

Notice that when the bus is early or "on time" then we count only the number of passengers. This implicitly assumes that each car has a single passenger and that a bus with n passengers has n times the weight of a car. Clearly we could divide the n_i number in the weight function by the average occupancy of a car if it is greater than one. Also, the above weight function becomes zero when there are no passengers on the bus. This implies that only current passengers on the bus are being considered in the objective function. With the inclusion of the bus passengers and the bus lateness in the computation of the objective function value, the $RHODES^{TM}$ strategy will tend to give higher priority for late buses with many passengers. We note that this weight function could be modified easily to account for expected delays of anticipated users downstream. These anticipated users could be forecasted given historical ridership data, or better estimated if real-time passenger information is being obtained from downstream bus stops.

Whether we use the *"phase constrained"* or the *"weighted bus"* approach we refer to our modified $RHODES^{TM}$ strategy as the $RHODES^{TM}/ BUS$-$BAND$ strategy. In the evaluation of the $RHODES^{TM}/BUSBAND$ strategy, one would expect that *(a)* the average delay of the buses would decrease, and *(b)* the average delay of all passengers (in buses and cars) in the network would also decrease, with, perhaps, car passengers incurring some additional delays. Given the real-time interactions between the signal phasing decisions and vehicle movements, the diversity of the traffic scenario, and the rich variety of driver behavior, we chose to evaluate $RHODES^{TM}/BUSBAND$ with a micro-simulation model, using different control strategies, including standard actuated control, $RHODES^{TM}$ without bus priority, and $RHODES^{TM}/BUS$-$BAND$.

4 Evaluation Using a Simulation Platform

It is clear that any type of real-time traffic control algorithm needs to be tested in the "laboratory" before it is implemented and evaluated in the field. The most appropriate method to do this "laboratory" testing is to *(1)* have a realistic simulation model of traffic flow at an intersection, *(2)* emulate the (loop) detection of the traffic flow, and *(3)* observe the resulting changes that would come about if the algorithm was implemented in place of the current control system.

The ability to represent dynamic recurrent and non-recurrent congestion, as well as other non-congested traffic conditions, is needed for measuring the algorithm's capability to respond to real-time traffic conditions. Also, simulation models used for testing must provide the same surveillance and detection information as that available in the field. The frequency of surveillance and detector system output and the frequency of the signal control input dictate the minimal resolution, and hence the responsiveness, of the signal control logic. The simulation model must be able to represent rates that will be achievable when the control logic is implemented for field-testing.

The simulation model requirements from a development and testing perspective differ from the requirements for performance evaluation. Clearly the most important requirement of a simulation model is that it accurately represents the dynamics of traffic flow and its response to dynamic signal control. This requirement dictates that the simulation model chosen for development and testing not be based on a macroscopic flow model that assumes constant cycle length and deterministic traffic flow characteristics. Rather, the model should include microscopic flow characteristics, such as car following and overtaking, and include an ability to simulate real-time traffic controls (not necessarily constant cycle lengths) and attendant vehicle response to actual traffic signals.

During the development and testing phase it is essential to have access to both traffic and signal control variables so that detailed behavior can be studied. One may distinguish between traffic simulation information/data that is needed for validation and testing and that information/data which is available as traffic surveillance/detection data for the signal control algorithms. For example, for the purpose of testing a traffic model used in an optimization routine, it may be desirable to compare the traffic model's state-of-the-traffic measures, such as queue length, to the corresponding measures in the simulation model. This form of testing requires that the traffic simulation model provides accurate measurements of queue lengths despite the fact the existing traffic surveillance technology may not provide this information.

Another important consideration is the frequency at which required testing data is available. For example, the average queue length for a simulation period is insufficient for testing a routine that estimates real-time queue lengths. This information must be available as frequently as possible, at least as frequent as estimated queues are generated.

Another major consideration in the selection and/or development of a simulation model is from the realization that we need to code the *BUSBAND* algorithms and test it via a simulation model. That is, we need to represent, identify and monitor buses in the simulation, and either track them throughout the region or detect them at specific points (e.g., at bus stops) to measure bus movement performance.

Based on the requirements and considerations discussed above, we developed a *CORSIM*-based[2] simulation model to implement and test our *RHODES TM/BUSBAND* approach. Fixed-time, semi-actuated and actuated signal control strategies (internal to *CORSIM*) were implemented and animations were observed to confirm if the traffic was indeed moving appropriately. Having fine-tuned the actuated timing parameters within the simulation model so that traffic performance was as good as can be expected, *RHODES TM/BUSBAND* was interfaced with the simulation model and evaluated.

An essential element of external real-time signal control logic is the traffic surveillance system. In our simulation experiments, we utilized the internal surveillance detector logic of *CORSIM* for the placement and processing of detector events, but we utilized an external *RHODES TM/BUSBAND* control logic for processing this detector data. Methods internal to *CORSIM* allowed

---

[2] *CORSIM* is a software package for modeling and simulating traffic on a network. It has been developed by the Federal Highway Administration of the U.S. Department of Transportation (FHWA).

us to estimate any necessary traffic parameters such as travel times, delays, and queue lengths, in addition to the standard count and occupancy values that were used by the external control logic.

4.1   "Phase-constrained" *RHODES* ^{TM}*/BUSBAND*

We simulated a subnetwork (an arterial with several intersections) within the City of Tucson for the first set of tests using the simpler *phase-constrained BUSBAND* logic. Effectively, this implies that the weight of each bus is set high so that *RHODES* ^{TM} tries and gives it a green phase when it approaches the intersection and the lateness and the number of passengers on the bus does not change priorities. See Knyazyan (1998) for details of the simulations. In these tests we generated buses on the *main street* at given times and, using the performance measures internally generated by *CORSIM*, we compared the following measures for the three scenarios *(1)* SAC: semi-actuated control within *CORSIM*, *(2)* *RHODES* ^{TM} without bus priority, and *(3)* RHODES-BP: *RHODES* ^{TM} with bus priority:

- Bus travel times and car travel times
- Bus delays at intersections and car delays at intersections
- Total person delays

The simulation results are summarized in Table 1. First it can be observed that *RHODES* ^{TM} with or without bus priority significantly reduced travel times and delays at intersections for all vehicles over SAC. As would be expected, inclusion of bus priority on the main street further decreased travel times and delays for the main street, but with a slight increase in these measures in the cross streets which did not have buses on them. This is to be expected, but one should realize that cross street delays are still substantially lower than the delays in the SAC baseline case.

Person delays that included passengers on buses as well as cars are also given in Table 1. Again, note the significant reductions with *RHODES* ^{TM} (with and without bus priority) as would be expected with real-time traffic-adaptive control. Comparing the two *RHODES* ^{TM} scenarios, the person delays were not changed much with bus priority on the main street. On the other hand, person delay was slightly increased on the cross streets when bus priority was used on the main street. This is to be expected because *RHODES* ^{TM}*/BUSBAND* provides bus priority to main street at the expense of some delays for the cross street vehicles. Again, these delays are much less than the baseline SAC case.

One also needs to compare *RHODES* ^{TM} bus delays and travel times, with and without bus priority, to see the effect of bus priority. Observe again that *RHODES* ^{TM} improved these measures over SAC. And, as would be expected because of its design, *RHODES* ^{TM} with bus priority shows considerable further improvement for bus delays and travel times.

	SAC	RHODES	RHODES-BP
All vehicles			
Average travel time on main street link (sec)	60.95	56.11	55.79
% reduction over SAC		7.94%	8.47%
Average travel time on cross street link (sec)	80.11	60.62	63.11
% reduction over SAC		24.33%	21.22%
Average intersection delay on main street (sec)	26.99	22.15	21.83
% reduction over SAC		17.90%	19.13%
Average intersection delay on cross street (sec)	47.42	27.95	30.43
% reduction over SAC		41.06%	35.82%
All passengers			
Total person-delay on main street (person-min)	1259	1030	1026
% reduction over SAC		17.90%	19.13%
Total person-delay on cross street (person-min)	698	444	500
% reduction over SAC		41.06%	35.82%
All buses			
Total travel time for all buses (minutes)	45.62	43.23	39.67
% reduction over SAC		5.24%	13.04%
Total stop delay for all buses (minutes)	30.75	28.29	25.29
% reduction over SAC		8.00%	17.76%

Table 1. Summary of results from simulation experiments with phase-constrained $RHODES^{TM}/BUSBAND$.

4.2 "Weighted bus" $RHODES^{TM}/BUSBAND$

In the second set of tests, we used a simulation model that is being developed for an FHWA-sponsored field-test of the $RHODES^{TM}$ traffic control strategy on an arterial with some cross streets in a Seattle suburb. The model is being developed by a contractor for FHWA and is based on real data. In our simulation experiments the scenario consisted of a single $RHODES^{TM}$ controlled intersection and for all practical purposes can be thought of as an isolated intersection being fed by streams of realistic car and bus streams.

For baseline conditions against which to evaluate $RHODES^{TM}/BUS$-$BAND$ we generated several buses at a bus stop upstream of the intersection, say Stop A. For each bus, we generated a "lateness" which was positive if the bus was late and negative if it was early; the distribution we used was from a uniform distribution with range [-30s, +30s]. We also generated a passenger count, from a uniform distribution with range [0, 30]. The baseline case was standard actuated control (SAC). At a bus stop downstream from the intersection, say Stop B (illustrated in Figure 1), we measured arrival times. Assuming in the baseline situation that, on the average, some buses arrive early, some arrive on time and some late, we let the average of these arrival times correspond to zero delay. Hence we added a fixed travel time component to the average arrival time at upstream Bus Stop A so that this holds and the average scheduled delay is zero. This same travel time component was used for the corresponding case with $RHODES\,^{TM}$ traffic control. Two cross street traffic volumes were used for the baseline case (SAC), while the main street volume was kept constant at 1074 vehicles per hour. Buses were generated upstream from Stop A. We compared $RHODES\,^{TM}$ with no bus priority and $RHODES\,^{TM}/BUSBAND$ (RHODES-BP) with these baseline conditions. Table 2 summarize the results.

For low cross street traffic volumes, totaling 550 vehicles per hour in both directions, $RHODES\,^{TM}$ reduced average travel times and intersection delays over SAC, as expected. $RHODES\,^{TM}/BUSBAND$ further reduced average bus delays over $RHODES\,^{TM}$ (without bus priority), but only slightly. For high cross street volumes, with demand of 1100 vehicles/hour, $RHODES\,^{TM}$, with and without bus priority, significantly increased travel speeds and reduced delays for cross street traffic, while decreasing bus delays on the main street (Table 2).

An impressive result in these experiments is the significant reduction in the variance of the bus delays at the downstream bus stop when $RHODES\,^{TM}$ is implemented. For example, the standard deviation for the delay decreased from 23.17s (SAC) to 17.4 ($RHODES\,^{TM}$ with no bus priority) and 17.04s (RHODES-BP) at high cross street volumes.

When passenger count is included, $RHODES^{TM}/BUSBAND$ decreased average passenger travel time over SAC and $RHODES\,^{TM}$ with no bus priority. For example, at high cross street volume, average bus passenger travel time decreased from 80.3 seconds to 75.5 seconds, while the passenger car speeds and delays on the cross street either were unaffected or slightly decreased.

5   Conclusions

This paper introduced a new approach to bus priority using the $RHODES\,^{TM}$ traffic adaptive logic. Simulation results showed that $RHODES\,^{TM}$ significantly increases average travel speeds and decreases the total traffic delays, for both passenger vehicles and buses. In addition, average and variance of bus delays are decreased with $RHODES\,^{TM}$. Since a major objective for a bus transit system is to reduce passenger delays by providing on-time performance, then $RHODES\,^{TM}$ bus priority logic, $BUSBAND$ results in some additional decrease in bus delays and passenger travel times with little effect on the rest of the traffic.

One can assume that if $RHODES^{TM}/BUSBAND$ has been implemented in the city, the overall standard deviation in bus travel times have been reduced. Reduction in standard deviation addresses another objective for a bus transit system: to improve schedule reliability and hence increase the bus ridership.

	SAC	RHODES	RHODES-BP
All vehicles - low cross street volume			
Average travel time from Stop A to Stop B (sec)	73.69	73.89	71.79
% reduction over SAC		-0.27%	2.58%
Std. Dev. travel time from Stop A to Stop B (sec)	9.4	8.1	7.7
% reduction over SAC		13.83%	18.09%
Average speed on cross street link (mph)	11.1	14.6	14.2
% increase over SAC		31.53%	27.93%
Average delay on cross street link (sec)	29.75	18.25	19.5
% reduction over SAC		38.66%	34.45%
All vehicles - high cross street volume			
Average travel time from Stop A to Stop B (sec)	80.65	80.30	76.46
% reduction over SAC		0.43%	5.2%
Std. Dev. travel time from Stop A to Stop B (sec)	13.19	10.32	9.5
% reduction over SAC		21.76%	27.98%
Average speed on cross street link (mph)	9.9	15	14.9
% increase over SAC		51.52%	50.51%
Average delay on cross street link (sec)	33.8	16.5	16.7
% reduction over SAC		51.18%	50.59%
Bus delays - low cross street volume			
Average passenger travel time from A to B (sec)	73.69	73.40	70.30
% reduction over SAC		0.39%	4.6%
Std. Deviation of bus delay at Stop B (sec)	21.55	19.41	18.51
% reduction over SAC		9.93%	14.1%
Bus delays - high cross street volume			
Average passenger travel time from A to B (sec)	80.65	80.30	75.50
% reduction over SAC		0.43%	6.39%
Std. Deviation of bus delay at Stop B (sec)	23.17	17.4	17.04
% reduction over SAC		24.9%	26.46%

Table 2. Summary of results from simulation experiments with weighted bus *RHODES ^{TM}/BUSBAND*.

Also, if *RHODES^{TM}/BUSBAND* has been implemented in the city, then the corresponding travel times used in the baseline case are no longer valid in the bus schedule. In that case, we need to examine the effects of decreased travel times on the schedule and the utilization of buses. First, given lower travel times, it can be argued that fewer buses will be used in the development of the bus schedule for the given level of service. This will allow the transit system to be profitable and/or economical.

Second, now the ($RHODES\,^{TM}$) baseline delay at downstream bus stop will average to zero since the bus schedule will include the consideration of lower travel times. Furthermore, if arrival times at bus stops are available using $RHODES\,^{TM}$ processed data ($RHODES\,^{TM}$ includes bus travel time predictions in its algorithmic logic) through an advanced traveler information systems, then this will also increase the desirability to use the transit system and hence increase transit system ridership and profitability.

Bibliography

Bretherton, D. (1996). *SCOOT Current Developments: Version 3*. Technical Report no. 960128, Transportation Research Board, 75th Annual Meeting, Washington, DC.

Chang, G.-L., M. Vasudevan, and C.-C. Su (1995). Bus-preemption under adaptive signal control environments. *Transportation Research Record 1494*, 146–154.

Cornwell, P.R., J.Y.K. Luk, and B.J. Negus (1986). Tram priority in SCATS. *Traffic Engineering and Control 27*, 561–565.

Gartner, N.H. (1983). OPAC: A demand-responsive strategy for traffic signal control. *Transportation Research Record 906*, 75–81.

Gartner, N.H., P.J. Tarnoff, and C.M. Andrews (1991). Evaluation of the optimized policies for adaptive control (OPAC) strategy. *Transportation Research Record 1324*, 105–114.

Head, K.L. (1995). An event-based short-term traffic flow prediction model. *Transportation Research Record 1510*, 45–52.

Head, K.L., P.B. Mirchandani, and D. Sheppard (1992). Hierarchical framework for real-time traffic control. *Transportation Research Record 1360*, 82–88.

Hunt, P.B., D.I. Robertson, R.D. Bretherton, and R.I. Winton (1981). *SCOOT: A Traffic Responsive Method of Coordinating Signals*. Lr 253, Transport and Road Research Laboratory, Crowthorne, Berkshire, U.K.

Khoudour, L., J.-B. Lesort, and J.-L. Farges (1991). PRODYN: Three years of trials in the ZELT experimental zone. *Recherche - Transports - Sécurité, English Issue 6*, 89–98.

Knyazyan, A. (1998). Application of RHODES to Provide Transit Priority. Master's thesis, Systems and Industrial Engineering Department, University of Arizona, Arizona, USA.

Luk, J.Y.K. (1984). Two traffic-responsive area traffic control methods: SCAT and SCOOT. *Traffic Engineering and Control 25*, 14–19.

Mauro, V. and D. Di Taranto (Eds.) (1990). *UTOPIA: Proceedings of the 6th IFAC/IFIP/IFORS Symposium on Control and Communication in Transportation, Paris, France*. Pergamon Press, Oxford.

Mirchandani, P.B. and K.L. Head (2001). *RHODES: A Real-Time Traffic Signal Control System: Architecture. Algorithms, and Analysis.* To appear in Transportation Research C.

Robertson, D.I. (1969). *TRANSYT: A Traffic Network Study Tool.* Lr 253, Transport and Road Research Laboratory, Crowthorne, Berkshire, U.K.

Sen, S. and K.L. Head (1997). Controlled optimization of phases at an intersection. *Transportation Science 31*, 5–17.

Sunkari, S.R., P.S. Beasley, T. Urbanik, and D.B. Fambro (1995). Model to evaluate the impacts of bus priority on signalized intersections. *Transportation Research Record 1494*, 117–123.

Yagar, S. (1993). Efficient transit priority at intersections. *Transportation Research Record 1390*, 10–15.

An Optimal Integrated Real-time Disruption Control Model for Rail Transit Systems

Su Shen and Nigel H.M. Wilson

Massachusetts Institute of Technology, Department of Civil and Environmental Engineering, Cambridge, MA 02139, U.S.A.
{sshen,nhmw}@mit.edu

Abstract. Rail transit systems are subject to frequent minor disruptions caused by random disturbances. Although these minor disruptions usually last no longer than 10-20 minutes, they can degrade the level of service significantly on a short headway service. This paper describes an integrated real-time disruption control model, formulated as a mixed integer program, for rail transit systems, which includes holding, expressing and short-turning strategies. Although the model is capable of dealing with multi-branch systems, the model was applied to a disruption scenario on a single line system as a first step. Two cases with 10 and 20 minute disruption duration are tested. The results showed that holding strategies combined with short-turning strategies reduced the mean passenger waiting time by 35% in the former case and 57% in the latter case, compared with not applying any control strategies. Expressing provided only modest additional benefits. Sensitivity analysis was used to investigate the impact of the deterministic disruption duration assumption. The results showed that holding and expressing solutions were fairly robust, but the effectiveness of short-turning solutions was quite sensitive to the accuracy of the disruption duration estimate. In one scenario, the passenger waiting time increased by 14% under an estimate 50% less than the actual disruption duration, compared with the result with correct estimate. Problem instances without expressing can be solved in less than 30 seconds of computation time with the branch-and-bound algorithm proposed to solve this mixed integer problem.

1 Introduction

Minor disruptions lasting up to 10-20 minutes occur frequently in rail transit systems due to random disturbances such as medical emergencies, car problems or signal malfunctions. To reduce the impact of these disruptions on passengers, transit agencies often employ real-time control strategies such as holding, expressing and short turning. In the holding strategy, trains ahead or downstream of the disruption are held after normal passenger boarding and alighting, and, therefore, the long headways created by the disruption can be reduced and the capacity of the held trains may be used to reduce the crowding on trains behind or upstream of the disruption. In the expressing strategy, trains behind the disruption may skip a few stations after the disruption is cleared. Hence, long headways ahead of the express trains are reduced. If a disruption is serious, the first few trains may be full after picking up passengers right after the disruption clearance. It may make sense to express these trains to avoid unnecessary delays and reduce passenger waiting time beyond the express segment. In the short-turning strategy, trains are short-turned from the reverse direction into the blocked direction ahead of the disruption. Short-turning can significantly reduce long headways. However, it is obviously restricted by the availability of crossover tracks in rail transit service.

The effectiveness of these control strategies relies upon a bird's-eye-view of the whole system. Unfortunately, it is difficult for human dispatchers to assess situations and make good decisions in real-time, even with the aid of advanced information technologies such as automatic vehicle location (AVL) and automatic vehicle monitoring (AVM) systems. On the other hand, with the rich real-time information from AVL or AVM systems, computer based decision support systems can help dispatchers to make better control decisions.

Barnett (1974) considered a simplified transit line with two terminals and one control point. A simple two-point discrete distribution is used to approximate the distribution of the vehicle arrival headways. The control scheme was holding the vehicles at the control point, and the decision variable was the optimal holding time. Turnquist (1989) discussed the recovery of schedule deviations with real-time vehicle location information. The strategy considered was controlling vehicle speed on multiple segments. Van Breusegem et al. (1991) designed a discrete-event traffic model and state feedback control algorithms to improve system stability. Li (1994) conducted research on the real-time bus dispatching problem with the terminus being the only control point. With different levels of real-time bus location information, the arrival times of buses at the terminus were estimated and dispatching time and route operating pattern selected to minimize passenger-waiting time. Furth (1995) considered using holding strategies to minimize the sum of passenger waiting time and delay after a delay occurred. The decision variables were the number of trains over which to spread the delay and the delay time for each train. Murata and Goodman (1998) used recursive optimization to select holding time and minimize passenger inconvenience, which is a function of the difference between actual waiting time and the expected waiting time.

Eberlein et al. (1999) presented research on real-time routine control problems with holding and expressing strategies. A deterministic model was developed to describe single-loop transit systems, which provided a good foundation for further research on this problem. O'Dell and Wilson (1999) presented formulations for disruption control problems with holding and short-turning strategies for systems with more than one branch. That model was applied to two scenarios on the Massachusetts Bay Transportation Authority (MBTA) Red Line, and solved with CPLEX 3.0 on a Sun SPARC 20 workstation

(CPLEX (1995)). Most problem instances took less than 30 seconds to obtain the optimal solution, making it feasible for real-time control decisions.

This paper describes an integrated model to deal with minor disruptions. Integrated is used because the model includes holding, expressing and short-turning strategies together. The objective is to minimize the sum of total platform passenger waiting time and weighted in-vehicle passenger delay. Although the model is capable of dealing with multi-branch systems, it was applied to a disruption scenario on a single line system as a first step. Two cases with 10 and 20 minute disruption duration are tested. We assume that the disruption duration is known when applying the formulation. The mixed integer formulation is solved with the CPLEX 4.0 MIP Solver on a Micron Pentium II 300MHz PC with 64M RAM using a branch and bound algorithm. All problem instances except the two with expressing in the 20 minute disruption case can be solved in less than 30 seconds. Sensitivity of different control strategies to the accuracy of the disruption duration estimate is then investigated.

2 Model Description

In this section, the assumptions, features, and data requirements of the model are first described (Shen (2000)). Modeling methods when the train order can vary and several other important concepts are introduced. The model is then described in detail. Since the original model is nonlinear, linearization of the model is also discussed.

2.1 Assumptions, Features and Data Requirements

The following assumptions are made:

- Passenger arrival rates and alighting fractions are deterministic, station-specific constants, which is realistic for short-headway transit service in which passenger arrivals are independent of train arrivals.
- Running times between stations are approximated by their expected values. According to empirical data, the coefficients of variation are usually not large in rail transit systems.
- The short-turning time is also approximated by its expected value, although its coefficient of variation could be large.
- The safe separation between trains is ensured by imposing minimum train departure-arrival intervals at each station.
- Dwell time is approximated by a piecewise linear function with respect to the number of passengers boarding and alighting and the crowding condition on the train. Although the dwell time function is expected to be nonlinear, the data collected shows linearity with respect to the number of passengers boarding and alighting.

- The duration of the disruption must be estimated to apply the model and it is assumed to be constant. This assumption is fairly strong. In practice, when a disruption occurs, its duration is usually unknown. This assumption is investigated later through sensitivity analysis.

The model has the following features:

- The transit line can have more than one branch.
- The capacity of trains is considered. Passengers will be left behind if a train is full.
- Trains can be short-turned at one, or more, locations, based upon the availability of crossover tracks.
- Many types of disruption can be considered, including a temporary closure or speed restriction on a track section.

The following input data are required:

- Track configuration, including branches and available crossover tracks.
- Train schedule.
- Passenger arrival rate and alighting fraction at each station for the time period of interest.
- Minimum train departure-arrival interval at each station, which may be obtained from the train control system specification or from simulation experiments.
- Running time between stations, which includes acceleration and deceleration time.
- Parameters for the dwell time function.
- Train capacity.
- Last station departure time for trains when the disruption occurs, which can be obtained from AVL or AVM data.
- Disruption location and duration estimate.
- Short-turning time for each possible crossover track.

2.2 Basic Concepts

Notation:

$a_{i,k}$ arrival time of train i at station k
$ah_{i,k}$ arrival headway for train i at station k
$d_{i,k}$ departure time for train i at station k
$dh_{i,k}$ departure headway for train i at station k
$dw_{i,k}$ dwell time of train i at station k
$h_{i,k}$ maximum platform waiting time for train i at station k
$ht_{i,k}$ holding time of train i at station k
H_k minimum separation at station k

Headway: Headway is a common concept used to represent the interval between two successive train arrivals or departures and is represented by $ah_{i,k}$ or $dh_{i,k}$ in Figure 1.

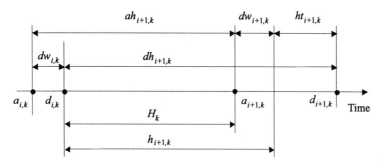

Figure 1. Headway, platform and in-vehicle waiting time, and minimum train separation at a station.

Maximum Platform Passenger Waiting Time and In-Vehicle Passenger Delay: Maximum platform passenger waiting time is defined to be the time from the departure of a train to the time when the next train has finished loading passengers and is ready to go, denoted by $h_{i+1,k}$ in Figure 1. As shown in Figure 1, dwell time, $dw_{i+1,k}$, is taken as part of platform waiting time. If a train is held as part of the control strategy, the time between the end of platform waiting time and actual departure time is taken as in-vehicle passenger delay, denoted by $ht_{i+1,k}$ in Figure 1. Note that in our definition, the dwell time and the holding time do not overlap and that the sum of the dwell time and the holding time is the time period the train is stopped at a platform.

Minimum Separation: Minimum separation at a station is the minimum train arrival-departure interval at each station, denoted by H_k in Figure 1.

2.3 Control Set and Impact Set

The control set is the set of trains and stations for which we might apply control actions. Depending on the disruption duration, operating headway, short-turning time and control system, the control set may consist of 2-4 trains ahead of the blockage to consider for holding, 1-2 trains to consider for expressing and 1-3 trains to consider for short-turning. For several reasons, it is desirable to limit the size of the control set. First of all, it is unrealistic to project train movements far into the future with any deterministic model due to system stochasticity. Instead, the formulation can be solved repeatedly with updated real-time information. Secondly, the marginal benefit from controlling trains decreases with the number of trains controlled (O'Dell and Wilson (1999)). Thirdly, the impact of the disruption will presumably be diminished down the line with proper control actions. Finally, larger numbers of trains and stations in the control set will increase computation time, while the small benefit from considering controlling additional trains and stations may not justify the loss from a less responsive decision making process.

The impact set is the set of trains and stations that are affected by the disruption and control strategies. The impact set obviously must be at least as large as the control set. Several factors influence the choice of impact set: disruption duration, crossover tracks locations, passenger flow profile, and location of disruption. The longer the disruption, the larger the number of affected trains and the farther down the line impacts will occur. If crossover tracks are available, short-turning strategies and the associated impact on the reverse direction should be considered. Passenger arrival rates and alighting fractions also affect the choice of control/impact set. At stations where the passenger arrival rates are high, the associated passenger waiting costs can be large, and the impact from the disruption and control strategies should be considered. If the blockage is close to the terminal station, the impact on the reverse direction may also have to be considered. Behind the blockage, enough trains should be included in the impact set so that no passengers will be left beyond the impact set.

2.4   Modeling Train Order in the Rail Transit System

In a rail transit system, the order of trains can vary for any of the following reasons: *(1)* track junction; *(2)* a train entering or leaving service; *(3)* a train short-turning. To model this behavior, we define a segment as a track section within which the order of trains cannot change, while across segments the order of trains may change. The end of a segment will be a terminal, a crossover track or a junction point. As shown in Figure 2, the train sequence does not change within any of the six segments identified. However, from segment G_1 to segment G_2, trains may branch off the main line and the sequence of trains may change. From segment G_4 to segment $G_{2'}$, if we decide to short-turn train T4, the train preceding T4 will be T1 instead of T3 on segment $G_{2'}$, and the train preceding T2 will be T4 instead of T1. We use m to denote a segment. If the end of m is a crossover track, we use m' to denote the reverse segment. For example, segment $G_{2'}$ is the reverse segment of segment G_2 in Figure 2.

Across segments, a set of potential candidate predecessors for a train is defined. For example, the set of potential predecessors for train T2 on segment $G_{2'}$ in Figure 2 includes trains T4 and T1. Within this set, we use predecessor binary variables $y_{j,i,m}$ to indicate the predecessor. $y_{j,i,m}$ equals 1 if train j precedes train i on segment m, and 0 otherwise. To simplify the model, we use segment occupancy binary variables $so_{i,m}$ by train and segment to indicate whether a certain train operates on a given segment.

2.5   Dwell Time Function

Dwell time is important in determining the departure time for trains at stations. It is related to the number of passengers alighting and boarding and the train crowding conditions. We expect as more people board and alight, at some point the marginal time for a passenger to alight or board will increase

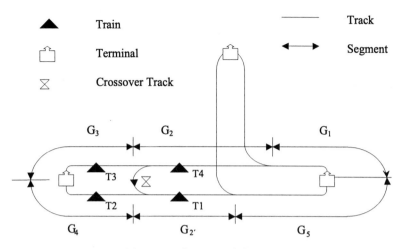

Figure 2. Segment definition.

because of increasing crowding level. Therefore, we expect dwell time to be nonlinear with respect to the number of passengers boarding and alighting, as shown by the dotted line in Figure 3. Rail transit dwell time data (Wong (1999); Lin and Wilson (1993)), however, suggests that linearity exists in the range where the number of passengers boarding and alighting is significant. Since there were not enough data to estimate the linear function for the non-crowded range, the following function is used to approximate the dwell time (the solid line in Figure 3):

$$dw = Max (C_0, C_1 + C_2 \cdot Alightings + C_3 \cdot Boardings)$$

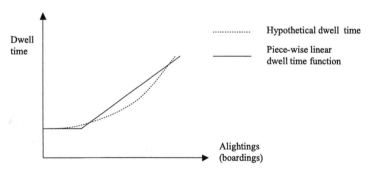

Figure 3. Dwell time function.

A reason to support this simplification is that under light flow conditions, the number of alightings and boardings may not have a strong effect on the dwell time. For example, provided other conditions are the same, the dwell time when there are three passengers boarding may not be much different from when there are six passengers boarding.

2.6 Disruption Control Problem Formulation

Notation:

$a_{i,k}$ arrival time of train i at station k

$dp_{i,k}$ departure time for the train preceding train i at station k

$dw_{i,k}$ dwell time of train i at station k

dw_k^0 typical dwell time at station k in that time period

$h_{i,k}$ maximum platform waiting time for train i at station k

$ht_{i,k}$ holding time of train i at station k

i_{BL} blocked or disabled train

$l_{i,k}$ passenger load on train i at station k

$l_{i,k}^0$ approximate passenger load on train i at station k

$p_{i,k}$ the number of passengers left behind by train i at station k

$p_{i,k}^e$ the number of passengers left by express train i at station k

$p_{i,k}^p$ the number of passengers left behind by predecessor of train i at station k

$r_{i,k}'$ the number of potential riders for train i at station k if it will not skip any later station

$r_{i,k}$ the number of potential riders for train i at station k considering possible expressing

$se_{i,k}$ 1 if train i starts expressing from station k, 0 otherwise

$so_{i,m}$ 1 if train i operates on segment m, 0 otherwise

$t_{k,k'}^s$ short-turning time from station k to k'

t_{BL} earliest time at which the blocked train or disabled train can move

$v_{i,k}$ 1 if train i is loaded to capacity at station k, 0 otherwise

$z_{i,k}$ variable to approximate the quadratic term of platform waiting time for train i at station k

$zt_{i,k}$ variable to approximate the quadratic term of holding time for train i at station k

A_k passenger arrival rate at station k

C_0 dwell time approximation for non-crowded condition

C_1 dwell time function constant for crowded condition

C_2 marginal alighting time for dwell time function under crowded condition

C_3 marginal boarding time for dwell time function under crowded condition

H_k minimum train separation at station k in time units

L train capacity

M sufficiently large number

Q_k passenger alighting fraction at station k

R_k non-inter-station-stopping running time from station $(k\text{-}1)$ to station k, including acceleration and deceleration time

S the set of stations in the impact set

$Sch_{i,t}$ scheduled dispatching time of train i at terminal t

S^t the set of terminal stations in the impact set

G the set of segments in the impact set

T the set of trains in the impact set

$T^p_{i,m}$ the set of trains that can be the predecessor of train i on segment m

$T^s_{i,m}$ the set of trains that can be the successor of train i on segment m

U^{iw} weight for in-vehicle waiting time

Decision Variables:

- Short-turn decision variables
  $st_{i,m}$: 1 if train i is short-turned at the end of segment m, and 0 otherwise.
- Express decision variable
  $sk_{i,k}$: 1 if train i skips station k, and 0 otherwise.
- Departure time of train i at station k
  $d_{i,k}$: continuous variable.
  Note that once the train departure time at each station is determined, the train holding time, $ht_{i,k}$, at each station is also determined.
- Predecessor binary variable
  $y_{j,i,m}$: 1 if train j precedes train i on segment m, and 0 otherwise.

Segment occupancy binary variables $so_{i,m}$ are used to indicate whether trains operate on certain segments. These variables are equivalent to the short-turning decision variables, because once we determine the short-turn decision variables the segment occupancy variables are also determined. They are only used to simplify presenting the formulation. Similarly, binary variables $se_{i,k}$ indicating starting an express section are completely constrained by express decision variables, and, therefore, are not decision variables.

Objective Function: In the context of real-time disruption control, operating cost is usually not a primary concern due to its short duration and because most of the operator's costs are already fixed. The top concerns may be to reduce customer complaints and to return to normal operation quickly. In this paper, the objective used is to minimize total passenger waiting time, including both platform passenger waiting time and in-vehicle passenger delay. Alternative objectives could be to minimize the maximum passenger waiting time or the number of passengers who can not board the first arriving train. The general objective function to be minimized is given by function (1).

$$\sum_{i\in T}\sum_{k\in S}\left\{\frac{A_k}{2}h_{i,k}^2 + p_{i,k}(d_{i+1,k} - d_{i,k})\right.$$

$$\left. + U^{iw}\left[\frac{A_k}{2}ht_{i,k}^2 + (l_{i,k} - A_k ht_{i,k})ht_{i,k}\right]\right\} \qquad (1)$$

We assume that the passenger arrival rate at any station is constant over the time period of interest. The first term is the platform waiting time for passengers arriving prior to train arrival and during the dwell time but before holding starts. The second term is the additional waiting time for passengers who are left behind or skipped by a train and have to wait for another train. The number of passengers left behind, $p_{i,k}$, is not a decision variable. Once the control decision variables are determined, $p_{i,k}$ is determined by the constraints. The third term is the waiting time for passengers arriving during the holding time. The last term is in-vehicle delay due to holding for passengers who are on-board when holding starts. Since passengers may perceive in-vehicle delay to be less onerous than platform waiting time, we weight these last two terms by $U^{iw}(\leq 1)$.

Formulation:[1]

$$\text{Minimize} \quad \sum_{i\in T}\sum_{m\in G} so_{i,m}\sum_{k\in m}\left\{\frac{A_k}{2}h_{i,k}^2 + p_{i,k}(d_{i+1,k} - d_{i,k})\right.$$

$$\left. + U^{iw}\left[\frac{A_k}{2}ht_{i,k}^2 + (l_{i,k} - A_k ht_{i,k})(ht_{i,k} + dw_{i,k})\right]\right\} \qquad (2)$$

subject to

$$d_{i,k} - d_{i,k-1} - R_k - dw_{i,k} \geq M(so_{i,m} - 1) \qquad \forall\, i \in T, k \in m, m \in G \quad (3)$$

$$d_{i,k} - dp_{i,k+1} - H_{k+1} + R_{k+1} \geq M(so_{i,m} - 1) \quad \forall\, i \in T, k \in m, m \in G \quad (4)$$

$$dp_{i,k} - d_{j,k} \geq M(y_{j,i,m} - 1) \qquad\qquad \forall\, i \in T, j \in T_{i,m}^p, k \in m \quad (5)$$

$$dp_{i,k} - d_{j,k} \leq M(1 - y_{j,i,m}) \qquad\qquad \forall\, i \in T, j \in T_{i,m}^p, k \in m \quad (6)$$

$$ht_{i,k} - d_{i,k} + d_{i,k-1}$$
$$\qquad + R_k + dw_{i,k} \geq M(so_{i,k} - 1) \qquad \forall\, i \in T, k \in m, k \notin S^t, m \in G \quad (7)$$

$$h_{i,k} - dp_{i,k} + ht_{i,k} \geq M(so_{i,k} - 1) \qquad \forall\, i \in T, k \in m, k \notin S^t, m \in G \quad (8)$$

$$d_{i,t} - Sch_{i,t} \geq M(so_{i,m} - 1) \qquad\qquad \forall\, i \in T, t \in S^t, t \in m \quad (9)$$

$$d_{i,k'} - d_{i,k} - t_{k,k'}^s - dw_{i,k'} \geq M(st_{i,m} - 1) \quad \forall\, i \in T, k \in m, m \in G, \text{emc} \quad (10)$$

---

[1] For segment m emc means: The end of m is a crossover track.

$$ht_{i,k'} - d_{i,k'} + d_{i,k}$$
$$- t^s_{k,k'} - dw_{i,k'} \geq M(st_{i,m} - 1) \qquad \forall\, i \in T, k \in m, m \in G, \text{emc} \quad (11)$$

$$dw_{i,k} \geq Max \ \{C_0, C_1 + C_2[(r_{i,k}(1 - v_{i,k}) + (L - (1 - Q_k)l_{i,k-1})v_{i,k}]$$
$$+ C_3\, Q_k\, l_{i,k-1}\}(1 - sk_{i,k})so_{i,m} \qquad \forall\, i \in T, k \in m, m \in G \quad (12)$$

$$r'_{i,k} - A_k\, h_{i,k} - (1 - Q_k)l_{i,k-1} - p^p_{i,k} = 0 \qquad \forall\, i \in T, k \in S \quad (13)$$

$$p^e_{i,k} - r'_{i,k}\big[Q_{k+1}\, sk_{i,k+1}$$
$$+ \sum_{u=k+2}^{S} Q_u \big(\prod_{v=k+1}^{u-1} (1 - Q_v)\big)sk_{i,u}\big] = 0 \qquad \forall\, i \in T, k \in S \quad (14)$$

$$r_{i,k} - r'_{i,k} + p^e_{i,k} = 0 \qquad \forall\, i \in T, k \in S \quad (15)$$

$$r_{i,k} - Lv_{i,k} \geq 0 \qquad \forall\, i \in T, k \in S \quad (16)$$

$$r_{i,k} - Mv_{i,k} \leq L \qquad \forall\, i \in T, k \in S \quad (17)$$

$$p^p_{i,k} - p_{j,k} \geq M(y_{j,i,m} - 1) \qquad \forall\, i \in T, j \in T^p_{i,m}, k \in m \quad (18)$$

$$p^p_{i,k} - p_{j,k} \leq M(1 - y_{j,i,m}) \qquad \forall\, i \in T, j \in T^p_{i,m}, k \in m \quad (19)$$

$$\{l_{i,k} - [l_{i,k-1}\, sk_{i,k} + (L\, v_{i,k} + r_{i,k}(1 - v_{i,k}))$$
$$(1 - sk_{i,k})]\}so_{i,m} = 0 \qquad \forall\, i \in T, k \in m, m \in G \quad (20)$$

$$p_{i,k} - [(p^p_{i,k} + A_k\, h_{i,k})sk_{i,k} + [(r_{i,k} - L)v_{i,k}$$
$$+ p^e_{i,k}(1 - v_{i,k})](1 - sk_{i,k})]so_{i,m} \geq 0 \quad \forall\, i \in T, k \in m, m \in G \quad (21)$$

$$p_{i,k} - (p^p_{i,k} + A_k\, h_{i,k} + l_{i,k-1}(1 - Q_k))st_{i,m} \geq 0 \quad \forall\, i \in T, k \in m,$$
$$(k+1) \in (m+1), m \in G, \text{emc} \quad (22)$$

$$st_{i,m} - so_{i,m} \leq 0 \qquad \forall\, i \in T, m \in G, \text{emc} \quad (23)$$

$$so_{i,m+1} + st_{i,m} \leq 1 \qquad \forall\, i \in T, m \in G, \text{emc} \quad (24)$$

$$so_{i,m} - so_{i,m-1} \leq 0 \quad \forall\, i \in T, m \in G$$
$$\text{if train } i \text{ is not likely to be short-turned onto segment } m \quad (25)$$

$$y_{i,i+1,m} + st_{i,m} + st_{i+1,m} \geq 1 \qquad \forall\, i \in T, m \in G, \text{emc} \quad (26)$$

$$y_{i,j,m'} - st_{i,m} \leq 0 \qquad \forall\, i \in T^p_{j,m}, m \in G, \text{emc} \quad (27)$$

$$\sum_{m \in G} st_{i,m} \leq 1 \qquad \forall\, i \in T \quad (28)$$

$$\sum_{j \in T^p_{i,m}} y_{j,i,m} = 1 \qquad \forall\, i \in T, m \in G \quad (29)$$

$$\sum_{j\in T^s_{i,m}} y_{i,j,m} = 1 \qquad\qquad \forall\, i \in T, m \in G \quad (30)$$

$$y_{i,j,m} + y_{j,i,m} \le 1 \qquad\qquad \forall\, i \in T^p_{j,m}, j \in T^p_{i,m}, m \in G \quad (31)$$

$$y_{j,i,m} - so_{i,m} \le 0 \qquad\qquad \forall\, j \in T^p_{i,m}, m \in G \quad (32)$$

$$v_{i,k} - v_{i,k-1} - sk_{i,k} \ge -1 \qquad\qquad \forall\, i \in T, k \in S \quad (33)$$

$$v_{i,k} - v_{i,k-1} + sk_{i,k} \le 1 \qquad\qquad \forall\, i \in T, k \in S \quad (34)$$

$$se_{i,k} + sk_{i,k-1} - sk_{i,k} \ge 0 \qquad\qquad \forall\, i \in T, k \in S \quad (35)$$

$$\sum_{k\in S} se_{i,k} \le 1 \qquad\qquad \forall\, i \in T \quad (36)$$

$$d_{i_{BL},k_{BL}} \ge t_{BL} \qquad\qquad (37)$$

$$y_{i,j,m}, v_{i,k}, v'_{i,k}, sk_{i,k}, se_{i,k}, st_{i,m}, so_{i,m} \in \{0,1\} \qquad \forall\, i \in T, k \in S, m \in G$$

$$d_{i,k}, dw_{i,k}, dp_{i,k}, h_{i,k}, p_{i,k}, pp_{i,k}, r_{i,k}, r'_{i,k}, p^e_{i,k}, l_{i,k} \ge 0 \;\; \forall\, i \in T, k \in S, m \in G$$

Because of the difficulty of solving the formulation with objective function (1), as discussed below, the objective function (2) is used. The stations in the impact set are divided into segments. If a train does not operate on a segment, there will be no cost associated with it on that segment.

Due to the structure of the dwell time function used, it is difficult to impose constraints to ensure that the dwell time is exactly equal to the time for the passengers to alight and board. The solution may potentially over-estimate the dwell time while under-estimate holding time if objective function (1) is used. Suppose we want to increase the departure time by Δ to $(d_{i,k} + \Delta)$. This can be achieved by increasing either the holding time $ht_{i,k}$ to $(ht_{i,k} + \Delta)$ or the dwell time $dw_{i,k}$ to $(dw_{i,k} + \Delta)$. (In this case, the dwell time is greater than the actual time for passengers to alight and board the train). In the following, we compare the changes in objective function (1) due to the change of holding time or dwell time. Since the departure time is $(d_{i,k} + \Delta)$ in either case, the terms related only to the departure times are not considered.

Suppose the holding time increases by Δ, (38) gives the corresponding change in objective function (1).

$$U^{iw}\left\{ \left[\frac{A_k}{2}(ht_{i,k} + \Delta)^2 - \frac{A_k}{2}ht^2_{i,k}\right] \right.$$
$$+ \left. [(l_{i,k} - A_k(ht_{i,k} + \Delta))(ht_{i,k} + \Delta) - (l_{i,k} - A_k ht_{i,k})ht_{i,k}]\right\}$$
$$\approx U^{iw}(l_{i,k} - A_k ht_{i,k})\Delta \qquad (38)$$

In contrast, if the dwell time is over-estimated by Δ, the change is

$$\frac{A_k}{2}(h_{i,k} + \Delta)^2 - \frac{A_k}{2}h^2_{i,k} \approx A_k h_{i,k}\Delta \qquad (39)$$

The value of (39) can be smaller than that of (38) if there is a significant passenger load on the train after alighting. For this reason, objective function (2) is used where the last term has $(ht_{i,k} + dw_{i,k})$ instead of $ht_{i,k}$. That is, the same cost is imposed for dwell time as for holding time. We can show that the overall cost by increasing $dw_{i,k}$ to $(dw_{i,k} + \Delta)$ is now larger than that of increasing $ht_{i,k}$ to $(ht_{i,k} + \Delta)$. If $dw_{i,k}$ is increased by Δ, then $h_{i,k}$ is also increased by Δ and the cost change is

$$A_k h_{i,k} \Delta + (l_{i,k} - A_k ht_{i,k})\Delta \tag{40}$$

The cost increase due to increasing $ht_{i,k}$ to $(ht_{i,k} + \Delta)$ is given by

$$U^{iw}(l_{i,k} - A_k ht_{i,k})\Delta \tag{41}$$

The value of (40) is no smaller than that of (41) because $U^{iw} \leq 1$. Therefore, the solution will never over-estimate the dwell time if holding is an option. In addition to ensuring solution validity, there are other advantages to including the dwell time in the last term. If objective function (1) is used, for passengers on-board the train after alighting (passengers staying on the train whose destinations are downstream), only the in-vehicle delay due to holding time is considered. However, during disruptions, dwell time is also increased. It theoretically makes sense to consider the in-vehicle delay due to increased dwell time for those passengers. The disadvantage of objective function (2) is that it double-counts the dwell time for passengers boarding the train before holding starts. The platform waiting time of these passengers, the first term of (2), as explained in § 2.2, already includes the dwell time. The last term of (2) again counts the dwell time as in-vehicle delay. However, since the dwell time is small, the double-counting should not have a significant impact on the solution.

Constraint (3) ensures that the departure time of train i at station k is no earlier than its departure time from previous station $(k-1)$ plus running time between these two stations and dwell time at station k. Constraint (4) ensures the minimum safe separation of trains at each station. Only if train i can run nonstop from station $k-1$ to k and the arrival headway at k is no less than the minimum separation at k, can it depart station $k-1$. Constraints (3) and (4) apply only if train i operates on segment m, i.e., if $so_{i,m} = 1$.

Constraints (5) and (6) determine the departure time of the train preceding train i with the predecessor indicator $y_{j,i,m}$. Constraint (7) determines the holding time and (8) determines the platform waiting time. Both of them apply only if $so_{i,m} = 1$. Constraint (9) ensures that the departure time of train i is no earlier than its scheduled departure time if it is at a terminal.

Constraint (10) is a short-turning constraint. If a train is to be short-turned via the crossover track on segment m, its departure time at the short-turning destination station can be no earlier than its departure time at the short-turning origin station plus the (constant) short-turning time and the dwell time at the destination station. Constraint (11) determines the corresponding holding time after short-turning.

Constraint (12) is the dwell time constraint. Only if a train operates on the segment and does not skip the station, does it incur a dwell time.

Constraint (13) determines the number of potential riders for a train at a station assuming it will not skip any station. It includes the passengers on the train and the passengers on the platform. Constraint (14) determines the number of passengers that will be left at a station because the train will skip their destinations. It uses the alighting ratio at each station to estimate the potential riders whose destinations are at each station, then uses the expressing decision variables to determine whether those people have to wait for the next train. For example, suppose there are r' potential riders at station k if the train does not skip any station. Based upon the alighting ratio, we can estimate how many of those r' potential riders will stay on the train beyond station $(k+q-1)$, and how many people will alight at station $(k+q)$. Then, if the train skips station $(k+q)$, those people whose destination is $(k+q)$ will have to wait at k for the next train. Constraint (15) determines the potential riders excluding those who cannot board the train due to expressing.

Constraints (16) and (17) determine the binary load variable value, so that $v_{i,k} = 1$ if and only if the train is loaded to capacity. Constraints (18) and (19) use the predecessor index to determine the number of passengers left by the preceding train at a station. Constraint (20) is the load constraint. If the train skips station k, its departure load at k will be the same as at $(k-1)$. If the train stops at the station, and if it is loaded to capacity, the load will be equal to the maximum load, otherwise, it will be equal to the number of riders. This constraint is valid only if the train operates on segment m.

It is assumed that the train to be short-turned must stop at the station immediately before the crossover track. Constraint (21) determines the number of passengers left by a train. It includes those people who cannot board the train due to either expressing or train capacity. Constraint (22) determines the number of passengers left by the short-turned train at the station immediately before the crossover track. All passengers on-board have to alight the train and wait for the next train.

Constraint (23) ensures that a train can be short-turned on segment m only if it operates on segment m. Constraint (24) ensures that a train will not operate on segment $(m+1)$ if it is short-turned on segment m. Constraint (25) ensures that any train not operating on segment $(m-1)$, will not operate on the following segment m, either. It does not apply to the situation where two segments are separated by a crossover track because trains may be short-turned from the other direction. At the merge point, this constraint means that if any train from one of the branches does not operate on the last segment on that branch (trunk portion), it cannot operate on the first segment of the trunk portion (that branch). Constraint (26) ensures that if neither train i nor its successor train $(i+1)$ is short-turned on segment m, train i will be the predecessor of train $(i+1)$ on segment m.

Constraint (27) states that only if train i is short-turned on segment m, may it be the predecessor of train j on the opposite segment m' in the reverse direction. Constraint (28) ensures that a train can be short-turned at only one of the available crossover tracks. Constraint (29) ensures that there can be only one predecessor of a train on a certain segment. Constraint (30) ensures a train can be the predecessor of only one train. Constraint (31) prevents any train that is the predecessor of another train also being the successor of that train. Constraint (32) states that any train not operating on a segment can have no predecessor on that segment. Constraints (33) and (34) state that if a train skips a station, its load must be the same as the load before that station, since there is no boarding and alighting.

Constraint (35) determines the express starting binary variables, which is used to define express segments. If the train stops at station $k-1$ but skips station k, there must be an express segment starting from station k. Generally speaking, multiple express segments may cause confusion to the passengers, complicate the announcements, and waste needed capacity. Therefore, Constraint (36) ensures there is only one express segment per train. Constraint (37) ensures that the blocked train can depart from the point of disruption only after the blockage has been cleared.

2.7   Model Simplification

The model above has a nonlinear objective function and nonlinear constraints. To solve it using a linear solver, we need to transform it into a linear formulation.

Piece-wise Linear Approximation of the Quadratic Terms: The objective function contains quadratic functions with respect to the platform waiting time and holding time. The following piece-wise linear functions can be used to approximate these quadratic functions:

$$h_{i,k}^2 \approx z_{i,k} = \max_{n=1,\ldots,N} \{a_n\, h_{i,k} + b_n\},$$

$$\text{or } z_{i,k} \geq a_n\, h_{i,k} + b_n, \quad \text{for } n = 1, 2, \ldots$$

$$ht_{i,k}^2 \approx zt_{i,k} = \max_{n=1,\ldots,N} \{a_n'\, ht_{i,k} + b_n'\},$$

$$\text{or } zt_{i,k} \geq a_n'\, ht_{i,k} + b_n', \quad \text{for } n = 1, 2, \ldots$$

Simplification of the Non-Separable Terms: The additional waiting time for passengers left by trains, which is the product of the number of passengers left and the departure interval, and the in-vehicle delay for passengers on-board, which is the product of the passenger load when holding starts and the holding time, are both non-separable terms. Applying the following reasoning, approximate values can be applied to one of the two variables in each of these terms in order to linearize the model.

When a disruption occurs, trains initially ahead of the disruption may be held to even out headways. However, it will never make sense to hold these trains for so long that they will not leave passengers behind at stations. If a train is to leave passengers behind after holding, a better solution can always be achieved by holding the train for less time resulting in reduced in-vehicle delay. Therefore, if the system meets the capacity requirement under normal operating conditions, no passengers would be left behind by trains ahead of the disruption. We only need to consider passengers left by trains behind the disruption. Since the headways of trains behind the disruption are usually constrained by the minimum headway, the minimum departure-arrival interval plus the typical dwell time, the minimum headway is a good approximation of the additional waiting time for passengers left behind by trains initially behind the disruption.

The other non-separable term, in-vehicle delay, is the product of the passenger load when holding starts and the holding time. For trains initially ahead of the disruption, normal operation is interrupted only when holding strategies are applied. Hence, the typical passenger load in that time period can be used as an approximation for the number of passengers on-board when holding starts. Of course, once trains are held, the typical load will be an underestimate of passengers on-board at following stations. However, O'Dell and Wilson (1999) showed that there are typically little marginal benefits from holding at more than one station for any train. Therefore, the product of the typical passenger load in that time period and the holding time can be used to approximate the in-vehicle delay for passengers on-board trains initially ahead of the disruption when held.

Unlike trains ahead of the disruption, passenger load for trains initially behind the blockage is usually high. Therefore, typical passenger load in that time period will be an underestimate of the true load. However, before the disruption clearance, passengers who are on-board blocked trains at stations behind the disruption have to wait for the whole disruption period. Hence, their in-vehicle delay is a constant and does not affect the decisions. After the disruption is cleared, trains behind the disruption are clustered together and there is a gap ahead of the first blocked train. Thus, there is no inherent reason to hold these trains unless there is another gap behind, and these blocked trains should move ahead as quickly as possible even without holding cost. Therefore, any underestimate of their passenger load should have little impact.

After these approximations, the objective function (2) becomes

$$\sum_T \sum_S \left\{ \frac{A_k}{2} z_{i,k} + p_{i,k}(H_k + dw_k^0) \right.$$
$$\left. + U^{iw} \left[\frac{A_k}{2} zt_{i,k} + l_k^0(ht_{i,k} + dw_{i,k}) \right] \right\} \quad (42)$$

where $z_{i,k}$ is used to approximate the quadratic term of normal waiting time for train i at station k, $zt_{i,k}$ is used to approximate the quadratic term of holding time for train i at station k, l_k^0 is the typical passenger load at station k in that time period, H_k is the minimum safe separation (minimum departure-arrival interval) between trains at station k, and dw_k^0 is the typical dwell time at station k in that period.

Transformation of Nonlinear Constraints: The formulation also includes some nonlinear constraints ((12), (14), (20), (21), (22)) to ensure that variables have different values under different control interventions. All these constraints can be transformed into linear constraints with binary variables and sufficiently large numbers M.

These large numbers are the upper bounds for different variables and can be determined heuristically. The tighter the upper bound, the smaller the feasible space will be, and the less time it will potentially take to solve the problem. For example, for the load constraint (20), train capacity is a good upper bound. For departure times, assuming we do not apply any control strategies, we can obtain the heuristic upper bounds on departure times for trains at different stations based upon the running time between stations, maximum dwell time and the duration of the disruption.

Constraint (14) determines the number of passengers left behind by the express trains. To simplify this problem, the following assumption is made: *Express announcements are made at the station immediately before the express segment.*

Under this assumption, the station immediately before the express segment is the only station where the express train dumps passengers. For disruption control, the possible trains that we may want to express are trains behind the disruption after the blockage is cleared. Since most of the benefit from expressing is the reduced waiting time at stations down the line, we want to express the train as early as possible. Hence, the express announcement can be made before the disruption is cleared at stations behind the disruption to save time, and the express segment can begin with the next station the train arrives at after the blockage is cleared. Therefore, the above assumption is realistic.

In addition to this assumption, a predetermined value is used to approximate the proportion of passengers dumped by the express train and the portion of passengers whose destinations are within the express segment, and thus, cannot board the express train at the station immediately before the express segment. Since this group of passengers is usually a small portion of all passengers and the train following the express train usually has a very small headway, this simplification should not have significant impact on the control strategies. To determine this proportion, we first heuristically define the maximum possible express segment based on alighting fractions. Intuitively, it is "expensive" to skip stations with large alighting fractions because large numbers of passengers will have to wait for the next train. Corresponding to the maximum express segment, the maximum portion of passengers dumped is used as the approximate value. This may overestimate the number of passengers dumped, but the impact should not be significant.

3   Model Application

The formulation is applied to a disruption scenario on a single linear transit line. A simplification of Massachusetts Bay Transportation Authority
(MBTA) Red Line, the heaviest ridership rail line in the Boston (Mass, USA)
region, was used to obtain a realistic representation. The MBTA Red Line is
first introduced and the simplifications made are addressed. Two disruption
durations are tested: 10 minutes and 20 minutes. To investigate sensitivity of
the control strategy effectiveness to the accuracy of the disruption duration
estimate, two scenarios with initial 10 minutes disruption duration estimates
but actual durations of 5 and 15 minutes are tested. The branching sequence
used in the branch-and-bound algorithm to solve the mixed integer problem
and solution times are also discussed. Although the test scenario is based on
the MBTA Red Line, the model should be applicable to other rail transit
systems in general.

3.1   MBTA Red Line and Simplified System

The MBTA Red Line is a heavy rail system consisting of two branches (Ashmont and Braintree), and a common Trunk portion. A simple representation
of the Red Line is shown in Figure 4. There are 5 stations on the Ashmont
branch, 6 stations on Braintree branch, and 12 stations on the trunk portion.
For modeling purposes, the two directions at a station are treated as different stations. Trains are dispatched from the branches. The headway on the
Ashmont branch is 8 minutes, and is 6 minutes on the Braintree branch with
a resulting effective headway on the trunk portion of 3.4 minutes. Between
the two branches, the Braintree branch is operating closer to capacity in the
morning peak period, and even a minor disruption can cause serious problems. The disruption scenario we analyzed is a temporary blockage close to
the terminus in the peak direction on the Braintree branch in the AM peak
period.

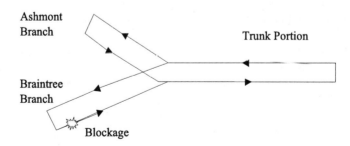

Figure 4. MBTA Red Line simple representation.

Since the disruption in the scenario occurs on one of the two branches, there are several associated difficulties. First, there may be required (forced) control actions at the junction point to ensure that trains enter the trunk portion with a safe separation. Whether holding is in response to the disruption or "forced" by the junction may not be clear. Second, since a blockage on one of the two branches does not completely disable the whole system, benefits from different control strategies may be confounded due to interaction effects with the other branch at the junction point.

In order to filter out the "noise" due to the junction and highlight the performance of the control strategies, only the Braintree branch is considered. Correspondingly, the following modifications were made:

- Trains and stations on the Ashmont branch are not considered.
- The passenger arrival rate at each station on the trunk portion is scaled by the ratio of Braintree trains to all trains. During peak periods, the ratio of Ashmont and Braintree trains is about 3:4. Therefore, 4/7 is used to scale the passenger arrival rate for stations on the trunk portion.
- At the same time, the minimum headway at each station on the trunk portion is scaled by 7/4 to accommodate trains from the Ashmont branch. Therefore, a solution feasible to the modified system is also feasible to the original system.

After these modifications, the same disruption is more serious in the simplified system. The solution to the simplified system will not be optimal but will be feasible for the original system. In terms of the solution speed, since the order of trains entering the trunk portion is not considered, fewer binary variables are needed to determine the preceding train on the trunk portion. Therefore, it potentially takes less time to obtain the optimal solution for the simplified system.

3.2 Input Data

Dwell time: Dwell time data was collected at one of the stations on the Red Line during the PM peak period. The estimated linear dwell time function is as follows, all times in seconds (Wong (1999)):

$$dw = \underset{(1.48)}{5.52} + \underset{(3.32)}{0.12} \cdot Alightings + \underset{(5.69)}{0.12} \cdot Boardings \quad (R^2_{adj} = 0.79)$$

The constant term has a small value and the t-statistic is only marginally significant. The reason for this result may be that the data collected are most concentrated in the area where the volume of boarding and alighting passengers is high. Little data was collected in the low passenger volume range. Therefore, a constant value is used to approximate the dwell time in the range where the volume of boardings and alightings is low. Based on the data, a value of 20 seconds was selected, which is close to the lowest dwell time in the data set. The final dwell time function (in seconds) is as follows:

$$dw = Max\ (20, 5.52 + 0.12 \cdot Alightings + 0.12 \cdot Boardings)$$

Such approximation in the low passenger boarding/alighting range can potentially result in overestimates of the dwell time. However, the slope of the dwell time function is expected to be small in the low alighting and boarding volume range. Therefore, the simplification should not have a significant impact on system performance.

Train Location: In the model, train location information is reflected in the latest departure time at stations when a disruption occurs. In this model application, this data item is estimated according to the running time between stations and the location of trains when the disruption occurs. In a real implementation, train arrivals at stations would be monitored and the latest train arrival time recorded.

Passenger Arrival Rate and Alighting Fraction: The disruptions tested occur at 8:15 AM during the height of the morning peak period. To maintain comparability with the prior analyzes by O'Dell and Wilson (1999), the same passenger arrival rates and alighting fractions are used. The data set was developed by O'Dell based on a data set collected by the Massachusetts Central Transportation Planning Staff (1991).

Normal Passenger Load at Each Station in the Time Period of Interest and Current Load: The product of the holding time and the normal passenger load at that station in that time period is used to approximate the non-separable passenger on-board delay. Based on the normal headway, arrival rates and alighting fractions, the typical load at each station in that time period is calculated. The passenger load of all trains when a disruption occurs is also required and is approximated with the typical load.

Train Running Time and Minimum Separation: With the control line information for the MBTA Red Line, Heimburger et al. (1999) developed a simulation model to determine the minimum safe separation of trains at each station and the corresponding maximum non-inter-station-stopping running time.

Weight for In-vehicle Delay: Since passengers may perceive in-vehicle delay to be less onerous than platform waiting time, a weight less than 1 is used for in-vehicle delay. According to the research by Abdel-Aty et al. (1995), 0.5 is a reasonable weight for in-vehicle travel time. In-vehicle delay should be more onerous than in-vehicle travel time but less onerous than platform waiting time. Therefore, a weight between 0.5 and 1 might be reasonable for in-vehicle delay. In this research, 0.5 is taken as the weight for in-vehicle delay.

Portion of Passengers Who Have to Leave the Express Train: As discussed in § 2.7, a predetermined value is used to approximate the proportion of passengers "dumped" by the express train and the portion of passengers whose destinations are within the express segment, and cannot board or have to alight from the train at the station immediately before the express segment. Based on the location of the disruption and passenger arrival rates and alighting fractions, the maximum potential express segment is pre-defined, and 0.1 is used to approximate this proportion.

3.3   Comparison of Control Strategies

Two disruption durations are tested: 10 minutes and 20 minutes. For each disruption duration, different control strategies are applied incrementally. First, no control strategies (NC) are applied, then only holding strategies are applied ("Holding Only" (H)), then both holding and expressing strategies can be applied ("Holding and Expressing" (HE)), and finally, holding, expressing and short-turning can all be applied ("Holding, Expressing and Short-turning" (HET)). To investigate the impact of expressing on solution time, holding combined with short-turning is also tested ("Holding and Short-turning" (HT)).

3.3.1   Group of Passengers for Impact Evaluating

To compare the effectiveness of different control strategies, we need to establish a common group of passengers to evaluate their waiting time. Ideally, this group should include all passengers who are affected by the disruption or the control strategies. In this research, passengers initially on-board controlled trains and trains behind the disruption in the impact set when the disruption occurs and passengers arriving within a certain time window at each station in the impact set are taken as the common group. The starting and ending time of the time window are explained below.

Precisely speaking, the starting time of the evaluation time window at each station should be the time after which arriving passengers board a blocked or a controlled train, or in other words, the departure time of the last uncontrolled train at that station. At a given station, if trains passing before the disruption occurs are not controlled under any control scheme, the starting time of the time window can be defined as the departure time of the last uncontrolled train at that station. This time should be the same across different control schemes because it is associated with an uncontrolled train.

However, if some trains passing the station before the disruption occurs are controlled at some point, the group of passenger affected can be over-estimated if the departure time of the last uncontrolled train is used as the starting time of the time window. For example, consider Figure 5, which shows a scenario when a disruption occurs.

Suppose the impact set includes stations A through I and A' through C', and trains T_1 through T_6 and T_1' through T_3'. The control set includes T_1 through T_6 and T_2'. T_2' is the only short-turning candidate. Suppose the first train controlled is train T_2, which is held at station E for a certain time. At station D, the latest departure time of an uncontrolled train is then the departure time of train T_1. That is, after this time, passengers arriving

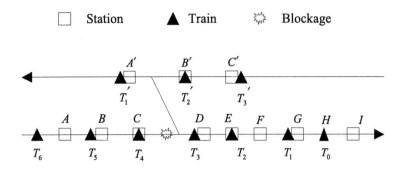

Figure 5. Determination of evaluation time window.

at station D may board a controlled train, e.g., train T_2. However, not all passengers arriving at D after the departure time of T_1 and boarding T_2 are affected by the disruption, e.g., passengers who alight at E. Only passengers on-board T_2 when it is controlled are affected.

Thus, if any train passing the station before the disruption occurs are controlled at some point under any control scheme, the starting time of the time window at that station is defined as the latest departure time before the disruption occurs. That is, the starting time may be the departure time of a controlled train, and, therefore, passengers arriving before the starting time may be affected by the control decisions. To deal with the problem, we include passengers on-board the train when control occurs. For example, in the above case, passengers on-board T_2 at E when T_2 is held are included in the impact evaluation set.

The ending time of the time window, should be the time after which no passenger perceives a delay, i.e., a waiting time longer than the normal headway. There are usually trains clustered behind the blockage with small headways. If a train initially behind the disruption does not leave passengers behind after the blockage is cleared, trains following it do not leave passengers behind, either. It is thus guaranteed that passengers arriving after the departure of this train have waiting time no longer than the normal headway because they do not need to wait for more than one train before boarding. Therefore, the departure time of the first train behind the blockage that does not leave passengers at any station can be used as the ending time. We define this train as the *margin train*. Under different control schemes, the corresponding departure time of this margin train may be different. For example, the departure time of the margin train in the "No-Control" case may be later than in the active control case. To establish a common ending time across control schemes, the worst-case time, the departure time of the margin train in the "No-Control" case, is used as the ending time of the time window. The waiting time for passengers arriving between the departure time of the margin train under other control schemes and the ending time of the time window is assumed to be the mean typical waiting time, which is half the normal headway. Since the headways for trains behind the blockage are usually smaller than the normal headway, this assumption should give a conservative estimate of benefits.

3.3.2 Results

After the problems for different control strategies are solved, the results are processed based on the evaluation group so that the number of passengers affected is fixed. Thus, the mean platform waiting time and in-vehicle delay can be used as measures of control strategy effectiveness. The results are summarized in Tables 1 and 2, which also include the no disruption (zero disruption duration) case (ND). Mean weighted waiting time is the sum of mean platform waiting time and weighted mean in-vehicle delay. Since passengers may perceive in-vehicle delay to be less onerous than platform waiting time, 0.5 is used as the weight.

According to the results, holding provides about 10-18% savings over no control. The benefit increases with the disruption duration in the 10-20 minute range. The benefit from holding is expected to stabilize eventually because train capacity is limited and this will limit holding times and benefits. In addition, in-vehicle passenger delay increases with holding time. The results show that expressing only provides modest savings beyond holding, which intuitively makes sense because skipping a station saves less than a minute and skipping multiple stations risks wasting needed capacity. For short-turning, one train is short-turned in the 10 minute disruption case, and the saving in passenger waiting time is 35%. In the 20-minute disruption case, two trains are short-turned, which results in a close to no-disturbance situation in the peak direction and provides the greatest saving. It should be noted that expressing is not used in the optimal solution for HET in either the 10 or 20 minute cases, though it is allowed. In other words, the solutions in the HT case are the same as HET. This suggests that short-turning combined with holding can provide significant benefits.

Control Strategy	Mean Platform Waiting Time (min)	Mean In-vehicle Delay (min)	Mean Weighted Waiting Time (min)	Saving
ND	3.00	0.00	3.00	-
NC	5.70	0.15	5.78	-
H	4.53	1.39	5.23	10%
HE	4.59	0.83	5.00	13%
HET	3.55	0.39	3.74	35%

Table 1. Comparison of strategy effectiveness for 10-minute disruption scenario.

In the scenario investigated, the disruption is close to the terminal. Hence, there are not many people on-board the held trains and the in-vehicle delay

Control Strategy	Mean Platform Waiting Time (min)	Mean In-vehicle Delay (min)	Mean Weighted Waiting Time (min)	Saving
NC	9.11	0.19	9.20	-
H	6.57	1.98	7.56	18%
HE	6.23	1.75	7.10	23%
HET	3.79	0.35	3.97	57%

Table 2. Comparison of strategy effectiveness for 20-minute disruption scenario.

is not very significant. If the disruption occurs at a different location, the saving from holding may be less. In addition, since the passenger flow in the off-peak direction is quite low, the increase of passenger waiting time in the off-peak direction due to short-turning has a very small effect, which favors short-turning. Nevertheless, the benefit from short-turning is expected to be significant because short-turning can dramatically reduce the gap. For expressing, since the skipped stations all have high passenger arrival rates, the cost of expressing is high. If the disruption occurs where the potential skipped stations have low passenger arrival rate, the benefit may be higher. However, considering the limited improvement in headway, the saving from expressing is still not expected to be significant in most cases.

3.4   Sensitivity Analysis

The most critical assumption in this model is that the disruption duration is known and deterministic, which is not true in general. In practice, there are many types of disruptions. When a disruption occurs, the time it will take to clear is usually unknown. The controller can only provide an estimate based on his/her experience and information at hand. To check the impact of the disruption duration assumption, we investigate the sensitivity of the control strategy effectiveness to the accuracy of the disruption duration estimate.

The initial disruption duration is assumed to be 10 minutes, but the actual disruption duration is 5 minutes and 15 minutes, respectively. In the 5-minute disruption case, it is assumed that the initial control strategies based on the 10-minute estimate are carried out and control strategies are re-optimized after 5 minutes, when the disruption is cleared. In Table 3, the "5 Minutes" column shows the results with the exact 5-minute disruption duration estimate, and the "10 Minutes" column shows the results with 10 minutes initial disruption duration estimate and re-optimization after 5 minute. ("Holding Only" and "Holding and Expressing" have the same re-optimization results). According to the results, there is only a slight increase under inaccurate estimates in the small disruption duration case.

In the case when the actual disruption duration is 15 minutes but the estimate is 10 minutes, it is assumed that the actual disruption duration is known 10 minutes after the disruption occurs, i.e., at the end of the estimated

10 minutes disruption duration, and the control strategies are re-optimized based on the remaining 5-minute additional disruption. This simplification assumes perfect information in re-optimization, which may potentially under-estimate the passenger waiting time. On the other hand, the re-optimization could be done earlier in practice and the results may be better. In Table 4, column "15 Minutes" shows the results under the correct disruption dura-tion estimate and column "10 Minutes" shows the results under the 10 minute initial estimate with re-optimization after 10 minutes.

Again, expressing is not used in the HET case based on the 10 minute disruption duration estimate. It is not used in the 5 minute actual disruption case after re-optimization, but it is used in the 15 minute actual disruption case after re-optimization to reduce the gap due to underestimating of the disruption duration.

Blockage Duration Estimate	5 Minutes	10 Minutes	
Control Schemes	H, HE, HET	H & HE	HET
Mean Weighted Waiting Time (min)	3.76	3.76	3.79
Increase due to Wrong Estimate	-	-	+0.8%

Table 3. Effect of over-estimating the disruption duration.

Blockage Duration Estimate	15 Minutes			10 Minutes		
Control Schemes	H	HE	HET	H	HE	HET
Mean Weighted Waiting Time (min)	6.34	5.97	3.77	6.37	6.21	4.31
Increase due to Inaccurate Estimate	-	-	-	+0.5%	+4.0%	+14.3%

Table 4. Effect of under-estimating disruption duration.

The increase of mean weighted waiting time is higher in the "Holding and Expressing" case than in the "Holding Only" case. This is because holding time is less after including expressing, and some trains have passed stations with high arriving passenger flows where holding can provide significant ben-efits.

According to these results, the effectiveness of holding and expressing strategies are not highly sensitive to the accuracy of the estimate. However, the effectiveness of short-turning strategies can be sensitive to the accuracy of the disruption duration estimate.

3.5   Solution Time

The mixed integer formulation is solved with the CPLEX 4.0 MIP Solver on a Micron Pentium II 300MHz PC with 64M RAM using a branch and bound approach. The order of branching on the binary variables affects the solution speed significantly. It is important to branch on variables that are associated with major decisions, and thus have significant impact on both feasibility and the objective function value. We propose the following branching sequence.

First branch on variables determining the order of trains, especially for multiple-branch rail systems. Thus, variables such as short-turn decision variables and predecessor binary variables should have high priority. Following short-turning variables and predecessor variables, express decision variables and load binary variables are the next ones to branch on. Since the passengers benefiting from expressing are primarily those at stations beyond the express segment downstream, the earlier the express train starts expressing, the larger the benefit may be. This suggests a logical way to branch on the expressing decision binary variables: first, branch on variables for the first express candidate following the sequence of stations it arrives at; after branching on the first candidate, branch on the second and third and so on. Trains behind the disruption are unlikely to be held after disruption clearance. After determining the expressing decision variables, the departure load of the trains can also be determined. Therefore, after branching on expressing decision variables, the corresponding load variable can also be branched on.

Tables 5 and 6 compare solution times with the default CPLEX MIP solver branching sequence and with the proposed sequence. The column entitled "S" shows the solution time with the proposed branching sequence.

Control Strategy	H		HE		HET	
	S	Default	S	Default	S	Default
Time (sec)	2.91	3.17	5.60	16.81	11.28	47.01
Nodes	79	82	253	643	84	489
Iterations	2895	3061	5056	14260	6121	26756

Table 5. Solution times for the 10-minute disruption.

In both 10 minute and 20 minute scenarios, the "Hold Only" problem takes the least time to solve. This is because the problem does not involve expressing and short-turning decision variables. With the default branching sequence, the 20 minute "Holding, Expressing and Short-turning" problem cannot be solved on the test platform due to insufficient memory. The solution time of the 20-minute "Holding, Expressing and Short-turning" problem with proposed branching sequence is shorter than for the "Holding and Expressing" problem. This is because short-turning provides large benefits in this scenario and good bounds can be obtained at an early stage. Since the

Control Strategy	H		HE		HET	
	S	Default	S	Default	S	Default
Time (sec)	12.10	21.60	155.01	793.47	68.32	-
Nodes	551	787	5085	28241	1162	
Iterations	11143	19918	122590	614187	24066	

Table 6. Solution times for the 20-minute disruption.

disruption occurs near the terminus, and few passengers are on the trains in the reverse direction, short-turning can provide significant benefits. If the disruption occurs in the middle of the line, branching on the short-turning decision variable may not provide us with nearly as sharp a bound.

Comparing the results of different control strategies, the expressing strategy only provides modest benefit beyond holding. As expected, skipping one more station or choosing another station to skip makes little difference in the objective value. Thus, many nodes have to be kept and explored to obtain the optimal expressing decisions. Table 7 compares the solution time with and without expressing using the proposed branching sequence.

Scenario	H	HE	HET	HT
10-Minute	2.91	5.60	11.28	12.06
20-Minute	12.10	155.01	68.32	24.72

Table 7. Solution times with and without expressing (in seconds).

We can see there is substantial improvement in solution time when fixing expressing variables in the 20-minute instances. In the 10 minutes scenario, since the disruption is not very long, it is less likely to be beneficial to skip many stations or express two trains. Therefore, it does not take much time to solve the "Holding and Expressing" and "Holding, Expressing and Short-turning" cases. In the 20 minutes scenario, expressing multiple trains and skipping multiple stations may be beneficial. In addition, more trains are likely to be loaded to capacity. Therefore, more nodes may need to be explored.

4   Conclusion

A deterministic model for real-time disruption recovery has been presented, which includes holding, expressing and short-turning strategies. The model was applied to a scenario with different disruption duration on a simplified (non-branching) version of the MBTA Red Line. Two disruption lengths were tested: 10 and 20 minutes. Sensitivity analysis was conducted to investigate the impact of the assumption that the disruption duration is known with certainty when applying the model. A branching sequence for the branch and bound process was also proposed, which produced significant decreases in solution time. According to the results:

- Holding strategies may provide 10-18% saving in passenger waiting time over the no-control case. Expressing only provides modest additional saving beyond holding. Combined expressing with holding, the saving was increased to 13-23%. Short-turning may provide substantial savings. Combined with holding, short-turning provided 35-57% saving in passenger waiting time. Though the test scenario may be favorable for short-turning and holding, it is expected that the cost saving from short-turning combined with holding may still be significant in other disruption scenarios.
- The effectiveness of holding and expressing is not sensitive to the accuracy of disruption duration estimate, which is an important parameter used in the model, but the effectiveness of short-turning can be sensitive.
- One of the bottlenecks of the solution process is to determine expressing decision variables. Since expressing may only provide modest benefit over holding, expressing may be taken as a secondary strategy in order to save solution time.
- With the proposed branching sequence, all the problem instances were solved to optimality within 30 seconds, except strategies involving expressing.

The model presented here has not yet been implemented in practice but the authors are continuing to work with the MBTA to develop simple decision support tools based on the concepts presented in this paper. Further research is also underway to relax the assumption of a known deterministic disruption duration.

Bibliography

Abdel-Aty, M.A., R. Kitamura, and P.P. Jovanis (1995). Investigating effect of travel time variability on route choice using repeated-measurement stated preference data. *Transportation Research Record 1493*, 39–45.

Barnett, A. (1974). On controlling randomness in transit operations. *Transportation Science 8*, 102–116.

Central Transportation Planning Staff (1991). *1989 Passenger Counts: MBTA Rapid Transit and Commuter Rail.* http://www.ctps.org/bostonmpo/mpo/ctpsn.htm, 07.04.2001.

CPLEX (1995). *Using the CPLEX Callable Library, Version 4.0.* CPLEX Optimization, Inc.

Eberlein, X., N.H.M. Wilson, and D. Bernstein (1999). Modeling real-time control strategies in public transit operations. In N.H.M. Wilson (Ed.), *Computer-Aided Transit Scheduling*, Lecture Notes in Economics and Mathematical Systems, 471, Springer, Berlin, 325–346.

Furth, P. (1995). A headway control strategy for recovering from transit vehicle delays. In *Proceedings of ASCE Transportation Congress, San Diego*, 2032–2039.

Heimburger, D.E., A.Y. Herzenberg, and N.H.M. Wilson (1999). Using simple simulation models in the operational analysis of rail transit lines: A case study of the MBTA red line. *Transportation Research Record 1677*, 21–30.

Li, Y. (1994). *Real-Time Scheduling on a Transit Bus Route*. Ph.D. thesis, Affiliee a l'Université de Montreal, Montreal, Canada.

Lin, T. and N.H.M. Wilson (1993). Dwell time relationships for light rail systems. *Transportation Research Record 1361*, 296–304.

Murata, S. and C.J. Goodman (1998). Optimally regulating disturbed metro traffic with passenger inconvenience in mind. In *International Conference on Developments in Mass Transit Systems*, London, UK, 80–85. April Conference Publication No. 543.

O'Dell, S. and N.H.M. Wilson (1999). Optimal real-time control strategies for rail transit operations during disruption. In N.H.M. Wilson (Ed.), *Computer-Aided Transit Scheduling, Lecture Notes in Economics and Mathematical Systems*, 471, Springer, Berlin, 299–323.

Shen, S. (2000). Integrated real-time disruption recovery strategies: A model for rail transit systems. M.sc. thesis, Department of Civil and Environmental Engineering, MIT, Cambridge, USA.

Turnquist, M.A. (1989). Real-time control for improving transit level-of-service. In *Proceedings of Conference on Applications of Advanced Technologies in Transportation Engineering, ASCE*, San Diego, USA, 217–222.

Van Breusegem, V., G. Campion, and G. Bastin (1991). Traffic modeling and state feedback control for metro lines. *IEEE Transactions on Automatic Control 36*(7), 770–784.

Wong, J. (1999). *Dwell time estimation on the MBTA Red Line*. Technical report, Department of Civil and Environmental Engineering, MIT, USA.

Design of Customer-oriented Dispatching Support for Railways

Leena Suhl, Claus Biederbick, and Natalia Kliewer

Decision Support & OR Laboratory, University of Paderborn,
Warburger Str. 100, D-33098 Paderborn, Germany
{suhl,biederbick,kliewer}@upb.de

Abstract. Traditionally, dispatching strategies in railways mainly concentrate on maintaining timeliness of trains and ensuring passenger connections. Although customer-orientation is getting more and more important today, little is explicitly known about the effects of various dispatching strategies into customer satisfaction. In this paper, we discuss the design of dispatching support systems for railway passenger traffic from the viewpoint of passenger orientation. We have implemented simulation and optimization based tools and validated them using extensive data from German Rail. We report on three systems: A coarse simulator based on global waiting time rules, a detailed agent-based simulator, and an exact optimization system. The system environment can be used offline, and partially online as well, to test and evaluate dispatching strategies. The focus of the paper is on the system design and its validation for the purposes of railway dispatching. The numerical results are still preliminary and have to be extended in subsequent studies.

1 Customer-oriented Dispatching

In many countries, railways are currently in the process of becoming privatized and operating in competition with other railway companies as well as with other modes of traffic. Although it has always been important to provide good service to customers, this becomes a central issue in the increasingly competitive situation.

Customer acceptance is positively influenced by providing high quality trips and passenger connections, but negatively influenced by delays and missed connections at operations. Because scheduled traffic is always subject to external factors, many types of disturbances cannot be avoided which make short-term changes to a given schedule necessary. When a disturbance such as a technical defect, accumulated vehicle lateness, missing crew members, or congestion, occurs, a dispatcher has to react within a few minutes, and decide about changes in the schedule, such as delaying connecting trains and reallocating resources (vehicles and crews). In railway traffic it often takes several days to reconstruct the planned schedule or a consistent one after ad hoc changes.

Although customer orientation is getting more and more important, little is known about how to reliably measure customer satisfaction in railways and how to design dispatching strategies that maximize customer satisfaction. There are several established simulation tools available for railway traffic (see, e.g., Siefer and Hauptmann (2001)), however, they are usually technology-oriented aiming at optimal dispatching of trains and other equipment. In this paper, we report on our activities in simulating railway traffic in a complex network, in order to test various dispatching strategies to maximize customer satisfaction. We present decision support tools which can be used to derive and evaluate explicit dispatching strategies that can be taken over by the responsible dispatchers.

Dispatchers of railway traffic must be continuously connected to the trains: The incoming information from the railway network is usually graphically represented on one or more computer screens in order to facilitate decision making. The most important decisions of a dispatcher have to do with wait-ing or rescheduling of trains in case of connecting passengers. If an incoming train is late, the connection train either has to wait for the incoming train (thus generating further delays), or leave on time (thus leaving the connect-ing passengers to wait for another connection). Because in a dense network, such as that of German Rail, passengers sitting in a certain train may have numerous destinations involving several different connecting procedures, it is non-obvious how to measure the overall impact of a certain decision into passenger satisfaction.

Four stages of customer-orientation are shown in Figure 1. Stage one is the status quo: Until now, it is only possible to inform passengers about the situation in the net through departure/arrival-boards and loudspeaker-announcements in stations. In trains, this is handled by conductors and train attendants. Passengers are forced to "pull" information relevant for their trip from these sources, i.e., they have to read boards and ask attendants. The major disadvantage about this is that neither in stations nor in trains the interests of a certain customer (to get from his origin to his destination) can really be considered. There is no instance being able to observe the complete journey.

Stage two brings a fundamental change: Customer specific on-trip infor-mation is not longer pulled by the voyager himself, but is pushed through modern mobile communication devices like cellular phones via Short Mes-sages Service (SMS). Because of the rapid development and spreading of new standards it is easy to forecast that these techniques will be redeemed by inexpensive mobile internet devices over the next years which enables in-teraction between dispatching systems and certain passengers. This is stage three of Figure 1. Customers can actively influence decisions, e.g., by just transferring an urgently needed connection directly to the dispatcher or one of his digital assistants.

Finally, stage four focuses solely on customer needs while making dis-patching decisions. This, of course, is restricted by technical feasibility and

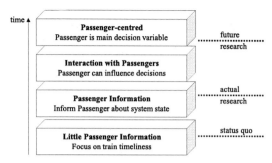

Figure 1. Stages of customer-oriented dispatching.

overall cost of a decision. "Technically feasible" means, that a redirection plan must be physically possible and secure, e.g., it is not possible to employ special trains when there are no free ones close-by. "Overall cost" confers to additional expenses that a customer-oriented solution would bring. For example, it may be too costly to charter a new crew and a new train for only a few passengers, instead of letting them take an overloaded train and give them certain bonuses.

In Suhl and Mellouli (1999) we used total passenger waiting time to compare the quality of different dispatching decisions. For a more detailed simulation approach, only one global decision variable is not sufficient. Some possible criteria for customer satisfaction from a customer's point of view are:

- Hard (measurable for dispatcher)
  - waiting times at stations
  - waiting times in trains
  - missed connections
  - trip properties as booked (e.g., dining-car, preferred train type, smoker/non-smoker, etc.)
  - on-trip-information quality
- Soft (not exactly measurable for dispatcher)
  - tidiness of trains and stations
  - security
  - good service, friendly staff
  - comfort of journey
  - quality of catering

However, among the criteria characterizing customer satisfaction, we usually emphasize timeliness (of customers, not trains (!)) to be the most important one in our studies. The main reason for this is that we are not able to modify most other decisive factors easily as they cannot directly be influenced by dispatching decisions.

In order to properly weight a customer's waiting times, we have to consider the circumstances, under which the customer has to wait. Therefore, we may

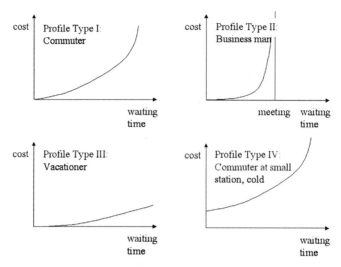

Figure 2. Cost functions for different passenger types.

define a parameter called waiting time quality: Does the customer wait in the train or at the station? If the latter, what type of station is it, i.e., is there any pastime, or is he or she forced to wait at the perhaps cold and windy platform? A second criterion influencing the quality of waiting time is the type of a trip. Waiting a few minutes longer may not be arduous if the customer is on a holiday trip, while it could be crucial for reaching a business appointment. This leads to waiting time costs for each passenger, which we represent with utility functions. Some typical utility functions are shown in Figure 2.

2   Basic Approaches and Tools

2.1   Simulation as a Decision Support Tool

Computer-based simulation has gained more and more importance for planning problems arising in logistics and transportation, especially for scheduling and dispatching problems (see Manivannan (1998), as an example). Its ability to handle even very complex systems makes it well-suited for this type of special-purpose network-based applications. Thus, we propose the use of simulation not only for visualization and real-time monitoring of transportation systems, but also for analyzing and, with a growing knowledge, improving customer satisfaction. In the first place, we use the simulator as an offline tool in order to test various strategies and derive conclusions for dispatchers.

Because complex simulation systems are computationally intensive, a detailed simulation can usually not be carried out in real-time to support dispatcher decision making immediately when a conflict occurs. However, if simulation of railway traffic is performed on several granularity levels, involving

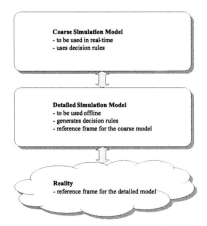

Figure 3. Granularity of simulation models.

more and less detailed models that are compatible with each other, then the less detailed models can be used in real-time as well. We propose a two-level simulation, consisting of a coarse model capable to compute a fast analysis in real-time according to given rules, as well as a detailed one in order to run accurate analyses and to generate the rules for the coarse level (Figure 3).

The coarse simulation model may be based on Waiting Time Rules (WTR) which define a general and fixed framework for dispatching decisions in case of delays.

Waiting Time Rule: A connection train has to wait for a delayed incoming train at least a given number of minutes - if the delay exceeds this number, a conflict occurs, and the dispatcher has to decide whether to wait or not.

The simplest case would be one global waiting time rule to be applied for all delays, independently from location and train type. However, the railway company may wish to use different rules depending on the type of the train (this is the case at German Rail currently). It would be possible to differentiate further, thus making the rules dynamically dependent on the amount of connecting passengers, time of the day, amount or frequency of the traffic, and so on. In any case, the coarse simulation is still mainly oriented on trains and connections. Its impact on the satisfaction of a given customer can be derived implicitly and is rather of general nature.

The detailed simulation approach adopts, on the contrary, an explicit customer-oriented view. It can be used to evaluate various sets of waiting time rules on a microscopic level involving each single passenger, as an extreme case. Thus, detailed information of a given dispatching strategy can be derived and used as an input to the coarse simulation - which is fast enough to be carried out in real-time. Thus, the detailed simulation may answer questions such as:

- Could the total delay time of all passengers be decreased by using a different set of waiting time rules? How do delay profiles of trains vary then?
- What is the effect of dynamic waiting time rules on the total passenger delay time?
- What are the net-wide effects of a disturbance if all global (and local) waiting time rules for trains are applied? How can these effects be measured in delay/waiting times for passengers?
- Are there typical delay profiles for some trains which could be changed by, e.g., varying buffer times or waiting time rules?
- Is there significant correlation between delay times of two trains which are directly or not directly connected? How can this correlation be formalized and represented?
- What are appropriate statistical distributions for delay times of trains in special stations and what are good estimates for their parameters?

Both the coarse and the detailed simulation can be implemented using the same basic simulation environment which involves the underlying railway network and time table. In the subsequent sections, we describe a basic distributed simulation environment that was developed at the University of Paderborn (see also Suhl et al. (2000)). The environment is capable of representing any (complex) railway network; for our purposes it was configured with data from German Rail.

To study the problems outlined above, it is necessary to represent the network itself and the events occurring in it correctly. Therefore, the simulation model has to be validated precisely using a - in a statistical sense - sufficient amount of data from the real system. In our study, this data was provided in form of log-files from the running system containing operating data.

2.2 Using Optimization in Decision Support

An adequately validated simulation model delivers valuable insight into the effects of dispatching decisions from the customer point of view. With simulation we can evaluate and compare given strategies, and thus make conclusions in order to identify the best ones. However, under certain assumptions we can go one step further and apply exact optimization techniques in order to determine a mathematically optimal dispatching strategy.

Additionally to the simulation models, it is possible to construct an optimization model that uses the same basic data about the network, timetable, and delays. In each case of a conflict - in the sense of the above definition of waiting time rules - there is a decision variable determining whether the departing train leaves on time or waits for delayed passengers. With the optimization model, we seek an optimal set of dispatching decisions in such a way that a given objective function is minimized or maximized, according to its definition, and that all relevant restrictions are fulfilled.

In subsequent sections, we report on a mathematical optimization model that is capable of including all long-distance trains of German Rail, and a representative amount of passengers per train. The optimization model was validated using data from German Rail, however, the results at the current stage can only give first impressions and determine directions to be studied in the future.

2.3   The Basic Simulation Environment

As noted earlier in this paper, dispatchers in railway systems cannot adequately integrate passenger wishes into their considerations, mainly because of two obviously interrelated reasons:

- lack of well designed information and decision support systems able to integrate customers, and
- lack of information about individual passenger behavior.

Even if reliable and extensive real-life data is missing, we can construct approximate passenger data and use it to simulate passenger behavior. In the following, we discuss the outline of our basic simulation environment that can be used for both the coarse and detailed simulation proposed above. The same basic data can be used in the mathematical optimization model as well.

The basic environment is principally capable of representing every single passenger. With other words, origin, destination, intermediate stations, and preferences of each passenger are given and stored within the system. Furthermore, a set of (simple) rules describing a single passenger's behavior has been deducted based on experienced passenger behavior. For the coarse simulation, it is possible to group passengers with similar itineraries, thus reducing the number of passenger units to be handled by the system and increasing its performance. This information is sufficient to develop and test methods that are capable of supporting each of the four stages of customer-orientation as shown in Figure 1. Except passenger behavior and data about the network and timetable, the basic environment also includes the simulation machinery, such as time propagation, event handling, and a random generator of delays with their location and length.

The technical realization of our simulation environment is event-oriented and distributed, i.e., one computer processes the timetable, one handles simulation time, and so on. This functional fragmentation was necessary because of the high complexity of the system. It makes the use of low-cost hardware (2-3 Intel Pentium/AMD Athlon processors at approximately 500 MHz) possible. The computers are connected with each other through Internet technologies. The implementation languages were Java and C++.

The environment includes simplified mechanisms for propagating delays, for recognizing conflict situations, and for performing what-if analyses to support estimating net-wide effects of a dispatcher's decision. The underlying object-oriented data management component contains a complete timetable of German Rail. It is able to manage the scheduled, actual, and expected state for all trips at all their intermediate stations, as well as minimum delays enforced by expert decisions.

3 Macroscopic View: Waiting Time Rule Simulation

3.1 General Framework

The coarse (or macroscopic) simulation level includes constant or variable
Waiting Time Rules, as described above. After generating simulated passenger flows for the main connections, and propagating a certain delay within
the network, we are able to estimate the accumulated number of passenger
waiting minutes induced by the delay. We use the total delay time of passengers as a central measure as it directly influences customers' behavior.
It can be roughly divided into two parts: the total of delay times affected
at the moment of analysis and the total of delay times induced by delaying
connection trains, either using waiting time rules or by dispatcher's decisions
in case of induced conflicts. Therefore, when we consider induced conflicts in
other stations and try to minimize their number, this will indirectly decrease
passengers' total delay.

   In the coarse simulation, we use waiting time rules, as described above,
that differentiate between train types. Thus, it is possible that fast trains,
such as ICE or IC/EC of German Rail, do not wait for slower trains, such
as SE or SB, but slower trains wait a certain number of minutes for faster
trains (see Table 1).

Type		Remarks
ICE	Inter City Express	Long distance train between major cities
IC/EC	Inter City/Euro City	Long/middle distance between bigger cities
IR	Inter Regio	Regional long distance train
SE	Stadt(city) Express	Short distance city connector
SB	Stadtbahn	City rail
Other	(including D, RB, RE)	less important middle distance trains

Table 1. Train types of German Rail.

   If we use constant waiting time rules all over the network, we speak of
global waiting time rules. Typical global regular waiting times are: SE-trains
for local traffic wait for EC-trains up to 10 min but EC-trains do not wait
for SE-trains. The ICE/EC/IC/IR-trains wait for each other up to 5 min.
SE-trains and other local trains also wait for each other up to 5 minutes.

   Furthermore, the rules in the macroscopic simulation consider Minimum
Transit Times (MTT) needed by passengers to proceed from one railway
platform to another within a station. Global minimum transit times are set
to 2 (or 3) min, if the connection train leaves on the same platform (or that on
the opposite side), and to at least 5 min otherwise, since passengers have to
change platform by using a tunnel. Results from our simulation study are an

evaluation of the quality of regular waiting time rules, suggestions for global or local corrections of these rules, and the development of simulation-based methods for these corrections to reach a better robustness against delays.

3.2   Underlying Assumptions

The macroscopic simulation makes assumptions about passenger flow that are realistic enough to give general indication about passenger satisfaction. We initialize each trip with a random number of passengers, e.g., an IC-train i is initialized with $150 + z_i$ passengers, where z_i is a normally distributed random variable. At each station, some passengers will *leave* train i, some will *change* to connection trains r, and some will *get into* train i. All the necessary (deterministic) calculations are based on simple functions which take size and reach of trains and the size of stations they meet at into account.

In case of passengers leaving the system their delay has to be added to the accumulated waiting time. It has to be increased as well, if passengers cannot reach their connections caused by a dispatchers' decision. We assume that passengers will wait for the next trip on this line rather than search for an alternative connection, if it is close enough in time (e.g., smaller than M minutes in the future). Nevertheless, if the time to wait is too long, they will take other trains on different routes (we suppose that there *is* one) and arrive at their destinations earlier than the delayed train would arrive. At last, the number of passengers entering a train equals the sum of passengers leaving it and those changing to another train. This number is also varied by a normally distributed random variable.

Every train can lose or regain time during a trip. To model this, we define normally distributed random variables for each type of train reflecting the relative change of delay on a trip between two stations. Extensive analysis of data from the real system led to this assumption.

In the study, we generated disturbances randomly according to time and place. We did not consider the type (the reason) or topology of disturbances, although this implies slightly unrealistic behavior of the system. We assume that all possible disturbances result in delays of the concerned trains. Because these delays will propagate from one train to some of its connectors, a single delay could have net-wide effects. Therefore, we are able to conduct ceteris-paribus-studies to compare different dispatching scenarios described below.

"Artificial" disturbances in a simulation run can be generated automatically or manually. We used two exponentially distributed random variables, one for the inter arrival times for disturbances and one for their delay length. The train to be delayed was chosen randomly from all trains or only from the driving/standing ones, i.e., dense parts (tracks and stations) of the network of course get more delays than sparse parts. Additionally, we introduced train-independent delays at each station to model congestion on outgoing tracks.

A conflict occurs, if the delay of a given train exceeds the regular waiting time defined by the waiting time rule. In the simulation system the computer instead of the dispatcher has to decide, whether a connecting train waits or not. Since there is no fixed set of rules a dispatcher follows, we are not able to model dispatcher behavior correctly in all cases. Dispatchers consider many fuzzy parameters in parallel, e.g., passenger requests submitted through the conductor, service rate on this trip, number of connecting trains in the following stations, track occupancy, etc.

For the simulation we had to simplify this complex task. We tested different dispatching strategies by using fast heuristics. We only distinguished between two classes of trains: The first class (class 1) includes the fast, high priority trains of types (1), (2) and (3) mentioned above, and the second one (class 2) contains all other types.

3.3   Impact of Waiting Time Rules

With the coarse simulation model we tested the impact of waiting time rules into the total number of conflicts, so that we randomly induced a number of conflicts and monitored their propagation within the network, as well as the number of successor conflicts induced by trains waiting for each other. Two conflict types are distinguished: minor conflicts, exceeding regular waiting time up to λ minutes, and major conflicts, exceeding regular waiting time more then λ minutes. We consider four different scenarios:

- A – a train will not wait for incoming trains at all.
- B – a connecting train will wait until all its regular feeder trains arrive.
- C – a connecting train will wait for its regular feeder trains, but only until regular waiting times are reached.
- D – a connecting train will wait for its regular feeder trains within the regular waiting times and in case of minor conflicts. Here, we modified the general waiting time rules by adding a parameter λ: if a delay exceeds the regular waiting time an amount that is less than λ, we have a minor conflict, otherwise we have a major conflict.

Scenarios A and B reflect extreme dispatching strategies (no waiting at all, always wait), scenarios C and D are intermediate strategies.

Two classes of trains are distinguished in the simulation: Trains of class 1 mentioned above, wait for each other for up to 5 min and do not wait for trains of class 2, which wait for each other for up to 5 min. Thus, RWT reduces to a 2×2-matrix, which is set to (5, 0; 10, 5). This corresponds to the global regular waiting times currently in use at German Rail. In the study, scenario D helps in drawing some conclusions about the effect of small changes of these regular waiting times. Logging all conflict situations that occur after propagating artificial disturbances helps in gaining insight into local and regional instabilities of the schedule. In this paper, we concentrate on conflict behavior and propagation in the railway network. Preliminary results from the above scenarios are stated in the next section.

Table 2 shows the results of a series of simulation runs under the above assumptions. We conducted 30 runs for each simulation experiment. For each run over approximately one and a half days, 30-40 disturbances are distributed in the average with an expected delay between 3 and 7 minutes. For each of the scenarios, the same disturbances are distributed for $\lambda = 2; 3; 5$.

For the extreme scenarios A and B, Table 2 shows the effect of propagating the disturbances generated by the distribution. Since connecting trains do not wait for incoming trains at all for scenario A, the number of (minor and major) conflicts simply reflects the number of primary delays induced at subsequent stations of those lines where the disturbances are generated (one level). The second extreme (scenario B) shows the dimension of delay propagation and of secondary conflicts throughout the network over several levels. Here, all passenger connections have been guaranteed, but the price is an extremely large number of induced secondary conflicts at subsequent stations after primary conflicts.

The number of conflicts per generated disturbance (shown in the second column of each conflict type) can be taken as a unifying normative indicator. Thus, for $\lambda = 2$ a disturbance induces in the average 1.395+0.067 delays on the first level (concerning connections on the same line) and 106.09+0.3958 delays over all levels within the considered trains of German Rails network. The first number in the summations indicates the number of major delays.

Scenario	λ	MAJOR CONFLICTS			MINOR CONFLICTS		
		Total #	Confidence Interval: $\alpha = 0.05$	Average per Disturbance	Total #	Confidence Interval: $\alpha = 0.05$	Average per Disturbance
A	2	50.03	[48.05; 52.02]	1.3950	2.40	[1.85; 2.95]	0.0669
	3	47.97	[46.05; 49.88]	1.3374	4.47	[3.79; 5.14]	0.1245
	5	44.57	[43.08; 46.05]	1.2426	7.87	[6.90; 8.83]	0.2193
B	2	3681	[3474; 3889]	106.09	13.73	[7.78; 19.60]	0.3958
	3	3675	[3468; 3881]	105.90	20.47	[13.08; 27.85]	0.5898
	5	3655	[3449; 3860]	105.32	40.27	[29.74; 50.79]	1.1604
C	2	52.50	[49.16; 55.84]	1.4100	3.70	[2.75; 4.65]	0.0994
	3	50.07	[47.05; 53.09]	1.3447	6.13	[4.59; 7.67]	0.1647
	5	46.33	[43.44; 49.22]	1.2444	9.87	[7.94; 11.79]	0.2650
D	2	141.4	[56.88; 225.85]	3.9488	126.6	[0.00; 280.3]	3.5363
	3	173.1	[17.88; 328.3]	4.8388	242.2	[0.00; 484.5]	6.7717
	5	1309	[783.2; 6374]	34.904	3266.	[1818; 4713]	87.083

Table 2. Results for extreme dispatching strategies A and B, and for intermediate dispatching strategies C and D.

A conclusion from Table 2 (scenario B) is that, when trains always wait, both the number and the length of delays increase drastically. This is indicated by the number of major delays which increases by a factor of 105-106 (for λ-values of 2, 3, and 5). This factor varies only slightly in the simulation, for instance, between 100 and 112 for a confidence of $\alpha = 0.05$ in the case of $\lambda = 2$. This shows the necessity of the waiting time rules and of dispatching decisions which have to reduce the amount of overall delays and at the same time minimize missed connections for passengers.

Now, it is interesting to see the impact of waiting time rules for reducing the propagation of delays within the net. Table 2 shows also results for scenarios C and D with different values of λ. Recall, that scenario C corresponds to dispatching decisions solely by the waiting time rules, i.e., trains do not wait in case of conflicts. For scenario D, trains wait in case of minor conflicts, this is equivalent to relaxing the waiting time rules by adding λ minutes to the maximum waiting times given by the rules.

The results for scenario C in Table 2 show that the waiting time rules currently in use by German Rail only slightly increase the numbers of major and minor conflicts relatively to the scenario A where trains do not wait at all. Because the network is dense, a slight increase of waiting times by 2 or 3 minutes (equivalent to the scenario D with $\lambda = 2; 3$) approximately triples numbers of major conflicts.

The "astronomical" increase in number of major conflicts for scenario D, $\lambda = 5$, shows the dimension of induced delays, if trains wait 5 minutes more than the regular waiting times: From $\lambda = 3$ to $\lambda = 5$ (only 2 min), the number of major conflicts increases by more than seven times. This already corresponds to more than one third of the conflict dimension of the extreme strategy B. This is mainly because a single major delay in the morning may propagate during the whole day and cause numerous major delays at other stations. Thus, an increase of regular waiting times may only be acceptable at the end of the day for latest connections, or only locally if the local network is not dense.

Therefore, it appears convenient to rather decrease the regular waiting times globally by a few minutes and perhaps increase these waiting times only locally in order to reduce missed connection at certain stations. To found conclusions in this direction, we have to extend simulation results by local considerations and to integrate passenger waiting times.

The coarse simulation, as described in this section, can be understood as a preliminary stage in order to evaluate customer-oriented dispatching strategies. It is possible to record the accumulated total waiting time for groups of customers following the same itinerary. Although our current numerical results do not give explicit numbers of passenger waiting time, some trends can be identified for given parameter values. The total measure of success strongly depends on the special data constellation, especially on the number of connecting passengers per conflict, and on the total waiting time of passengers that miss their connections because the connecting train left before they reached the connection station.

4 Microscopic View: Agent-based Simulation

4.1 Outline of the Agent-based System Architecture

Software agents are (broadly speaking) computer programs which are able to behave autonomously in a certain sense, interact with other software agents, thus building communities, and move within a digital network. Because there are usually several interacting agents, we often speak about multi-agent systems. It is conceptually appropriate to use software agents to implement a microscopic view of the railway system, because an agent is usually configured to represent a microscopic item like one specific train or station, even one specific customer.

From the computer science point of view, multi-agent technology can be understood as an advancement of object-oriented programming, thus presenting a new programming paradigm. Especially, the concepts of autonomous behavior and intentionality provide new dimensions to model objects with control functionality; this is not possible with traditional object-oriented programming techniques.

We generally distinguish between two basic variants of architecture models for multi-agent systems: *deliberative* and *reactive* architecture. Agents in the deliberative model behave analogously to expert systems as they are known in artificial intelligence. They possess an explicit symbolic model of their environment, together with logical inference mechanisms in order to be able to derive conclusions and evaluate possible actions. A well-known deliberative agent architecture model is the "Belief, Desire, Intention" model of Bratman (1987). Reactive agents react on certain stimuli from their environment with executing specified actions. The easiest way of implementation are if-then rules, e.g., *if* obstacle ahead *then* go left. Agent goals are not given explicitly, intelligence arises incrementally a non-centralized way, leading to a new modeling paradigm for distributed systems.

In our opinion, the agent concept is very well suited for modeling parallel and distributed simulation systems running on low-cost hardware. Furthermore, the agent concept enables us to integrate (autonomous) real world players, such as customers and dispatchers, into the simulation, thus supporting customer information and distributed real-time dispatching as well.

Seen from a bird's eye view, the microscopic system consists of three major agent servers, each of them having local subsidiaries (cf. Figure 4), together with a main simulation server.

- Passenger Server: Global generator of passengers in the system or interface to real-world-data, respectively.
- Topology Manager: Provides detailed information about the net, e.g., how many tracks are there between two major cities, what is the speed limit on certain tracks, where (geographically) are the stations, etc.

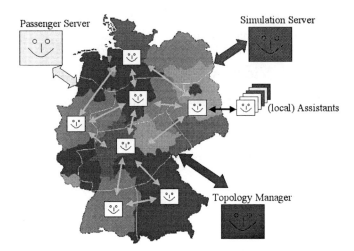

Figure 4. Region concept of the simulation system.

- Local Assistants of dispatchers: Although there are two important dis-patching instances in each region of the German Rail, transport direc-tion (to ensure passenger connections) and operations control (to en-sure timeliness of trains), there is only one instance per region modelled in the system, because this is a passenger's logical view upon the sys-tem. The dispatchers are supported by local computer-based assistants implementing special features, such as trouble seeker/predictor, trouble shooter/evaluator, and passenger navigator.

- The Simulation Server is responsible for all simulation tasks, e.g., event handling, mapping and scheduling, logging, analyses, synchronization and all kinds of parameter settings.

All system components are designed as deliberative agents. As an example, the Topology Server autonomously recognizes incoherence of the input-data, reacts on special events logged during simulation, is able to add this to its knowledge-base, and, therefore, learns systematically about the net. It is important to backup these central servers peripheral in their respective local mirrors.

From a detailed point of view, Figure 5 shows the system structure within the geographic regions. A human or automated dispatcher controls trains, passengers, and stations. When a conflict occurs, he or she searches for good solutions. It is possible to use the train agents for peripheral decision making in a reactive architecture as well: Trains negotiate with each other about guar-anteeing connections or not. Input data are passenger information (gathered from the actual and – if already known – future passengers), such as route, preferred train type, and transfer speed at stations, as well as the expected delay profile of the train, list of endangered connections, and so on.

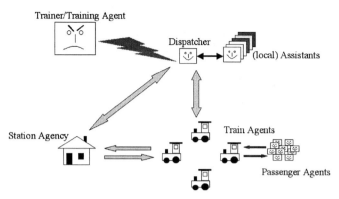

Figure 5. Agent communication within regions.

The agent-based environment is not only useful in simulations, but can be implemented for real moving trains as well. Therefore, the negotiation process takes place at the station where the trains involved have a feeder/connector-relationship. Agents meet at the station, because online-communication between driving trains may be difficult to ensure. Since agents themselves – instead of the negotiation content – move from train to station, only a short transmission window is needed. Online-negotiations would need longer availability of communication channels. This strategy reduces the dispatcher's workload dramatically, since he or she does not need to solve every conflict situation explicitly. However, if a dispatcher wants to keep control over each decision, he or she can use the result of the train negotiations as a supporting input.

4.2   Experiments with a Prototype System

We carried out a prototype implementation of the agent-based detailed simulation based on the IBM Aglets Software Development Kit, Version 1.1b3 (see www.trl.ibm.co.jp/aglets). The implementation exploits the aglet ("agent-applet") concept within a distributed computing environment connected over the Internet. An aglet is a mobile Java object being able to move between hosts over the Internet. Aglets communicate with each other by exchanging messages.

The main goal of the detailed simulation study was to find out, whether it is possible with current technologies to simulate a large real-life network of thousands of stations, trains, and passengers. The network of German Rail involves about 5000 stations and 30000 trains daily. Because a train passes 10 stations on the average, there are about 600000 departures and arrivals to be considered per day. The simulation was carried out within a distributed environment consisting of five regional dispatching units that correspond to five geographic regions of German Rail.

Characteristic	1:100	4:1	100:1
Number of passengers, who missed their connection	944	908	796
Passenger rate, who missed their connection	31.95%	30.73%	26.94%
Number of trains, which did not take at least one passenger	270	256	227
Trains rate, which did not take at least one passenger	4.64%	4.40%	3.90%

Table 3. Passenger statistics.

Major classes of agents describing objects in the real-world situation are: passenger agent, train agent, station agent. On top of these agent objects, the basic system includes several object types of more technical nature, such as database aglet, event handler, delay, and discussion.

In order to validate the agent-based system, we carried out experiments with real data. We were able to represent all connections and trains within the five-region network. Because of performance reasons within the aglets environment, it was not possible to generate an agent for each passenger (this would have meant millions of agents). However, a representative number of roughly 5000 passenger agents with realistic itineraries was generated.

In our experiments, approximately 50% of all trains were delayed, with an average delay length of sixty minutes. A delay does not remain constant in the course of a trip, because train time tables include some dwell time that can be exploited to reduce the delay lengths. If a connection is endangered by the feeder's delay, the two trains involved start a negotiation process, that can be described in the following way:

A delayed feeder train requests the connecting train whether a given connection can be ensured. The request contains information about the length of the delay and the number of transit passengers involved. Thus, the connecting train knows how long it would have to wait in order to ensure the connection, and how many passengers would miss their connection otherwise.

The two trains involved negotiate whether the connecting train should depart or wait. A negotiation strategy is characterized by two central parameters: The first parameter describes the customer-related profit per each additional passenger who will be carried further (instead of leaving to wait for a next connection). The second parameter presents the loss, or negative profit, for each additional minute of delay of the train.

Tables 3 and 4 present the results of our simulation runs with three different negotiation strategies. A 100:1 – relationship means that a connecting train waits for one individual passenger up to 100 minutes. Thus, as many passengers as possible reach their connections. If this ratio equals 1:100, then the feeder waits one minute only in the case that thereby at least 100 passengers would reach their connections. These two parameter settings present the two extreme strategies: Either trains wait for each other always (100:1), or they do not wait at all (1:100). Despite the high threshold of 100 minutes

Characteristic	1:100	4:1	100:1
Accumulated delay in minutes	160675	161664	163085
Number of delayed trains	2908	2929	2961
Trains rate with total delay	49.97%	50.34%	50.89%
Cumulated negotiation delay in minutes	0	44	1377
Number of trains having negotiation delay	0	6	39
Rate of Trains having negotiation delay	0.00%	0.10%	0.67%
Trains which were late, additionally to input delays	-15	6	38
Cumulated caught up delays in minutes	14813	13868	13780
Rate of cumulated caught up delays	8.44%	7.90%	7.79%

Table 4. Delay statistics.

missed connections are not significantly reduced because of the unrealistic number of the delays in our example as well as their extreme length. The third negotiation strategy (4:1) is an intermediate one. It implies that an additional delay of four minutes is weighted with as much loss as the profit of reaching one passenger connection would be.

5 Optimization for Customer-oriented Dispatching Support

The methods of exact mathematical optimization were applied to dispatching problem instances of different complexity, in order to get solutions optimal in the sense of customer orientation. Result of the optimization is not only a single decision recommendation, but a dispatching plan for a complete decision period.

A further aspect of our investigations is the real-time applicability of mathematical optimization in daily railway traffic dispatching. Additionally, the optimization system can evaluate decisions in training courses for dispatchers as well as the quality of heuristic approaches. Accomplished experiments for validation of the model are based on real timetable and topology data from German Rail.

5.1 Basic Model Definition

For a train $j \in Trains, TrainStations_j$ denotes all stations used by j. For short, the index set of stations is denoted by I, and the indices of $Trains$ are given as J and K, too. $L^{1/2}$ are numbers of linear segments in the approximated cost functions for in-vehicle and platform waiting times, respectively.

Variables:

$T_{i,j}^{a/d}$ $\forall\, j \in Trains, \forall\, i \in TrainStations_j$
ordered sets of arrival and departure times.

$Y_{i,j,k}$ $\forall\, k \in Trains, \forall\, i \in TrainStations_k, \forall j \in Feeders_{i,k}$
binary decision variables

$$Y_{i,j,k} = \begin{cases} 1 - \text{if } k \text{ departs not earlier as } T_{i,j}^a + mtt_{i,j,k} \\ 0 - \text{otherwise} \end{cases}$$

Model Parameters:

$t_{i,k}^{a/d}$ timetabled arrival and departure times

$\tau_{act_k}^{a/d}$ latest train arrival/departure times

$rwt_{i,j,k}$ regular waiting time, train k has to wait for train j

$mtt_{i,j,k}$ minimum transit time for passenger to change a platform

$mdt_{i,k}$ minimum dwell time of train k in station i

$d_{i-1,i,k}$ train running time from station $i-1$ to station i

$in_{i,k}$ passenger load of train k in station i (departure)

$ch_{i,j,k}$ passenger changing from train j to train k in station i

$takt_{i,k}$ waiting time for the next connection (approximately)

$rwtOn$ consider / not consider regular waiting time conditions

Objective Function: The total weighted passenger waiting time, including platform passenger waiting time (weighted by α) and in-vehicle passenger delays (weighted by β) is to minimize. An objective function is given by function (1).

$$\text{Minimize } C_1 = \sum_{IK} in_{i,k} * (T_{i,k}^d - t_{i,k}^d) * \alpha$$

$$+ \sum_{IJK} ch_{i,j,k} * (1 - Y_{i,j,k}) * takt_{i,k} * \beta \qquad (1)$$

Additionally, we integrate a nonlinear objective function to express the correlation between amount and duration of passenger delays. It is plausible to assume that a short delay for a large group of passengers is not as bad as a long delay for a few passengers, even if the accumulated waiting time is the same. The left side of Figure 6 visualizes the graph of the cost function, which could be approximated as piecewise linear function with arbitrary precision, which is shown in the right part of the figure.

$$C_2 = \sum_{IK} (in_{i,k} * \sum_{l^1=1..L^1} z_{l^1} * U_{l^1,i,k}) * \alpha$$

$$+ \sum_{IJK} (ch_{i,j,k} * \sum_{l^2=1..L^2} w_{l^2} * V_{l^2,i,j,k}) * \beta$$

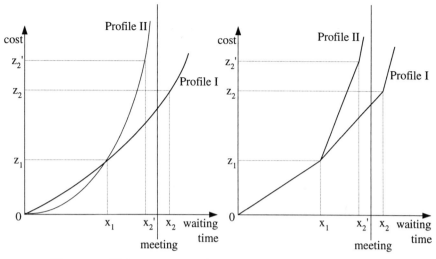

Figure 6. Waiting time cost function and linear approximation.

$U_{l^1,i,k}$ and $V_{l^2,i,j,k}$ are the special variables for modeling of piecewise linearity for in-vehicle waiting time modeling and for platform waiting time modeling, respectively. These variables underlay SOS2 restrictions (Special Ordered Set type 2).

Constraints:

$$T_{i,k}^a \geq t_{i,k}^a \tag{2}$$

$$T_{i,k}^d \geq t_{i,k}^d \tag{3}$$

$$T_{akt_k,k}^a = \tau_{akt_k}^a \qquad \forall\, k \in delayedTrains \tag{4}$$

$$T_{akt_k,k}^d = \tau_{akt_k}^d \qquad \forall\, k \in delayedTrains \tag{5}$$

$$T_{i,k}^d \geq T_{i,k}^a + mdt_{i,k} \tag{6}$$

$$T_{i,k}^a \geq T_{i-1,k}^d + d_{i-1,i,k} \tag{7}$$

$$T_{i,k}^d - T_{i,j}^a - mtt_{i,j,k} \geq M_1 * (Y_{i,j,k} - 1) \qquad \forall (i,j,k) \in IJK \tag{8}$$

$$T_{i,k}^d - T_{i,j}^a - mtt_{i,j,k} \leq M_2 * Y_{i,j,k} \qquad \forall\, (i,j,k) \in IJK \tag{9}$$

$$rwtOn * (rwt_{i,j,k} - (T_{i,j}^a + mtt_{i,j,k} - t_{i,k}^d)) \leq M_3 * Y_{i,j,k} \,\forall\, (i,j,k) \in IJK \tag{10}$$

The inequalities (2) and (3) limit arrival and departure times of the trains by the given timetable. The equations (4) and (5) initialize the arrival or departure times of all late trains in their current station. Condition (6) considers temporal relationship of train's arrival and departure times in a same station. Travel time to the next station consider equations (7).

If one of the feeder trains of the train k arrives late, the dispatcher decides whether the connection comes off. Let j and k be feeder and connector in the station i. Transition passengers can enter the connecting train k earliest at time $(t_{i,j}^a + mtt_{i,j,k})$. The variable $Y_{i,j,k}$ in conditions (8) and (9) describes whether this connection comes off.

M_1 and M_2 values must be selected large enough in such a way, that *(1)* the restriction (8) does not limit the range of the expression $(t_{i,k}^d - T_{i,j}^a + mtt_{i,j,k})$ for $Y_{i,j,k} = 0$, and *(2)* the restriction (9) does not limit the range of this expression for $Y_{i,j,k} = 1$.

The waiting time regulations are modelled as conditions (10). By setting the parameter $rwtOn$ to 0 these constraints can be switched off. Parameter M_3 in (10) should be selected similar to M_2 large enough.

Model extensions: Various model extensions which describe the planning problem more detailed were formulated, among other:

- Consideration of predicted data.
- Distance allocation scheme and capacities on distances.
- Limiting the number of dispatcher decisions.

5.2   Results

Several instances of the planning problem were solved with the help of the optimization program library MOPS (cf. Suhl (1994)). Table 5 shows some characteristic results selected from various runs carried out on an AMD-K6-2 processor with 350 MHz and 256 MB of memory. (The primary delays are summarized as disturbances.)

Impact of the Waiting Time Rules: The computations show that the optimization system can solve large mixed-integer problems within a short time. The system can also support the search for better waiting time rules. Heuristics for the planning strategies can be tested and evaluated.

The waiting time rules of the German Rail were introduced to facilitate the dispatching task by specifying a maximum time a connection train has to wait without need of dispatcher's analysis. Waiting times depend on the types of feeder trains. They are based on the assumption that it is more useful to let the connection train wait in case of small delays of its feeders. The Table 5 contains some examples of how the consideration of the waiting time rules affects the passenger delays. Test runs with and without consideration of the rule waiting periods were executed for different problem instances. Omitting restrictions means generally an extension of the solution space of a problem and often leads to better objective values. However, for many parameter sets and particularly for smaller problem instances (with modeling time shorter than 1 hour) omitting the regular waiting time restrictions does not cause improvement of the objective value. An example for a modeling time of 4 hours is specified in the parameter sets 1.5.1 and 1.5.2 in Table 5.

With a modeling time under one hour omitting the waiting time rules restrictions a 5% reduction of the total passenger delay was observed. The cost savings grow with increasing modeling time. A comparison of the parameter

parameter set	1.4.2	1.4.3	1.5.1	1.5.2	1.5.3	1.5.4
modeling time	1	1	4	4	4	4
region	F	F	F	F	F	F
apply WTR	yes	no	yes	no	yes	no
disturbances	14	14	5	5	10	10
restrictions	5125	5125	21479	21479	21479	21479
variables	4211	4211	17034	17034	17034	17034
nonzeros	11493	9007	48241	37675	48241	37675
LP time (sec)	7.47	6.87	32.02	29.71	30.16	27.91
IP time (sec)	46.36	27.07	145.5	135.17	134.01	168.34
B&B-Nodes	1263	1027	2087	1878	2976	4143
IP-iterations	369	184	178	182	206	177
waiting time (mins)	165000	161700	131600	131600	137100	116900

Table 5. Impact of the Waiting Time Rules.

sets 1.5.3 and 1.5.4 shows a 15% cost reduction by omitting the waiting time rules. The number of "wait"-decisions dispatcher have to make grows rapidly compared to the model with waiting time regulations. In both approaches model solution times were similar.

These results show that waiting time rules are meaningful in the rail operations control, as long as the decisions are made by a dispatcher, without computer-aided decision support. If the dispatcher receives support from an optimization system, the waiting time rules can be omitted and for each individual link, decisions after the optimizer recommendations can be made.

6   Conclusions

In this paper, we argued that railway dispatching should be performed primarily customer-oriented, not mainly equipment-oriented as the traditional way. We have developed a railway simulation environment, based on a distributed computing network, that is capable to evaluate passenger satisfaction according to given measures, such as total accumulated passenger waiting minutes. We propose a two-stage simulation system, whereby a fast coarse-grained simulation can be used in real-time with fixed waiting-time rules for trains. In a second stage, the use of software agent technology allows us to build a fine-grained simulation which is basically capable to model each passenger within the network. If the fine simulation model is connected into a real-time dispatching system, it can be used to generate individual passenger information in case of delays. More information can be provided through a mathematical optimization model. The preliminary results show that it is possible within our approach to represent even very complex networks and that the environment can deliver valuable insight for dispatchers.

Bibliography

Bratman, M. (1987). *Intention, Plans, and Practical Reason.* Harvard University Press, Cambridge.

Manivannan, M.S. (1998). Simulation of logistics and transportation systems. In J. Banks (Ed.), *Handbook of Simulation: Principles, Methodology, Advances, Applications, and Practice,* Wiley, New York, 571–604.

Siefer, T. and D. Hauptmann (2001). Computer aided planning of railroad operation. In M. Pursula and J. Niittymäki (Eds.), *Mathematical Methods on Optimization in Transportation Systems,* Kluwer, Dordrecht, 37–47.

Suhl, L. and T. Mellouli (1999). Requirements for, and design of, an operations control system for railways. In N.H.M. Wilson (Ed.), *Computer-Aided Transit Scheduling, Lecture Notes in Economics and Mathematical Systems,* 471, Springer, Berlin, 371–390.

Suhl, L., T. Mellouli, and J. Goecke (2000). Informationstechnische Unterstützung des Störungsmanagements im schienengebundenen Personenfernverkehrs. In J.R. Daduna and S. Voß (Eds.), *Informationsmanagement im Verkehr,* Physica, Heidelberg, 125–143.

Suhl, U. (1994). MOPS: A mathematical optimization system. *European Journal of Operational Research 72,* 312–322.

Determining Traffic Delays through Simulation

Penglin Zhu and Eckehard Schnieder

Institute of Control and Automation Engineering, Technical University of Braunschweig, Langer Kamp 8, 38106 Braunschweig, Germany
{zhu, schnieder}@ifra.ing.tu-bs.de

Abstract. This paper presents a simulative approach to determine train delays due to stochastic disturbances in railway traffic such as technical failures. The approach is based on the modelling of the railway traffic and the impacts of technical failures. A discussion about the modelling is given. The model is capable of simulating the traffic processes both under normal conditions and in case of technical failures. With respect to the determination of train delays and the delay distributions, the investigation possibilities with the model are demonstrated by an example. Some potential applications of the approach are summarised.

1   Introduction

In railway traffic the punctuality of train operations is one of the most important service quality measures both for travellers and for the railway company. Usually, railway traffic is regulated according to the timetable. If no disturbance happens during the traffic, trains run punctually. But that is often just the ideal picture. In real operations, the railway traffic suffers from different stochastic disturbances such as failures of technical devices, bad weather conditions, accidents, natural catastrophe etc. So the train delays are somehow not avoidable. In order to take some measures to ensure a high punctuality degree, it is very helpful and desirable to know the delay risks and delay extent in different situations and in case of disturbances.

Train delays can be classified into two types: the primary delay and the consequent delay. A train directly affected by a disturbance or an event is called primarily delayed. The primarily delayed trains may cause further delays of other trains, such delays are then called the consequent delays. The lengths of primary delays depend on the type, location, time point, expansion and duration of the disturbances, while the consequent delays are determined by the mutual interactions among trains and the limitations of railway infrastructure resources. To determine the extent of train delays one must consider the disturbance properties and all the train movements on the track and their mutual influences.

Some analytical methods were developed to calculate train delays in an abstract way (Higgins et al. (1995); Jochim and Schwanhäußer (1993);

Mühlhans (1990)). These methods are useful in planning phases or for some abstract and global observations, but they could not deal with the complicated traffic processes concretely in relation to a certain traffic plan (timetable) and specify the delays of the individual trains. An evaluation of a certain timetable is thereby impossible. On the other hand, simulative approaches enable us to take into account both the properties of disturbances and the mutual interactions among trains. It is thus suitable to investigate concrete traffic situations and train operations.

In this paper we discuss a simulative method to determine train delays, which is based on the modelling of the traffic processes in railway system and the influences of disturbances. The modelling is realised with Coloured Petri Nets (CPN, Jensen (1992)). At first the traffic processes without disturbance are modelled, then the influences of technical failures in railway traffic are integrated into the model. In the models the primary train delays are taken as stochastic events and their lengths are stochastic variables following certain distribution functions (e.g., the exponential distributions (Kraft (1981)). For a given primary delay having occurred at certain time and place, the corresponding consequent delays can be determined through simulations by means of the traffic operation model.

The paper is organised as follows: In Section 2 the modelling of a railway system is introduced. Section 3 discusses the influences of different technical disturbances and its modelling. An example will be presented in Section 4 to demonstrate the simulative investigations with the model in relation to the determination of traffic delays. Conclusions are given in Section 5.

2   Modelling of the Train Traffic

In literature some Petri Nets models for railway systems have been reported (van der Aalst and Odijk (1995); Montigel (1994); Hielscher et al. (1998)). But these models often concentrate only on parts of the railway traffic, either on the transportation process (van der Aalst and Odijk (1995); Hielscher et al. (1998)), or on the safety technology (Montigel (1994)). Because they make no clear separation between the transportation processes and the control system functions, the model will become very complex and difficult to understand when the whole system shall be modelled and investigated. Moreover, none of these models tells us the consequences when disturbances impact on the railway traffic.

In our model, a clear separation is made between train transport processes and the traffic control system (Figure 1). The model of transport processes describes the train movements on a railway line, while the algorithms and procedures of train operations are modelled in the control system model where dispatching criteria and constraints can be taken into account. Moreover, the impacts of technical failures in the railway traffic are analysed and modelled regarding their occurring places in stations, on tracks or on trains, respectively. The integration of failure impacts on the traffic into the model makes

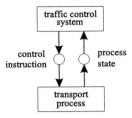

Figure 1. System structure.

it possible to determine train delays by simulation in case of technical disturbances in railway traffic. In the following subsections, the modelling concepts will be discussed. A more detailed introduction of the model can be found in Zhu and Schnieder (2000b).

2.1   The Traffic Processes

The train movements on railroads show some repeating behaviour, e.g., simplified the train movement in a track section consists of acceleration, running with constant speed and braking. The track topology may also be divided into some basic components. For instance, a railway track can be abstracted at the highest level as a chain of stations and the track sections between stations. A track section between two stations is usually further divided into blocks, each of which can only be occupied by one train at any time. So, when modelling the transportation processes, it is reasonable to build a class of basic models, which could be reused in the modelling to describe the entire train movements on tracks.

In our work a generic basic station model and a generic basic block model have been built, which can be used to model the railway transportation processes by simply connecting them (after parameterised) into a railway line (Figure 2). In this way, the modelling process is considerably simplified and the model complexity strongly reduced.

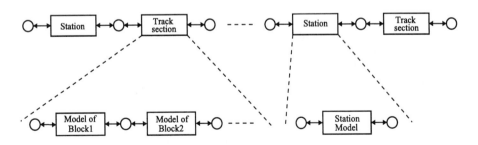

Figure 2. Modelling of the train traffic on railroad.

2.2   The Control System Functions

Beyond the train transport processes the train control system coordinates the train movements and manipulates the tracks (operation and control of signals, switches etc.). The control and safety processes governing the traffic processes are of discrete and causal properties. It is suitable for Petri Nets to model these causal relations. In the model, the functionality of the control system and the train operation algorithms on single line tracks are modelled, corresponding to the detail degree of the models of the subordinated traffic processes. The modelling principles can also be applied to double line tracks, but one needs to rebuild the model for different algorithms and procedures of train operations that are used there.

The modelling of a train control system involves the following aspects:

- Timetables are represented in the model with a CPN colour set (Jensen (1992)) which shall be assigned and initialised with actual data of the traffic process at investigation
- The procedure of train operations and communications between stations
- Disposition algorithms, taking into account the ranking of train, degree of delay and deadlock avoidance (Zhu and Schnieder (2000a))
- Safety conditions and processes
- Constraints and limitations of the track infrastructure
- The position of trains
- The state of tracks (free or occupied)

As an advantage of the clear separation between models of the control system and the traffic processes, very complicated and dynamic dependencies in train operations and disposition decisions can be taken into account in the control system model, while enabling a good model understanding at the same time.

3   Modelling of the Influences of Disturbances

If some technical devices in the railway system fail, the train operations will be affected. In general, the influences of technical failures can be reflected by the following scenarios all of which may lead to train delays:

(1) The stop times of some trains in the affected station and the capacity of the station to provide and prepare tracks for train traffic may change.
(2) As a result of temporarily reduced permitted speed, the travelling time of trains in the affected track section may be prolonged.
(3) Additionally, the train movements at some position may stop totally if the failure is very severe.

The technical failures differ from the defects of track side devices and the defects on railway vehicles. If a track side defect occurs in a station, its influences are of type (1). If the defect occurs on a track section between two stations, its influences are of type (2) or (3). Otherwise, if some parts of train devices fail, their influences may be type (1) or (2) or (3) depending on the current location of the train and the nature of the defect. The device failures on a train affect only this train and it is the only train which may be primarily delayed. In our work the influences of disturbances are modelled according to their breaking-in places, i.e., in railway station, on track section and on trains (Zhu and Schnieder (2000b)).

Technical Disturbance in Station: The impacts of a technical disturbance in station are described by the changes of train stop times. The scheduled stop time of a train in a station consists of the minimum stop time for necessary operations and of an additional buffer time to take into account the possible wait time in traffic. In the case of technical failures the necessary operations at station may be hindered so that the train's actual minimum stop time may be prolonged. If the prolonged portion of the minimum stop time exceeds the train's buffer time, the train will be delayed. On the other hand, the delay of a train can to some extend be compensated or reduced through the stop buffer time in its scheduled stop time, e.g., by shortening the actual stop time to the minimum for necessary operations.

In order to catch the influences of technical failures on the traffic, the extensions of the train's minimum stop times must be determined. The prolongation of minimum stop time depends on the type and location of the failure and is thus stochastic. In our model, the extension of the train's minimum stop time at a station is calculated with exponential distribution functions (other functions are also possible). The parameters of the distribution functions can be determined with the statistical data from train operations.

Technical Disturbance on Track Section: The failures on track section result in that either the affected track block must be blocked for traffic or all trains' speeds must be cut down during the disturbance. When a train is driving in that block at the occurrence of the failure, the train is directly affected and will be primarily delayed. Whether the train must stop or can still drive with a reduced speed, depends on the nature of the failure. If the traffic can run further but with a slower speed during the disturbance, the speed profile of the affected track block is accordingly changed in the model.

Technical Failure on Train: Like the failures of track side devices, the defects of train devices also differ. Defects could be those with which the train can still drive, or those with which the train must stop at once. In the first case some of the train characteristic data such as permitted speed, acceleration and braking capacity may be changed during the existence of the failure. In the second case the train is only allowed to move after the repair of the defect. In the model, all these changes have been considered and modelled, correspondingly.

By associating the models, which model the influences of technical failures, with the models of the railway traffic (see Section 2), it is possible to investigate the impacts of those failures on the entire train traffic. As a consequence, the final train delays (including consequent delays) can be determined, e.g., through simulations.

4 Investigation Examples

In this section an example is presented to demonstrate the applications of the discussed modelling. A fictive single line track in Figure 3 is taken as the object of the investigations. This track consists of four stations and ten blocks (B1~B3, B5~B7, B9~B12). It is assumed that the train traffics are operated according to the timetable in Figure 4 which is shown in the form of a time-distance diagram. The timetable is an input of the model and it is represented in the format of a CPN colour set. From the timetable in Figure 4 it can be seen that five trains run from station 1 to station 4 (train 2, 4, 6, 8, 10) and five trains from station 4 to station 1 (train 1, 3, 5, 7, 9). The ranking of the trains is assigned as follows: ICE > IC > SE > G.

At first the track infrastructure, the traffic operation algorithms and the influences of technical failures are modelled as discussed above. Then the characteristic data of the track have to be put into the model, before the simulations can be started. These data, besides the timetable data, include:

- Track data, such as block distribution, block length, speed profile.
- Train characteristic data, such as permitted speed, acceleration and braking capacity.
- Parameters of the relevant stochastic distribution functions, e.g., the distribution of primary delays.

With the traffic operation model of the track, different simulative investigations can be carried out.

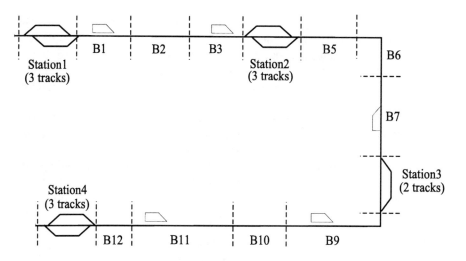

Figure 3. A single line track.

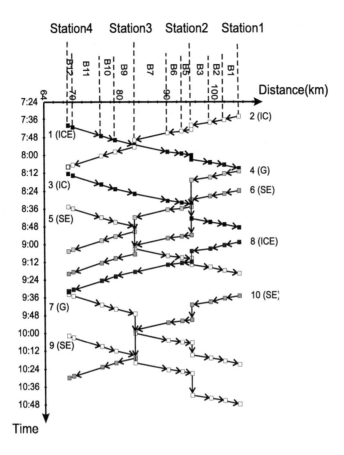

Figure 4. Timetable of the track in Figure 3.

4.1 Determination of Train Delays Caused by a Certain Failure

The influences of technical failures on the traffic can be investigated under different conditions such as different failure types, different occurring places, different occurring times, etc. For the sake of demonstration, three failure examples are given below. Table 1 shows the description of the failure scenarios. To each of the scenarios the train delays shall be determined, respectively. The simulation results of these failure scenarios are presented in various forms in Figures 5 – 7 (the trains are divided into two groups according to their running directions).

4.2 Investigation of Delay Distributions

If some parameters describing a failure are known not being deterministic but following certain stochastic distributions, e.g., the primary delays are usually

Failure 1	The failure occurs in B7 at 7:50, lasts 30min., block not available for traffic during the failure, primary delay = 10min.
Failure 2	The failure occurs in B7 at 8:10, lasts 90min., allowed speed $v = 40$ km/h during the failure, primary delay = 8min.
Failure 3	The failure occurs in B7 at 8:30, lasts 90min., allowed speed $v = 20$ km/h during the failure, primary delay = 10min.

Table 1. Failure scenarios.

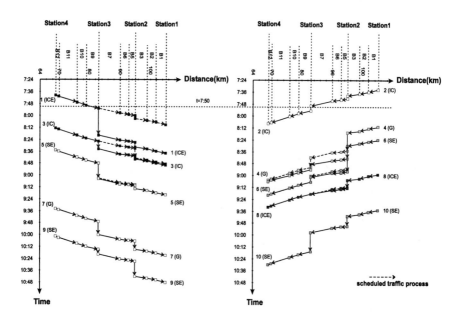

Figure 5. Traffic process in case of failure 1 in B7.

stochastic, then the distributions of the train delays at different stations can be determined through stochastic simulations by means of the traffic operation model.

Provided that a failure in block B7 at 7:45 leads to the primary delays following a negative exponential distribution with the mean value $\frac{1}{T} = 10$ min., the density function is given by $f(t) = 0.1e^{-0.1t}$. By random sampling of the density function, a series of primary delay values can be generated. Corresponding to these values, the final train delays are determined through simulations. Then the distributions of train delays can be calculated on the basis of evaluating the simulation results. Figure 8 shows the evaluated delay density functions of Trains 1 and 2 at their end stations (Stations 1 and 4) due to the mentioned failure in block B7.

In the above, a small model of a simple track and the investigation possibilities with the model were demonstrated. In the same way, a model for

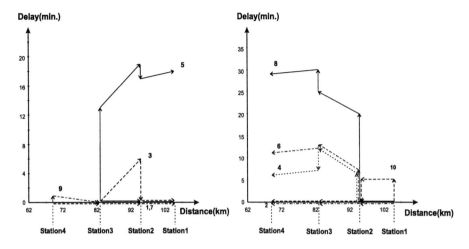

Figure 6. Special development of train delays in case of failure 2.

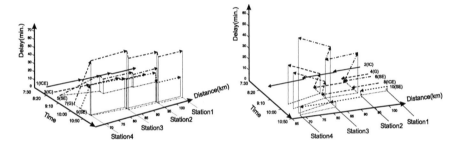

Figure 7. Delay-time-distance diagram in case of failure 3.

a quite long railway line can be constructed. Actually, the approach was developed aiming at modelling of larger railway networks.

The above investigations with the model may have different applications, such as:

- Determine train delays and delay risks due to technical failures
- Testing the robustness of a timetable by simulating the train operations under different disturbances and finding out the train delays
- Performability evaluation of railway traffic in conjunction with a dependability analysis of the technical devices in railway systems (Zhu and Schnieder (1999))
- Planning support of maintenance works on tracks, the influences of those works and the train delays can be determined by simulation before the works begin

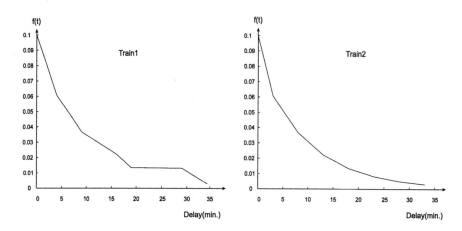

Figure 8. Density function of train delays.

5   Conclusions

In this paper an approach has been introduced to determine the train delays at railway traffic under stochastic disturbances. The approach is based on the modelling and simulation of the railway traffic and the influences of technical failures with Coloured Petri Nets. The train transportation processes and the control system functionality are separated clearly and modelled, respectively. In the modelling of traffic control systems the procedure of train operations and different ancillary conditions are considered. The train transportation processes on tracks are represented by means of the generic basic models of railway stations and blocks, whose reuse improves the modelling efficiency and reduces the model complexity. The influences of possible disturbances on tracks, in stations and on trains are analysed and modelled. The obtained traffic operation model is able to simulate the railway traffic processes both in normal case and in case of disturbances. In this way, the traffic delays under different situations can be determined.

Bibliography

Hielscher, W., L. Urbszat, C. Reinke, and W. Kluge (1998). On modelling train traffic in a model train system. In *Workshop and Tutorial on Practical Use of Coloured Petri Nets and Design/CPN, Aarhus, Denmark, June 10–12, 1998*.

Higgins, A., E. Kozan, and L. Ferreira (1995). Modelling delay risks associated with train schedules. *Transportation Planning and Technology 19*, 89–108.

Jensen, K. (1992). *Coloured Petri Nets: Basic Concepts, Analysis Methods and Practical Use.* Springer, Berlin.

Jochim, H. and W. Schwanhäußer (1993). *Neueichung der Qualitätsmaßstäbe.* 1. Zwischenbericht, Verkehrswissenschaftliches Institut der RWTH Aachen, Aachen, Germany.

Kraft, K.H. (1981). *Zugverspätungen und Betriebssteuerung von Stadtschnellbahnen in systemtheoretischer Analyse.* Ph.D. thesis, Technical University of Braunschweig, Braunschweig, Germany.

Montigel, M. (1994). *Modellierung und Gewährleistung von Abhängigkeiten in Eisenbahnsicherungsanlagen.* Ph.D. thesis, ETH Zürich, Switzerland.

Mühlhans, E. (1990). Berechnung der Verspätungsentwicklung bei Zugfahrten. *ETR – Eisenbahntechnische Rundschau 39*, 465–468.

van der Aalst, W.M.P. and M.A. Odijk (1995). Analysis of railway stations by means of interval timed coloured Petri nets. *Real-Time Systems 9*, 241–263.

Zhu, P. and E. Schnieder (1999). Performability-Modellierung von Bahnsystemen. In *Proceedings of International Symposium ZEL'99*, Zilina, Slovakia, 243–254.

Zhu, P. and E. Schnieder (2000a). Integrated modelling of railway traffic with Petri nets. In *Proceedings of IMACS/IFAC Symposium on Mathematical Modelling (3rd MATHMOD)*, Vienna, Austria, 453–456.

Zhu, P. and E. Schnieder (2000b). Modelling and performance evaluation of railway traffic under stochastic disturbances. In *Proceedings of 9th IFAC Symposium on Control in Transportation Systems*, Braunschweig, Germany, 330–339.

Optimization Approach to Support the Grouping and Scheduling of Air Traffic Control Sectors

Ana Paula Barbosa-Póvoa, Paula Leal de Matos, and Lúcio Rocha

Center of Business Studies of IST/SAEG,
Instituto Superior Técnico, Av. Rovisco Pais, 1049-001 Lisboa, Portugal
apovoa@ist.utl.pt, {pmatos,luciorocha}@alfa.ist.utl.pt

Abstract. Congestion in the European airspace calls for measures to make best use of the available capacity. The grouping and scheduling of sectors to minimize flight delays is one of these measures. This paper introduces the decision problem of grouping and scheduling air traffic control sectors and proposes a generic framework to represent the interactions between the air traffic sectors and the flight demands. The representation is a bipartite directed graph denoted as Sector-Flight Network (SFN) where two types of nodes are defined, the sector and the flight nodes. All the existing interactions between these two types of nodes are detailed in a simple and comprehensive form. Based on this representation a mathematical formulation is proposed to address the problem of grouping and scheduling Air-Traffic Control (ATC) sectors. The model provides the grouping and scheduling of sectors during a given time interval accounting for controller availability and sector capacity constraints so as to minimize flight delays. The possibility of optimizing the number of air traffic controllers is also considered. This approach leads to an integer programming model that is solved using a standard branch-and-bound method. The model is tested on two examples: a subset of French airspace and the Lisbon Area Control Center (ACC) airspace. Conclusions on the appropriateness of the model are taken and further improvements identified.

1 Introduction

Congestion in the air-transportation system has been plaguing air traffic both in the US and in Europe for nearly 20 years. To protect air traffic control from overloads, a planning activity called Air-Traffic Flow Management (ATFM) emerged during the seventies that compares the available capacity with forecast traffic sometime prior to the flights, tries to anticipate overloads, and takes control actions to prevent them.

The controlled airspace is divided into 3D geographic elements called ATC sectors. A team of controllers ensures that the traffic crossing each sector is safely separated. In simple terms, the capacity of an ATC sector is the number of flights that the ATC team of that sector is able to supervise per period of time. Sectors can be grouped or split depending on the number of controllers available and on the traffic. Splitting sectors will increase the total capacity of that part of the airspace.

When the traffic expected to cross the sector exceeds the capacity, traffic delays occur. For example, delays caused by congestion have affected approximately 14% of the traffic in February 2000, a low season month, and more than 20% of the traffic in the summer of 1999. The delays in the summer of 1999 were greater than 20 minutes per delayed flight (see EUROCONTROL (2000b)). ATFM tries to limit the extent and impact of those delays (see de Matos and Ormerod (2000)).

Delays increase the operating costs, e.g., the global costs of delays to airspace users of Air-Traffic Management (ATM) is estimated to be greater than 5.7 billion Euro in 1999 (EUROCONTROL (1999)). In addition, air traffic is expected to grow at an annual rate of 4.6% between 2000 and 2007 (EUROCONTROL (2000a)).

One of the flow management decisions that has to be taken two days before flights take place is, given the number of air traffic controllers available on that date, how to combine air traffic sectors in order to minimize delays caused by congestion. At present, there are practically no tools to address this problem. This paper, to our knowledge, contributes by providing a first model to support the grouping, and scheduling of ATC sectors, an urgent and relevant problem. To model this problem a new framework is developed in this paper. This is denoted as the *Sector-Flight Network*, a generic representation that describes all possible interactions between sectors, flights, and airspace usage during a given time interval.

Based on this representation an integer linear programming formulation is developed. The model supports grouping and scheduling of ATC sectors providing the sector combinations that minimize flight delays. The optimal solution results in the grouping and scheduling of sectors during a given time interval accounting for controller availability and sector capacity constraints.

The paper is structured as follows. The next section provides background information to the problem and reviews the literature. Section 3 describes the SFN framework and its use in defining the problem under study. Section 4 introduces the mathematical model that is applied to the two cases: *(1)* a set of four sectors from French airspace and *(2)* the Lisbon ACC. Finally, conclusions on the applicability of the model are drawn in Section 6 together with directions for future research.

2   Background

There are substantial differences between ATFM in Europe and in the US, de Matos and Ormerod (2000) address these differences. In the US, the Air-Traffic Control System Command Center coordinates flow management, but in the US most of the congestion is experienced at the airports. In Europe, on the other side, the coordinated air traffic control and flow management is more difficult due to the numerous countries with their own airspace. Many flights in Europe are short but have to cross several airspaces and, therefore, congestion affect not only airports, but also the en-route airspace. Thus, the thrust of air traffic management and control efforts in Europe has to be the integration and centralization of control activities. The Central Flow Management Unit (CFMU) in Brussels was created in 1989 to provide air traffic flow management for the countries of the European Civil Aviation Conference (ECAC).

In the US most planning is done during the hours before a flight departs, whereas in Europe planning starts six months before departure and involves not only flow managers but also different national administrations, area control centers, and representatives of aircraft operators. This range of stakeholders with divergent interests makes ATFM in Europe more problematic than in the US. US researchers tend to term planning before take-off "strategic" and after the flight takes-off "tactical." In Europe, strategic planning covers the period from six months to a few days before departure, pretactical planning occurs on the two days before departure, and tactical planning takes place on the day of departure until take-off. Measures affecting airborne flights are strictly in the realm of ATC rather than ATFM.

At pretactical and tactical levels, flow managers in Europe handle congestion by negotiating increases in capacity with ATC, by allocating slots to aircrafts (called ground-delays in the US), and by vertical or horizontal reroutings. A departure slot, usually at a later time than initially scheduled, is issued to flights heading for congested locations. Slots can be at airports, ATC sectors or just airspace junction points. A slot allocation program is called a regulation. Contrary to the US, a European flight is often subjected to several regulations.

CFMU is assisted by Flow Management Positions (FMP) based at the main European ATC centers. They supply the CFMU with local information on capacity and the traffic situation, and interface with the ATC center and the aircraft operators departing from the area. At pretactical and tactical levels, the FMP together with the ATC center have to decide how many and which ATC sectors have to be opened and closed during the day of operation. That is, they have to decide how to group and schedule the opening and closing time of the different sectors of each center. More opened sectors provide more capacity but also require more controllers. On the other hand, it is wasteful to open more sectors in an underused part of the airspace or at a time where the traffic is low. Typically, they open more sectors during peak hours, and group more sectors during low traffic periods. Note that sectors are open so that all airspace covered by the ATC center is controlled 24 hours a day. Their decision problem can be stated as, given the number of air traffic controllers available, they have to decide which sectors should be opened during a certain time interval so that flight delays are minimized. In this paper we address this problem.

Research on decision support for ATFM is over a decade old. Odoni (1987, 1994) defines the ATFM problem domain, identifies some of the major issues, and suggests decision support needs, mostly based on the US situation. de Matos and Ormerod (2000) provide similar European mapping work. Research initially concentrated on tactical optimization models for allocating ground-delays for the US case, with congestion limited to airports. Andreatta and Romanin-Jacur (1987) address the case of one airport where congestion lasts for a single time period. Terrab and Odoni (1993) present an exact solution method for one airport, several periods, and deterministic capacity. Richetta and Odoni (1993) provide a linear programming solution to a multi-period single airport case where capacity is stochastic. Vranas et al. (1994b) and Navazio and Romanin-Jacur (1998) present integer formulations for a network of airports, taking into account the interdependency between operations at different airports. Formulations exist to deal with dynamic situations (Richetta (1995); Vranas et al. (1994a)). Later, optimization models are developed where congestion also affects en-route sectors (Lindsay et al. (1993); Tošic et al. (1995)). Helme (1992) describes a multi-commodity network flow formulation. Vranas (1996); Nachtigall (2000) propose optimization models for allocating tactical ground-delays in Europe accounting for flights crossing several congested airspace elements while Bertsimas and Stock (1998) model the allocation of ground-delays and also speed control of airborne traffic.

ATFM is a planning activity that complements ATC. ATFM plans so that the number of flights crossing a sector does not exceed its capacity, whereas ATC ensures, at a more microscopic level, that the flights crossing the sector are safely separated. Bianco and Bielli (1993) propose different network models for air traffic control that include flow control measures both before and after flight departure, ranging from ground-delays to queues at holding points. More recently, the concept of "free flight" (airlines have more freedom to decide on flight route, altitude, and speed) which is not yet implemented, has led to research on new ATFM models. Andreatta et al. (2000) propose a model and an algorithm to assist a central authority in allocating arrival times of flights. This model is meant for a scenario where airlines decide on the departure time, route, and speed of their flights to meet the arrival times issued by the central authority.

The problem, faced by European FMP, of grouping and scheduling ATC sectors such that congestion delays are minimized, has, to our knowledge, never been addressed in the literature before.

3 Sector-Flight Network Representation

The *Resource-Task Network* representation proposed by Pantelides (1994) and used as one of the main frameworks for scheduling and designing industrial batch plants (see, e.g., Barbosa-Póvoa and Pantelides (1997)) is the basis to the generic framework SFN proposed in this paper. This framework

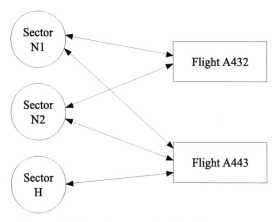

Figure 1. Sector-flight assignment for the airspace representation.

is used to model the problem of grouping and scheduling ATC sectors. Each sector that is used by a set of flights, is constrained by the associated capacity.

The sector and flight assignment describing the airspace utilization (as represented through the SFN framework) is a bipartite directed graph as illustrated in Figure 1. In this graph, two types of nodes are used, sector and flight nodes. In this example, two flights and three sectors are presented – Flight A432, which uses sectors N1 and N2, and flight A443 using sectors N1, N2 and H.

The SFN representation leads to a simple and elegant mathematical formulation for the problem of optimizing the grouping and scheduling of ATC sectors as shown in the next section. Assuming a time-uniform discretization and considering that each flight has a fixed duration τ_i, sectors are consumed[1] and produced[2] at a finite set of discrete times during the duration of the flight dictated by each flight plan.

Each flight i may have associated different types j describing the different routes that can be taken by the flight. Each of these routes involves different sector groupings and usage.

A single variable describes flight i of type j (e.g., characterizes flight i in different instances defined through the use of different sector groupings) taking off at time t. This variable is denoted by $N_{i,j,t}$ and determines the demand of flight i of type j on the various sectors included in the grouping of sectors of type j. In particular, the "production" (e.g., usage) of sector s at time θ relative to the take-off of flight i of type j at time t is given by $\mu_{ijs\theta}N_{ijt-\theta}$ where $\mu_{ijs\theta}$ for all i, j, s, θ, $\theta = 0, \ldots, \tau_i$ are known constants. Negative values for these constants indicate *consumption*, rather than *production*, of *sector s*.

---

[1] The term "consumed" can also be interpreted as "engaged" or "reserved."
[2] The term "produced" can also be interpreted as "disengaged" or "released."

Additionally, a set of variables, R_{st} is defined to represent the available capacity of *sector* s at time interval t (e.g., $R_{st} = 2$ means that two flights can pass through sector s at time t).

4   Problem Definition

The optimal combination of sector usage to minimize flight delays is subjected to a certain number of air traffic controllers. Limited sectors capacity can be obtained by solving the following problem: Given is a process description in terms of its SFN representation (sectors, flights), the sector capacities, the planning horizon, the flight plans, and the number of available controllers. The problem is to determine the sector combinations (or grouping) and the sector schedules to minimize the total flight delays.

4.1   Mathematical Model

Using the SFN representation, the model to describe the optimization of the grouping and scheduling of ATC sectors is developed discretizing time. This discretization is made fine enough so that all flights can take off or land at time interval boundaries. Sectors are produced and consumed at discrete times, thus sector changes can only occur at interval boundaries.

Indices, sets, and variables are defined as follows:

Indices:
i flight
s sector
j type of flight (e.g., since a flight can use different sector combinations, different types are given for the same flight using different combinations)
t time interval

Sets:
I set of flights
S_{ij} set of sectors used by flight i of type j
J_i set of instance of flight i with the type of flight i
S set of sectors
Ω_s set of sectors incompatible with sector s that cannot be opened if sector s is open

Variables and Constants:
d_i scheduled departure time of flight i
l_i scheduled landing time of flight i
τ_i duration of flight i
R_{st} available capacity of sector s for time interval t
N_{ijt} 1 if flight i of type j takes off at time t, 0 otherwise; $t \geq d_i$
X_{st} 1 if sector s is used at time interval t; 0 otherwise
$\mu_{ijs\theta}$ consumption ($\mu_{ijs\theta} = -1$) or production ($\mu_{ijs\theta} = 1$) of sector $s \in S_{ij}$ by flight i of type j at time θ relative to the take-off of flight i

In addition, the model comprises the following types of constraints.

Excess Sector Balance: These express the fact that sector availability (excess sector) changes from one time interval to the next due to the interactions of this sector with the active flights.

$$R_{st} = R_{s,t-1} + \sum_i \sum_{j \in J_i} \sum_{\theta=0}^{\tau_i} \mu_{ijs\theta} N_{ijt-\theta} \qquad \forall \, s,t \qquad (1)$$

Quantity R_{s0} corresponds to the initial availability of sector s (sector capacity) and is assumed to be given.

Excess Sector Capacity Constraints: The amount of excess sector at any given time is positive and has as an upper bound given by the sector capacity at each time interval. Therefore, we have

$$0 \leq R_{st} \leq C_{st} \qquad \forall \, s,t \qquad (2)$$

where C_{st} is the given sector capacity at each time interval within the planning horizon (e.g., number of flights that can pass through sector s at time interval t).

Activation of Sectors and Incompatibility Constraints: During the planning horizon, different sector combinations can describe the airspace. These combinations are a function of the requirements of the flight as well as the availability of controllers. Combinations may involve the grouping of individual sectors, which can cause incompatibility. For instance, a sector cannot be opened if another sector from the same group is already open. To know which sectors are active (e.g., open) the following constraints have to be fulfilled.

$$C_{st} - R_{st} \leq X_{st} C_{st} \qquad \forall \, s,t \qquad (3)$$

Thus, when the left side of (3) takes a value ≥ 1, meaning that the sector is active, the variable X_{st} takes the value of 1. Using this information the incompatibility between sectors is guaranteed as follows.

$$\sum_{s' \in \Omega_s} X_{s't} \leq 1 \qquad \forall \, s,t \qquad (4)$$

Sectors - Controllers Constraints: A fixed number of controllers is generally assigned to each sector. This number depends on the type and organization of airspace. We assume that one controller controls one sector. Note that each sector is crossed by several flights whose number is constrained by the sector capacity. Considering that the number of controllers A_t is given for each time interval t, we can use the following constraint:

$$\sum_s X_{st} \leq A_t \qquad \forall \, t \qquad (5)$$

Therefore, the number of active sectors at a time interval t must be smaller or equal to the number of controllers.

Flight - Types Occurrence Constraints: As defined above, we may have different types j for each flight i. This is represented in the SFN framework with different routes for the flight (different set of sectors – elementary or grouped). Based on this and knowing that only one of these types j may occur within the planning horizon, we can add the following constraint:

$$\sum_{j \in J_i} \sum_t N_{ijt} = 1 \qquad \forall\, i \tag{6}$$

Objective Function: The objective function is the minimization of flight delays. The delay is modelled by using a penalty factor for each flight associated with all the times different from the landing time of flight i. This penalty is defined as:

$$p_{it} = |t - l_i + \tau_i| \qquad \forall\, i, t \tag{7}$$

where t is the departure time of flight i. Therefore, the objective function is:

$$Minimize\ z = \sum_t \sum_i p_{it} \sum_{j \in J_i} N_{ijt} \tag{8}$$

Constraints (1) – (7) along with the objective function (8) define the integer model that is used to optimize the grouping and scheduling of ATC sectors for a given airspace, possible sector definitions and usages – represented in the SFN framework – and assuming a predefined availability of controllers per time interval. If the number of controllers becomes a variable, the model can easily be adapted by considering A_t as a variable and replacing (8) with

$$Minimize\ z = \sum_t \sum_i p_{it} \sum_{j \in J_i} N_{ijt} + \sum_t c_t^v A_t \tag{9}$$

where an additional variable cost (c_t^v) is introduced to account for the number of controllers. Thus, the optimal result would give us not only the grouping and scheduling of sectors but also the number of controllers that minimize the sum of flight delays for a given set of flights and the costs for the controllers.

5 Examples

Two examples are presented to show the model applicability. The Generic Algebraic Modelling System GAMS (see Brook et al. (1988)) was used combined with the CPLEX optimization package (see CPLEX (1995)).

5.1 Example 1

Four contiguous sectors of French upper airspace were chosen: LFBUN1, LF-BUN2, LFBUH1, and LFBUH2, which will be denoted by N1, N2, H1, and

H2. We assumed that these sectors could be grouped in some particular combinations, namely H and N, where H is the sector that results from grouping H1, and H2, and N that results from N1 and N2.

As input for the model, hundred flight plans were randomly chosen from the traffic filed on the 25th of April 1996, provided by EUROCONTROL. The 24 hours of this day were discretized into time intervals of five minutes. The constraints on the maximum number of available controllers and maximum capacity per time period were set considering a maximum traffic capacity of three flights per time interval for each elementary sector (N1, N2, ...), and two flights per time interval for each grouped sector. The number of controllers was defined allowing for the fact that there is more traffic during the day. Therefore, a maximum of three controllers was considered between 7h[3] and 19h (the busiest period in terms of traffic), and two controllers in the remaining intervals. Results are shown in Figure 2, where the sector usage and flight schedules are depicted. The optimal value was five time intervals of delay, corresponding to a total delay of 25 minutes. Five flights were delayed one time interval, flights v8, v19, v33, v48, and v53 (see Figures 2 and 3, delays are represented in grey before the actual flights).

Sector capacity shortage and lack of controllers explain the delay obtained. Flight v8 was delayed due to capacity shortage on sector H. This sector with a capacity of two was already being used by flights v36 and v62. The same happened to flight v19 since sector H2 with a capacity of three was being used by flights v9, v12, and v65.

The lack of controllers explains the delay of flight v48. If v48 departed on time, during time interval 254, three sectors, H, N2, and N1, would have to be open. This would have required three controllers but at this time of the day only two controllers were available.

5.2   Example 2

The second example is the application of the SFN model to a real situation: the grouping and scheduling of Lisbon ACC sectors. The purpose of this second example is to measure the effectiveness of the model and to compare the results obtained with the actual scheduling procedures. The flight plans recorded on April 14th were used as input data. The data was provided by EUROCONTROL.

The Lisbon ACC is composed of six elementary sectors, LPDEMOS, LP-VERAM, LPNOR, LPCEN, LPSUL, and LPMAD which can be grouped originating four combined sectors, LPOESTE (LPDEMOS + LPVERAM), LPLESTE (LPNOR + LPCEN + LPSUL), LPNORCE (LPNOR + LPCEN), and LPSULCE (LPSUL + LPCEN).

At each time interval, the Lisbon ACC airspace configuration is defined by a combination of the ten sectors listed above. The maximum number of active sectors at a given time is directly related to the number of controllers

---

[3] Times are given on a 24 hour time scale.

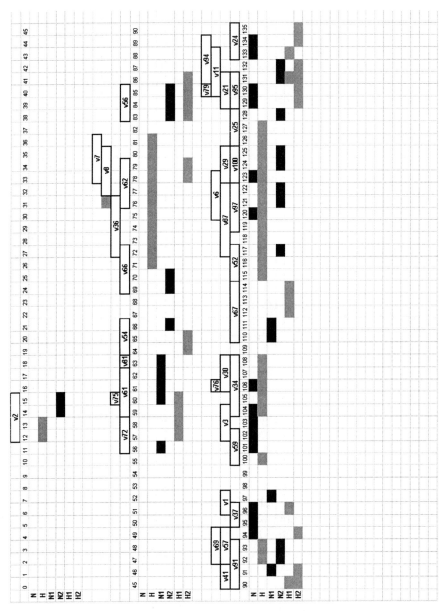

Figure 2. Usage of sector and flight schedule.

available, six for the periods from 7h until 22h and three for the remaining periods. Capacity per sector was provided by the Lisbon ACC: 12 flights per five minute interval.

The results presented in Figures 4 – 6 refer to three two hour periods on this date. For each 2 hour period, computation time varied between 35

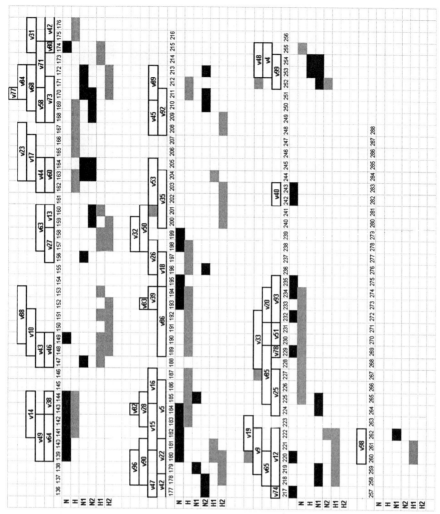

Figure 3. Usage of sectors and flight schedule (continued).

(20-22h) and 194 CPU seconds (14-16h) on a Pentium III, 500 MHz. The charts show the time intervals and the sectors that are active during each interval. Dark grey represents the solution produced by the SFN model while light grey corresponds to the real schedule used by the ACC on that date.

Naturally, the results produced by the SFN model do not differ from the solution used by the Lisbon ACC controllers in very busy or very quiet periods. This can be seen in Figure 4 between 14 and 16 hours (interval 169 to 192). The active sectors are the same in both charts and correspond to the configuration that maximizes airspace capacity by using all 6 elementary sectors available. The delay for both layouts is 5 units of time.

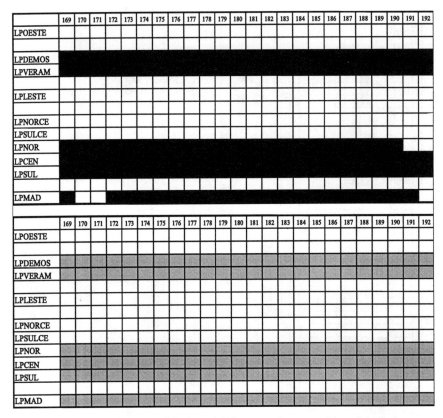

Figure 4. Utilization of Lisbon ACC sectors between 14 and 16 hours.

For periods of low traffic the results – not shown in this paper – are similar with the exception that, in this case, the configuration adopted corresponds to a smaller number of sectors: LPLESTE, LPOESTE, LPMAD. The major differences between the solutions adopted on that date and the ones produced by the SFN model occur over periods of intermediate traffic levels. For example, in Figure 5, between 18h and 20h the sector layout adopted is different despite the fact that the overall number of active sectors is the same. The solution used by the Lisbon ACC groups the sectors corresponding to the sector LPOESTE, leaving the elementary sectors that constitute sector LPLESTE active. The model solution splits sectors LPDEMOS and LPVERAM and collapses two other sectors, LPSUL and LPCEN, originating LPSULCE. This results in a value of 11 for the SFN solution. Compared to the value provided by the ACC solution, 75, it represents an improvement of 85%.

A similar situation occurs between 20h and 22h. While the Lisbon ACC keeps the same sectors open as before between 18 and 20h, the model changes from LPNOR/LPSULCE to LPNORCE/LPSUL (see Figure 6). The solution

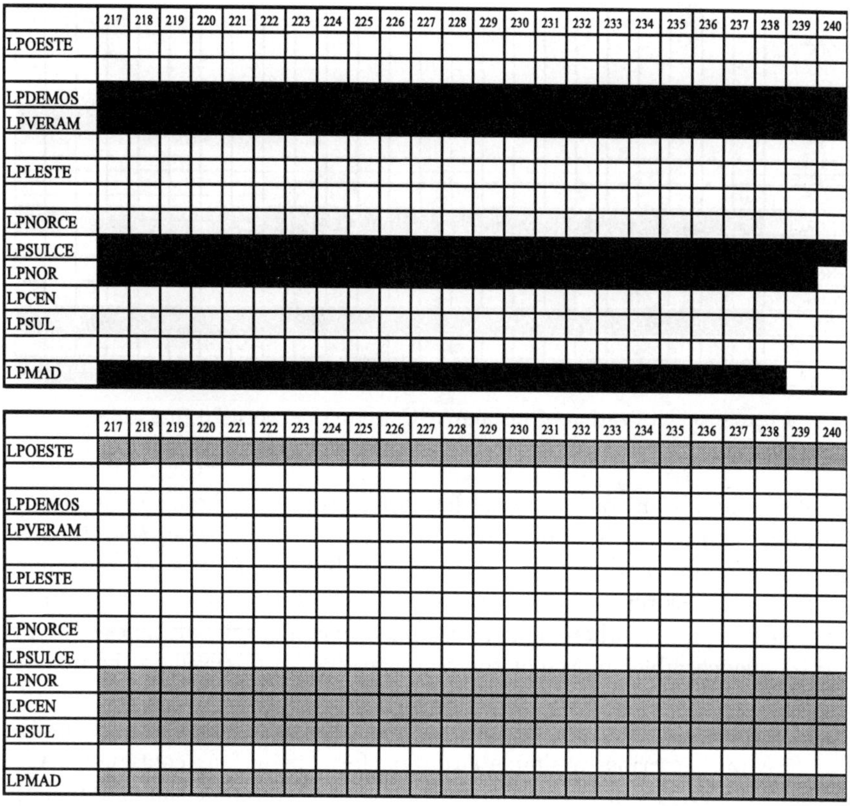

	217	218	219	220	221	222	223	224	225	226	227	228	229	230	231	232	233	234	235	236	237	238	239	240
LPOESTE																								
LPDEMOS																								
LPVERAM																								
LPLESTE																								
LPNORCE																								
LPSULCE																								
LPNOR																								
LPCEN																								
LPSUL																								
LPMAD																								

	217	218	219	220	221	222	223	224	225	226	227	228	229	230	231	232	233	234	235	236	237	238	239	240
LPOESTE																								
LPDEMOS																								
LPVERAM																								
LPLESTE																								
LPNORCE																								
LPSULCE																								
LPNOR																								
LPCEN																								
LPSUL																								
LPMAD																								

Figure 5. Lisbon ACC sectors utilization between 18 to 20 hours.

used by the Lisbon ACC corresponds to an overall delay of 22 time intervals, whereas the model solution provides a zero delay situation. Again the model leads to considerable improvement. In conclusion, these results show that the model can significantly reduce traffic delays.

6 Conclusions

This paper has introduced the decision problem of grouping and scheduling air traffic control sectors in Europe. A framework entitled Sector-Flight Network is proposed where all the existing sector-flight assignments during the planning horizon are represented in a simple and comprehensive way. Using this framework, a mathematical model has been developed to optimize the grouping and scheduling of sectors. The model provides the optimal grouping and scheduling of air traffic sectors accounting for capacity constraints of controllers and sectors so as to minimize total delay of flights.

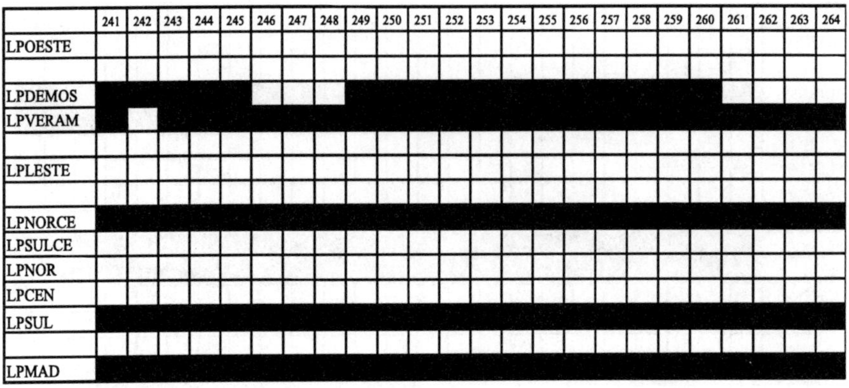

	241	242	243	244	245	246	247	248	249	250	251	252	253	254	255	256	257	258	259	260	261	262	263	264
LPOESTE																								
LPDEMOS																								
LPVERAM																								
LPLESTE																								
LPNORCE																								
LPSULCE																								
LPNOR																								
LPCEN																								
LPSUL																								
LPMAD																								

Figure 6. Utilization of Lisbon ACC sectors between 20 to 22 hours.

A second variant of the model has been proposed where the number of air traffic controllers is also a decision variable. In this case, the objective function accounts not only for delay of flights but also for cost of controllers.

The problems referred to have been formulated through an integer linear programming model where binary variables define the take-off times of flights and integer variables represent the sector availability as well as the number of controllers required.

The applicability of the model has been illustrated with two examples. The first one describes a simple situation formed by four contiguous ATC sectors from Southern France. The example indicates that the model is appropriate to this decision problem and leads to the identification of directions for further work. The second example describes a real situation, the Lisbon ACC on the 14th of April 2000 that has been solved quite effectively by the model. Considerable reduction of traffic delay was achieved by the model when compared to the actual grouping and scheduling of sectors on that date. Therefore, it can be stated that the model proposed is a promising decision aid for the grouping and scheduling of ATC sectors.

Other model generalisations are going to be object of future work. Considering that many flights follow practically the same route, the possibility of using flows of traffic instead of flights should be investigated. This possibility would substantially reduce the size of the optimization model.

Acknowledgements

The data used in this paper was provided by EUROCONTROL and NAV. We would like to thank Marcel Richard, Pierre Loubieres, Marc Bisiaux, and Serge Manchon from EUROCONTROL for their help in opening doors, defining the decision problem and obtaining the data. We are also grateful to José Caetano from the Lisbon ACC, and Michelle Le Guillou from the Paris FMP for all the information that they kindly provided.

Bibliography

Andreatta, G., L. Brunetta, and G. Guastalla (2000). From ground holding to free flight: An exact approach. *Transportation Science 34*, 394–401.

Andreatta, G. and G. Romanin-Jacur (1987). Aircraft flow management under congestion. *Transportation Science 21*, 249–253.

Barbosa-Póvoa, A.P. and C.C. Pantelides (1997). Design of multipurpose batch plants using the resource-task network framework. *Computers & Chemical Engineering 21S*, S703–S708.

Bertsimas, D. and S. Stock (1998). The air traffic flow management problem with enroute capacities. *Operations Research 46*, 406–422.

Bianco, L. and M. Bielli (1993). Systems aspects and optimization models in ATC planning. In L. Bianco and A.R. Odoni (Eds.), *Large-Scale Computation and Information Processing in Air Traffic Control*, Springer, Berlin, 47–99.

CPLEX (1995). *Using the CPLEX Callable Library, Version 4.0.* CPLEX Optimization, Inc.

de Matos, P.L. and R. Ormerod (2000). The application of operational research to European air traffic flow management. *European Journal of Operational Research 123*, 125–144.

EUROCONTROL (1999). Special performance review report on delays (January–September 1999). In *Performance Review Comission*.

EUROCONTROL (2000a). Air traffic statistics and forecasts: Number of flights by region 1974–2015. In *European Air Traffic Management Programme /Strategy Concept and System Unit – STATFOR*.

EUROCONTROL (2000b). *Delays in February 2000.* CODA, http://www.eurocontrol.be.

Helme, M. (1992). Reducing air traffic delay in a space-time network. *IEEE International Conference on Systems, Man and Cybernetics 1*, 236–242.

Lindsay, K., E.A. Boyd, and R. Burlingame (1993). Traffic flow management modeling with the time assignment model. *Air Traffic Control Quarterly 1*, 255–276.

Nachtigall, K. (2000). Luftverkehrssteuerung in Europa. In J.R. Daduna and S. Voß (Eds.), *Informationsmanagement im Verkehr*, Physica, Heidelberg, 145–165.

Navazio, L. and G. Romanin-Jacur (1998). The multiple connections, multi-airport ground holding problem: Models and algorithms. *Transportation Science 32*, 268–276.

Odoni, A.R. (1987). The flow management problem in air traffic control. In A.R. Odoni, L. Bianco, and G. Szegö (Eds.), *Flow Control of Congested Networks, NATO ASI Series, Series F: Computer and Systems Science*, 38, Springer, Berlin, 269–288.

Odoni, A.R. (1994). Issues in air traffic flow management. In *Proceedings of Conference on Advanced Technologies for Air Traffic Flow Management*, Deutsche Forschungsanstalt für Luft- und Raumfahrt e.V. (DLR), Bonn.

Pantelides, C.C. (1994). Unified frameworks for optimal process planning and scheduling. In D.W.T. Rippin and J. Hale (Eds.), *Proc. 2nd Conf. Foundations of Computer-Aided Operations*, CACHE Publications, 253–274.

Richetta, O. (1995). Optimal algorithms and a remarkably efficient heuristic for the ground-holding problem in air traffic control. *Operations Research 43*, 758–770.

Richetta, O. and A.R. Odoni (1993). Solving optimally the static ground-holding policy problem in air traffic control. *Transportation Science 27*, 228–238.

Terrab, M. and A.R. Odoni (1993). Strategic flow management for air traffic control. *Operations Research 41*, 138–152.

Tošic, V., O. Babic, M. Cangalovic, and D. Hohlacov (1995). Some models and algorithms for en route air traffic flow management. *Transportation Planning and Technology 19*(2), 147–164.

Vranas, P.B.M. (1996). Optimal slot allocation for European air traffic flow management. *Air Traffic Control Quarterly 4*, 249–280.

Vranas, P.B.M., D.J. Bertsimas, and A.R. Odoni (1994a). Dynamic ground-holding policies for a network of airports. *Transportation Science 28*, 275–291.

Vranas, P.B.M., D.J. Bertsimas, and A.R. Odoni (1994b). The multi-airport ground-holding problem in air traffic control. *Operations Research 42*, 249–261.

New Revenue Management Strategies for Railway Network Providers

Imma Braun[1], Karl Albrecht Klinge[2], Martin Schroeder[1], and Eckehard Schnieder[1]

[1] Technical University of Braunschweig,
Langer Kamp 8, D-38106 Braunschweig, Germany
i.braun@tu-bs.de, {schroeder,schnieder}@ifra.ing.tu-bs.de
[2] spektra Informationssysteme GmbH,
Lennéstr. 3A, D-39112 Magdeburg, Germany
klinge@spektra.de

Abstract. At present the calculation process to determine travel routes and times for railway operations, as well as to find available railway slots, is done manually by the network provider and is very time intensive. Consequently the network provider cannot react as flexible as necessary to short-term demand of slots. Therefore, the network provider is mainly interested in a reservation system allowing an easy and fast planning and marketing of free slots as well as in higher yields/revenues of slot assignments. In future the planning and booking of a railway slot must be supported by an information system which takes these aspects into consideration.

In this paper we will present a concept for a future European railway slot management system for slot-marketing, allowing reduced processing times for the assignment of slots as well as providing demand oriented pricing strategies.

1   Introduction

As an effect of the European Union (EU) directive 96/48 forcing the European railways to open their network for third parties, competition concerning passenger and freight transportation in guided traffic will increase. For this reason the assignment of slots will play an important role to European railway operators as well as to network providers. A slot (German: *Trasse*) is the main service product of a railway network provider and indicates the right for a railway operator to use a defined railway track within a defined time window by a defined train type. This service product includes the processing of railway slot inquiries, the construction of a timetable, the usage of the railway infrastructure with respect to the running movement as well as the provision of additional information during the train ride.

Nowadays the utilisation rates of railway lines are very heterogeneously distributed between the over-loaded main lines and the secondary lines, which are menaced by degeneration. The average rate of utilisation of a secondary line is far below 50%, which results in high average costs for the provision and utilisation of these lines and hence high prices (on the basis of full costing). For this reason expensive equipment such as radio based control systems (e.g., ETCS, European Train Control System) does not pay on secondary lines for network operators.

The current pricing system of railway slots is too inflexible. For instance, the present system does not consider the actual and expected demand for a railway slot, the flexibility of a customer (departure/arrival time window), the customer type (price elasticity of demand) and the time of booking (purchase in advance). In order to stimulate the demand on secondary lines and thereby to balance the rate of utilisation between primary and secondary lines, a new pricing system has to be introduced. This pricing systems must take the following aspects into account:

- Customer segmentation based on booking behaviour and their willingness to pay.
- Expected demand (forecasting) derived from historic data with respect to customer segmentation.
- Fixing of slot-prices, based on free capacities and prognostic demand.
- Assignment of slots, based on free capacities and prognostic demand.

The continuously growing competition in guided traffic, largely based on the EU directive 96/48, leads to an increased importance of the slot assignment for both the network provider and the railway operator. The demand for railway slots for freight and passenger traffic becomes more and more short-termed. Therefore, the railway operator and consequently the network provider has to react very quickly, but currently the majority of network providers is still planning slot allocations mostly manually which is very time intensive and results in a booking time period of up to six weeks. In order to prevent the loss of market shares to other transport modes (e.g., road traffic), the network providers need to accelerate the booking process. A fast and efficient slot planning requires new routing strategies considering:

- Strategic aspects, e.g., European Freightways, Trans-European Transport Network (TEN) strategy net 21, etc.
- Tactical aspects, e.g., technical network access criteria.
- Operational aspects, e.g., travel time, load, or cost optimised routing.

There are two main aspects to improve the situation of a network provider. On the one hand a network provider needs a computer based distribution system to determine free capacities and to plan railway slots. On the other hand he needs a demand-oriented pricing to increase revenues by marketing the free capacities of secondary lines to railway operators with a high price elasticity of demand (e.g., freight traffic) as well as by marketing the scarce capacities of primary lines to railway operators with low price elasticity of demand. Therefore, a Europe-wide coordinated distribution system for railway slots (slot management system) is needed, which allows network operators a customised and flexible demand-oriented price segmentation for the provision and utilisation of railway lines.

In the following we will give an overview of the taxonomy of different revenue management problems (also called yield management problems) and classify the given railway slot problem. Afterwards we will present our suggestion for the system architecture of a railway slot distribution system, which includes the pricing model. Finally we give a short conclusion and an outlook for further research aspects.

2   Current Pricing Mechanisms for Railway Slots

In this section we give an overview of the current pricing mechanisms for railway slots, based on the railway slot pricing scheme of Deutsche Bahn AG (DB AG) introduced in 1998 (cf. Haase (1998)). Reacting to the EU directive 91/440 forcing a cost separation between railway operator and network operator as well as to the national German law §14 AEG (Allgemeines Eisenbahngesetz) providing the right for all railway operators to use the existing German railway network, the railway network branch of German Railways (DB Netz AG) introduced its first track pricing scheme in 1994. This scheme was last updated in 1998 and is still in a continuous process of improvement. Concerning the characteristics of the railway infrastructure the new pricing scheme takes into account that:

- The depreciation period for railway infrastructure (e.g., tunnels, bridges, railway crossings, interlockings) is very long (up to 75 years).
- The usage of railway infrastructure is spatially fixed. After installation, railway infrastructure cannot be moved or used for other purposes (irreversibility of investment costs).
- The cost for the usage of the railway structure are predominantly fixed. Variable and marginal costs are unknown but considered to be low.
- Each railway network can be assigned a special system character with respect to the transportation modes.

In order to provide incentives for an increased usage of the railway network and considering the existing cost structures of the railway infrastructure a two-stage pricing scheme was chosen as already being used successfully by other network operators (e.g., power supply). Since DB Netz AG does not receive any subsidiaries for the operation of the railway network (in contrast to other European railway network operators like the Netherlands or Sweden) the pricing scheme has to allow a cost-covering operation.

The pricing scheme for railway slots consists of a fixed pricing component independent of the concrete use of infrastructure (InfraCard) as well as a variable pricing component mirroring the cost structure of the railway infrastructure. The InfraCard enables a railway operator to use a defined part of the railway network at lower variable costs. As an additional benefit, it offers customer related operation performance evaluation (e.g., punctuality evaluation) as well as support with respect to the planning of the timetable. The InfraCard only refers to a defined network with a minimum length of 1000km/500km (passenger/freight traffic) and the chosen railway tracks have to form a network. Besides only railway operators with regular transport operations can profit from the InfraCard. The pricing of the InfraCard considers the following aspects:

Size of network: The greater the size of the used network, the more expensive is the InfraCard. There are no price degressive elements in the pricing scheme, the price is only linearly related to the network length (price/network kilometer is constant).

Quality of network: Pricing is based on the quality of the infrastructure related to the supported speed and knot capacities. For instance, high speed lines require expensive control systems like LZB (continuous automatic train-running control) or ETCS. Regional passenger traffic at short intervals requires high knot capacities (several platform tracks, complex signaling). Therefore, the railway network is divided into six categories (K1-K6).

Contract period: The longer the period, the higher the discount on the InfraCard price. This discount will be justified by the long depreciation periods of railway infrastructure. A long contract period reduces the investment risk for the railway network operator. A discount of 2% will be given for a contract period of two years and up to 10% for a period of ten years.

The variable pricing component regulates the pricing of direct network usage after the InfraCard price has been paid. It takes the following elements into account:

Load balancing: Three different loading classes are formed in order to distribute total load evenly over the railway network. To every railway track a loading factor has been assigned (from 1, representing the lowest loading class, to 3, representing the highest loading class), to establish different pricing categories.

Flexibility of timetable: Flexibility with respect to the timetable (departure time window, arrival time window, possible routes) supports an efficient use of railway network capacities. Basic interval traffic is one of the most demanding products and, therefore, not discountable; flexible transportation modes are cheaper (e.g., the use of free already constructed slots).

Further discounts/surcharges: Further discounts and surcharges are – according to the pricing scheme from 1998 – possible, but until now not exactly evaluated. In future a discount may be provided with respect to the train infrastructure (e.g., use of innovative train control systems or ecologically friendly technologies).

Small railway operators, who cannot take advantage of the InfraCard (the railway network being too small) have access to the network, too, but are charged according to the VarioPrice pricing scheme (higher variable costs directly related to the use of the network but no fixed costs). Depending on the amount of rides of the railway operator the use of the VarioPrice pricing scheme can be advantageous as the VarioPrice depends on:

- Track kilometer (length of ride)
- Loading class of tracks/track segments
- Quality of railway network infrastructure

The current pricing system of the DB AG shows first signs of a revenue management system (cf. load balancing or flexibility of the timetable), but for introducing yield management strategies for the marketing of railway slots, the pricing system has to be more demand oriented. To achieve a high yield, the prices of the slots must refer to the demand of the slots, i.e., a high demand leads to a high price and a low demand leads to a low price, correspondingly. Therefore, on the one hand the future demand of the slots must be forecasted and on the other hand the prices must be dynamically adapted to the demand (cf. Section 3). These strategies are currently not supported by the DB AG pricing systems, but have to be implemented as the basic idea of the pricing module in the suggested slot management system (cf. Section 5).

3   Taxonomy of Revenue Management Problems

The production process of the railway network provider is characterised by the simultaneity of production and consumption of goods. The result of the production process – the slot – is an immaterial and perishable inventory, which has to be consumed on a special date and is no more available afterwards. Because of these distinguishing product marks, the network provider requires special marketing methods, such as revenue management. Revenue management is a technique that assists the allocation and pricing of perishable goods to different customer types at different times, in order to maximise both the revenue and the utilisation in a complementary manner (cf. Daudel and Vialle (1994); Vollmar (1994)).

By using revenue management techniques the capacity is split into several classes with different price levels (e.g., full price and discount classes) and the clients are segmented into different customer groups according to their demand profiles. For each customer group a prognosis of their future demand is made. Based on these prognoses about the booking process, the number of units to be sold in each fare class as well as the fare of each class are dynamically determined in progression of the booking process. These strategies build a basic revenue management model. In advanced revenue management systems, cancellations, overbooking and bumping strategies can be taken into account for optimizing the utilization and prices of the slots in order to maximise the total value of revenues for the railway network provider.

Revenue management problems can be characterised by the following (main) aspects (see, e.g., Friege (1996)):

- goods are perishable
- high fixed costs of providing the service and low variable costs by selling an additional good
- the capacity is inflexible
- considerable variation of demand
- the services are reserved in advance
- the customers can be differentiated by demand
- the price of the goods can be differentiated

These aspects are, e.g., given in airline, car rental and lodging industries, but also in the marketing of railway slots.

Weatherford and Bodily have introduced a taxonomy of revenue management problems which is shown in Table 1 (for detailed explanation of the taxonomy see Weatherford and Bodily (1992) or Weatherford (1998)). Many of the taxonomy aspects cannot be classified for the slot management problem yet, such as the customer's reservation demand profiles and, therefore, the number of fare classes, the willingness to pay or the turned down reservation, etc. These aspects are the subject of further investigations of the authors and will be presented in a subsequent paper. In the following the yield management problem will be described in respect to pricing and assigning railway slots to railway operators by using the taxonomy of Weatherford and Bodily:

Nature of resource: The nature of the units of the perishable asset can either be discrete (e.g., seats in a plane) or continuous (e.g., electrical power). Although usually slots are regarded as discrete units like airline seats, the duration of a slot can be variable. Due to the train type and train set power, the duration of the occupancy of a railway line is variable and can be varied continuously. For this reason, the nature of the resource of railway slots can be characterised as continuous.

Capacity: The capacity of the resource can be fixed or non-fixed. Most yield management problems are characterised by a limited capacity of the resources. This also applies for the railway slot problem, because the capacity of a railway line can only be increased, e.g., by installing an improved train control system or building a new parallel railway track, but this leads to a high time lag (and high costs) and can not be done to adapt short-term variations of the customer demand.

Prices: By using revenue management techniques prices are usually predetermined by a pricing group independently from the decision which is made in respect to the amount of discount units to be sold. Another alternative is to predetermine the amount of units at each fare level first, and then set the prices for each price level in an optimal manner. Lastly, pricing and allocating capacity to the fare levels can be done jointly. Currently, the prices for railway slots are set predeterminedly and independently of the forecasted and actual demand for slots. Corresponding to revenue management for the marketing of railway slots, one of the above mentioned strategies should be used in future, whereas the railway network operators are free to choose an appropriate method for the pricing of slots and assigning the railway capacity.

Elements	Descriptors
Resource	discrete/continuous
Capacity	fixed/nonfixed
Prices	predetermined/set optimally/set jointly
Willingness to pay	buildup/drawdown
Number of discount price classes	$1, 2, ..., I$
Reservation demand	deterministic/mixed/random-independent/ random-correlated
Show-up of reservations	certain/uncertain without cancellation/ uncertain with cancellation
Turned down reservation	lost/recaptured
Forecasting: 1. Seasonal data 2. Used method	no/yes moving average/exponential smoothing/ regression/quadratic spline/other
Group reservations allowed	no/yes
Diversion	no/yes
Displacement	no/virtual nest/bid price
Bumping procedure	none/full-price/discount/auction/other
Asset control mechanism	distinct/parallel nested/serial nested/other
Decision rule	simple static/advanced static/dynamic

Table 1. Taxonomy of revenue management problems.

Willingness to pay: Some customers are willing to pay more for the service as the availability date draws closer, for other customers the opposite applies. Today railway operators usually have to book their railway slots one year in advance so that the yearly timetable can be prepared. After the construction of the timetable, the resulting free line capacities are sold about 5 to 6 weeks in advance due to today's complex processes of slot construction. Up to now, the customer behaviour and the willingness to pay for railway slots has not yet been examined in detail. To apply adequate revenue management techniques to the marketing of railway slots, the customer behaviour has to be investigated first, but normally the willingness to pay of customers in the transport sector increases with the approaching date of the transport. Therefore, the discount fares for slots should be made available for those customers who are willing to book a slot early.

Discount price classes: Discount price classes refer to different groups of customers. The current pricing of railway classes is more technically oriented and (indirectly) considers only two classes of customers: railway operators of passenger trains who make use of higher train control standards of railway lines (which e.g., allow higher speeds) as well as railway operators of freight trains who use simple standards. In order to set up appropriate discount price classes in future, the composition of the customer groups has to focus more on the customer behaviour than on the technical aspects.

Reservation demand: Each fare class can have a different demand (deterministic, random, dependent on other classes). For the different customer classes of railway operators, there does not yet exist any information about the demand. Therefore, this point of the customer's behaviour has to be investigated as well.

Show-up of discount and full price reservations: The show-up of discount and full-price reservations describes the aspect whether a customer (discount-price or full-price customer) preoccupies the reserved good/service or not as well as if cancellations are permitted. Today, railway operators in Germany can reserve different slots, but they do not need to use them. In this case, a fee for the reservation is applicable but they do not have to pay for line usage. Cancellations are allowed from the day of booking up to the point of utilisation.

Turned down reservation: If a request for a special railway slot can not be fulfilled because the slot is already assigned to another customer, the revenue of this reservation is lost for the railway network provider. In some cases the request can be switched to another slot (e.g., another route or another time window), so that the revenue can be recaptured.

Forecasting: Forecasting the demand is a very important task in every yield management problem. The forecasting data for railway slot management has to be described as seasonal data. Both, short seasonal variation intervals (such as freight traffic operators' demands on weekdays in contrast to weekends) and long seasonal variation intervals (such as reinforced passenger traffic operators' demands on holidays) can appear in railway slot marketing (Kuhla (1998)). The type of the forecasting method should be as easy as possible but as sophisticated as needed and has to be determined appropriate to the customer's reservation demand.

Group reservations: Usually group reservations are not considered in most yield management problems. Group reservations by railway operators are applicable for two reasons: First railway operators offer their customers a certain service, which means a certain timetable. Therefore, railway operators usually do not ask for a single slot but for a certain number of slots. Second due to the increasing competition in guided traffic and the open access to the railway infrastructure for railway companies, smaller railway operators found alliances in order to enhance their negotiation power.

Diversion: The diversion refers to the booking behaviour of the customers who are willing to pay the full price for a service. A distinction is drawn between "full price customers" who also take a discount price unit in case it is available, and those who will not take a discount price unit. In the given railway slot problem, the customers can not be separated into full price customers and discount price customers. Railway operators willing to pay full prices for a slot, would also take a discount price slot if it is available.

Displacement: Yield management systems can take the displacement of different kinds of reservations into account. This criterion should be considered for the railway slot management, too. For the railway network provider this means, that he has to contemplate different origin-destination combinations, which bring different revenues. For instance, he has to decide whether to give away a slot (e.g., Hamburg to Hanover at a special time) at a discount price level to a customer taking a long route (e.g., Hamburg to Munich) and, therefore, needing more than one slot, or to prefer reserving this slot for a customer booking only this one slot for a short trip (e.g., Hamburg to Hanover), but willing to pay the full price.

Bumping procedure: This aspect handles the situation that sometimes the demand exceeds the capacity. Airlines, e.g., usually overbook their planes taking the expected no-shows into account. Today's timetable construction tools of railway network operators do not allow conflicts of slots. For this reason, an overbooking of slots is not possible in theory. In reality, some special situations in railway traffic (irregularly scheduled) occur, which cause slot conflicts, therefore, overbooking of railway lines in a very low extent is possible nowadays. Taking into account the new planned production concepts of railway operators like, e.g., modular freight train (cf. Zirkler (1998)), it is expected that no-shows of railway slots will increase and, therefore, overbooking becomes more and more important for the railway network provider. Further investigations will have to evaluate, which bumping procedure in case of overbooking should be used.

Asset control mechanism: The asset control mechanism describes the aspect, if the number of requested units of one fare class exceeds the defined limit of units to sell in this fare class. For the marketing of railway slots it seems to be reasonable that the units are serial nested. That means, that each fare class has a certain limit of units to be sold. If the limit of a higher fare class segment is reached before the limits of the lower fare classes are reached, the requests of high-fare-class-bookings are satisfied by using slots of the lower fare classes. Therefore, the full prices class has access to the whole number of slots of one railway track (the whole capacity). The other fare classes have only access to the determined number of units of their own class as well as the contingents of the lower fare classes.

Decision rule: In conjunction with the asset control mechanism, the decision rule determines in which way the entire yield management problem – i.e., the assignment of slots to prices and customers – is solved. Further investigations have to consider which decision rules are adequate for selling railway slots by using revenue management strategies.

Layer	Character of Management	Task	Time Horizon
SLOM 3	Long-Term Management	Creation of Yearly Timetable	1 – 2 Years
SLOM 2	Medium-Term Management	Marketing of Free Slots/Capacities	Days – Months
SLOM 1	Short-Term Management	Incident Management	Minutes – Hours

Table 2. Hierarchical structuring of slot management systems.

According to the classification of the revenue management problem, there are different algorithms and/or heuristics on forecasting, determination of optimal class size as well as optimal class fares to maximise the enterprise's revenue. Every special revenue management problem requires special algorithms. For most revenue management problems such algorithms do already exist in other industries (see, e.g., Weatherford (1998) or Subramanian et al. (1999)). At present these algorithms are examined, whether they can – in a modified form – be adapted to the railway slot marketing problem.

4 Hierarchical Structuring of Slot Management Systems

A slot management system can be hierarchically structured into different subsystems (see Table 2) depending on the temporal horizon of the respective management (cf. Erdmann et al. (1994); Fay and Schnieder (1999)):

- For the creation of the annual timetable, a long-term management system is needed, supporting the assignment of slots in the planning phase (SLOM 3). The time horizon for booking those slots, included in the annual timetable, is about one to two years. This comprises especially the assignment of slots for passenger traffic.
- After creating the annual timetable, the remaining available slots are marketed by a medium-term slot management systems (SLOM 2). This system considers a planning time horizon in the dimension of days to months. It manages the unscheduled traffic and can react to short-term demands of the railway operators.
- For the incident management of railway slots (time horizon of minutes to hours), a short-term slot management system (SLOM 1) is required. This system supports the planning manager in evaluating operational decisions of assigning or changing slots under economic aspects.

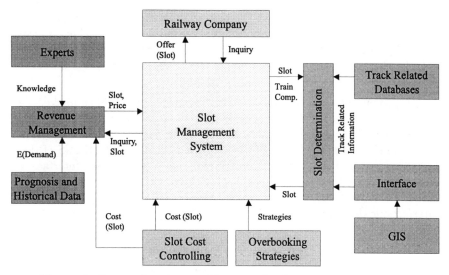

Figure 1. Proposed system architecture of a slot management system.

5   Architecture for a Future Slot Management System

In the following a concept for a slot management system is presented which supports the functions of SLOM 3 and SLOM 2. In a modified form the presented architecture can also fulfill the function of an online management system (SLOM 1). The structure of a future, modularly structured slot management system (cf. Schroeder et al. (1999)) is shown in Figure 1. It consists of three modules, which operate independently from each other and allow to apply revenue management strategies to the marketing of railway slots.

The central element "Slot Management System" receives the inquiries of railway companies concerning available slots and provides appropriate offers for the customers. It is responsible for the coordination, interaction and information exchange between the modules.

The module "Slot Determination" is responsible for the calculation of the travel time and for the determination of suitable slots according to the customer inquiry. All relevant track-related information for the calculation of travel time (e.g., static speed profiles in dependency of the train composition and geometry of the track) as well as for the assignment of different slots (topology of the railway network, track restrictions for certain train compositions, etc.) is provided by railway specific geographic information systems (e.g., DB GIS, Köthe and Schmitt (1993)). Other track-related databases can be used if necessary via appropriate interfaces.

After the transfer of the customer inquiry (consisting of the place of origin and destination, a date and time window as well as all relevant train specific parameters) the module supplies corresponding slot suggestions and the appropriate distance-time-diagrams taking into consideration the already reserved/booked slots. Since the construction and assignment of suitable slots is currently done manually with the help of different lists and by telephone among the timetable/slot editors (for this reason it is a very time-intensive process), the processing time for a customer inquiry will be significantly shortened by the application of this module.

At present DB AG is implementing its own computer supported slot assignment system for the generation of timetables (called DAVIT, Rauh (1998)). On the other hand there already exist further approved tools for an online slot assignment like SIMU++ (Simu (2000)) or FAKTUS (Faktus (2000)). Most of these tools do not provide an external open interface for communication. Thus they cannot be connected with other tools for the economic assessment or pricing of slots. Therefore, a standard for an external interface for the communication has to be developed and implemented.

The module "Slot Cost Controlling" supplies the relevant cost related information (fixed and marginal costs of the slot) for the pricing module. After the transfer of a slot to the cost controlling instrument for slots, marginal and fixed costs of this slot are determined and forwarded to the pricing module. At present no controlling tools exist which are able to provide fixed and marginal costs for railway slots. At the TU Braunschweig an appropriate concept was designed for the determination of these cost types (Braun (1998)). This concept will be implemented by using the controlling module of the standard software SAP R/3. For the external communication SAP R/3 supports different communication standards (e.g., CORBA (OMG (1998))), which can be used for the communication between the SAP R/3 controlling module and the slot management system.

The module "Revenue Management" is responsible for slot-pricing as well as for the assignment of priority figures to specific slots (as input for the online slot management concerning the assignment of slots in the case of deviations from regular traffic). All relevant information concerning customer type and slot is provided by the other modules "Slot Determination" and "Slot Cost Controlling" as well as the data bases. To improve the revenue of the network provider, the pricing has to be demand-oriented. Therefore, yield management strategies for pricing are used for the price calculation of slots. These strategies use information on the actual free capacities (they have to be provided by the slot assignment module), estimated demand for slots, marginal and fixed costs for the utilisation and allocation of slots, the customer type and behaviour (e.g., different types of freight and passenger traffic) as well as historical data (e.g., data about overbooking, reservation process or rejected inquiries). By the use of revenue management algorithms, network load on main lines and secondary lines can be distributed more evenly, free capacities of the network should be exploited commercially in a more efficient way than today and profit on highly demanded lines can be increased.

6   Conclusions

In future, railway slot management systems are required to achieve an effective marketing of slots. An appropriate system architecture for such a system has been presented. As shown, some modules are currently fully or partly implemented (e.g., Slot Determination). The main technical problem is the communication between these tools as well as the coordination of these tools. Appropriate external interfaces, a standard for the information exchange and a communication control unit have to be developed.

Revenue management algorithms for the pricing of services already exist (airline seats, hotel rooms). One main research field is the implementation of appropriate algorithms for the pricing of network provider services (in this context the assignment and use of a slot is understood as a main service).

For the migration of a slot management system as a global distribution system for slots we suggest the following way:

- The first step will be the implementation of an information system for railway companies as part of the global distribution system. By means of this information system railway companies will be able to plan their (Europe-wide) trips in a more efficient way than today.
- The second step will be the implementation of a reservation system. Inquiries/bids for slots can be priced by the reservation system and appropriate offers can be generated directly by the reservation system (such as airline reservation systems). Moreover, inquiries/bids can be passed over to the network providers' slot management system, where offers will be generated automatically or manually and then forwarded via the reservation system to the railway operator.

Our goal is that the reservation of a railway slot will be as fast and easy as the reservation of an airline seat via a computer reservation system. Efficient pricing algorithms such as those currently used in airlines reservation systems will result in a higher and more evenly distributed utilisation of the network.

Bibliography

Braun, I. (1998). *Möglichkeiten der Integration von Prozeßkostenmanagementsystemen in Verkehrsbetrieben.* Institut für Regelungs- und Automatisierungstechnik, Braunschweig. Masters thesis.

Daudel, S. and G. Vialle (1994). *Yield management: Applications to air transport and other service industries.* Presses de l'Inst. du Transport Aerien, Paris.

Erdmann, L., E. Schnieder, and A.G. Schielke (1994). Referenzmodell zur Strukturierung von Leitsystemen. *at – Automatisierungstechnik 42,* 187–197.

Faktus (2000, February). *Programm zur Fahrplan-Feinbearbeitung für Eisenbahnstrecken und -knoten.* Aachen: RWTH Aachen, Lehrstuhl für Verkehrswirtschaft, Eisenbahnbau und -betrieb. http://www.rwth-aachen.de/via/Ww/arbeiten/faktus.html.

Fay, A. and E. Schnieder (1999). Knowledge-based decision support system for real-time train traffic control. In N.H.M. Wilson (Ed.), *Computer-Aided Transit Scheduling, Lecture Notes in Economics and Mathematical Systems*, 471, Springer, Berlin, 109–125.

Friege, C. (1996). Yield - Management. *WiSt – Wirtschaftswissenschaftliches Studium 25*, 616–622.

Haase, D. (1998). Das neue Trassenpreissystem der Deutschen Bahn. *Internationales Verkehrswesen 50*, 460–465.

Köthe, K. and A. Schmitt (1993). Basisdaten der Vermessung für das Bahn-Geoinformationssystem. *ETR – Eisenbahntechnische Rundschau 6*, 401 – 408.

Kuhla, E. (1998). Frachtexpress. In *DEUFRAKO 1978 - 1998*, bmb+f, INRETS and Ministère de L'Équipment, du Logement, des Transports et du Tourisme, Bonn, Germany, 48–50.

OMG (1998). *The Common Object Request Broker: Architecture and Specification.* Object Management Group. http://www.omg.org.

Rauh, H.-F. (1998). Trassenmanagement bei der DB AG. *ETR – Eisenbahntechnische Rundschau 10*, 620 – 625.

Schroeder, M., E. Schnieder, and E. Kuhla (1999). Innovative slot management systems for guided traffic. In *WCRR 1999 – Proceedings of the World Congress on Railway Research.*

Simu (2000). *Interactive Timetable Construction and Simulation of Railway Operation.* Institut für Verkehrswesen Eisenbahnbau und -betrieb Hannover. http://www.ive.uni-hannover.de/engl/software/software.html.

Subramanian, J., S. Stidham, and C.J. Lautenbacher (1999). Airline yield management with overbooking, cancellations and no-shows. *Transportation Science 33*, 147–167.

Vollmar, T. (1994). *Yield-Management: Begriff, Inhalt und Einsatzmöglichkeiten im Dienstleistungsbereich.* Arbeitspapiere zum Marketing. Ruhr-Universität Bochum.

Weatherford, L.R. (1998). A tutorial on optimization in the context of perishable-asset revenue management problems for the airline industry. In G. Yu (Ed.), *Operations Research in the Airline Industry*, Kluwer, Boston, 68–100.

Weatherford, L.R. and S.E. Bodily (1992). A taxonomy and research overview of perishable-asset revenue management: Yield management, overbooking, and pricing. *Operations Research 40*, 831–844.

Zirkler, B. (1998). *Planung und Disposition eines Train-Coupling and -Sharing-Systems im Eisenbahngüterverkehr.* Technical Report Nr. 52, Institut für Verkehrswesen, Eisenbahnbau und -betrieb der Universität Hannover.

Impacts of Deregulation on Planning Processes and Information Management Design in Public Transit

Joachim R. Daduna

University of Applied Business Administration Berlin
Badensche Straße 50 - 51, D-10825 Berlin, Germany
daduna@fhw-berlin.de

Abstract. The deregulation in public transit has been advanced by the European Union in the last years. Limitations emerge from the obligation under public law of (local) municipalities to guarantee a basic mobility for all population strata and from the utilization of public transit as an instrument of regional structure development. Within the existing organizational structures of public transit, the conflict between the introduction of market structures and the perception of legal tasks cannot be solved. Therefore, basic changes are necessary that enable the creation of market structures with the help of bidding processes, the development of necessary new concepts for the (operational) planning processes, as well as the design of an efficient information management. A possible solution concept is introduced based on a vertical separation of organizational structures, subdividing centralized tasks (fulfilled by transit providers) and decentralized tasks (carried out by operating companies). The interconnected effects of the bidding processes are shown, as well as the influences on the design of software tools for planning and operational control.

1 Economical and Political Aspects of Public Transit

In recent years public mass transit in the Federal Republic of Germany as well as in other countries of the European Union (EU) had to undergo a fundamental change (cf., e.g., Costa (1996)). On one hand this is based on (internal) national effects, on the other hand there is a considerable (external) influence forced by the European Commission (EC), especially concerning the deregulation of (public) transit markets (see the Regulation (EEC) No 1191/69 of the Council of 26 June 1969). The structural and organizational concepts in public transit, that grew over a couple of years, must be questioned on the basis of the pending changes. Here conflicts occur from the clearly deviant objectives of the different interest groups, especially the public transit companies, (local) political decision makers, and the EC-policy with the political targets of a market orientation (cf., e.g., Ilgmann and Petzel (1998)).

A fundamental political question is whether to give a higher priority to the public (mass) transit in opposition to (individual) passenger-car traffic, while considering ecological as well as economical aspects. The objective is to especially reduce the traffic volume in the passenger-car traffic in inner-urban areas and in conurbations, in order to reduce the traffic-related environmental pollution (cf., e.g., Heymann (2001)). With the realization of this central target in transportation policy, however, the existing financial framework of the governmental budgets have be considered because the available means for subsidizing public transit decrease more and more. Therefore, suitable possibilities need to be established in order to guarantee a long-term funding. Since public transit forms an essential component of (local) community services and an important instrument of the local and regional development policy, the (local) governmental units still must comply with their legally tied-up obligations. For these reasons, public transit cannot be understood as an exclusively market-oriented commercial service business, offering cost-covering (or even profit-making) at (public) transit markets. It is rather a (governmental) obligation, where guaranteed service levels have to be defined by the responsible political and/or governmental decision makers. However, the operations to carry out the tied-up service level must take place under cost considerations in order to limit the necessary public subsidies.

A basic problem of the pending change processes results from the strained relations between deregulation policy and market-orientation forced by the EC as well as the obligation to perform the responsibility under public law. In this paper, a possible concept is introduced that bandages these two conflicting objectives in a suitable form, outgoing from changed organizational structures in public transit. The central point of this concept is a vertical separation with an assignment of the different responsibilities, that essentially refers to the question of establishing the service levels on the one hand and the operations on the other hand. In this context a suitable form of inviting (public) tenders for the provision of transportation services has to be considered.

Existing organizational structures in public transit are described in the following section. Afterwards the objectives and limitations of deregulation are shown and based on this framework an alternative organizational model is presented. On the basis of this solution, proposals for the bidding processes are discussed taking an efficient vehicle scheduling into account. In the last section, the most essential problems for an information management, that result from the modified organizational structures, are shown and possible solutions for future developments are presented.

2 Organizational Structures in Regulated Transit Markets

Apart from (long-distance) rail transit, the demand for public transit services has been focused on local areas in the past, so that the responsibility for public transit has concentrated on (local) governmental units (e.g., municipalities and counties). Organizational structures have evolved, that were geared with the local (administrative) community structures. Municipality and rural community owned public (mass) transit companies, built on this political framework, had a predominant position for many years in performing transportation services (cf., e.g., Heymann (2001)). Typically, these companies were completely responsible for strategic and operational planning as well as operating and control.

With an increase of mobility and the simultaneous spatial expansion in the last years, traffic flows changed so that they no longer coincide with the administrative structures. Therefore, the discrepancy between public transit service areas and demand structures increased considerably. In order to meet these changed requirements, two different cooperation models were developed:

Public transit working pools: Immediate cooperations between adjacent (operating) transit companies, mainly concerning tariff and/or timetable arrangements.

Public transit associations: Foundation of independent organizations through transit companies, which are operating together in a certain service area. These cases are usually accompanied with a transfer of different strategic and operational functions to the transit associations.

The spread of public transit associations leads to an organizational concept, which can be described as a 3-level-model, based on hierarchically structured and functionally oriented responsibilities (Daduna (1995)):

Local administration unions: These authorities (mainly affiliated to the public sector) have to guarantee the influences under public law. Their fields of competence are the coordination of regional planning and of basic agreements for the offered extent of mobility, the definition of targets for service development, and the financing of social shares of public transit.

Public transit associations: The organizations are mainly responsible for realizing the (political) targets and for carrying out centralized planning tasks (e.g., line network design and timetable construction).

Operating (public) transit companies: These organizational units are responsible for the (physical) production of the *public transit* service.

This model was a suitable solution for some years in order to react to the changed demand structures. However, it was based on structures in the public transit markets, that were essentially determined by governmental regulations as well as cross-subsidized companies owned by municipalities and rural communities. In most cases, these companies showed a low economic efficiency in operations (see, e.g., Ewers and Ilgmann (2000)). With the enforcement of deregulation and the introduction of market-oriented conditions through the EC, the need of considerable organizational changes and the abolition

of governmentally granted monopolistic positions arises. One of the critical
issues in this respect is the necessary separation of strategic planning and operations, in order to establish bidding processes as an unrenounceable basis
of non-discriminating competition. This cannot be achieved within a 3-level-
model, since the partners of a transit association, that has to organize the
invitation of tender, are (usually) the operating companies. Therefore, other
models, that are suitable for this modified framework, must be developed.

3   Deregulation and Impacts on Organizational Structures

Even nowadays public transit is a typical community service with a considerable demand for (public) subsidies. That is, up to now there was no
market-oriented competition between different providers since public transit
licenses were the only legal basis for offering transit services. These were usually possessed by municipality and rural community owned companies. One
possible attempt to remove the (organizational) connection between planning and operating is deregulation, which covers, in a wider sense (cf. Laux
(1993)), applied counter strategies to supervised, limited, ordered, and co-
organized influences of governmental institutions on private industries and
commerce, including public companies and non-profit organizations. Based
on this idea, the targets of deregulation in public transit are the removal of
existing competitive distortions, the creation of competition under the service
suppliers, including privately owned companies.

With the realization of these targets some limitations appear, since establishing the service levels does not represent a pure economic question but
mainly a political decision, i.e., under the responsibility of (local) governmental authorities. At first, the influences under public law must be seen here,
from which the obligations of municipalities and rural communities derive to
guarantee a basic offer of mobility for all population strata (cf. Gonenc et al.
(2000)). Therefore, minimum service standards that guarantee, e.g., the at-
tainability of public administration facilities and public utility services must
be fixed as well as the necessary school transport. Furthermore, as mentioned
above, the offered public (mass) transit has to be considered as an essential
instrument of regional and community structure development.

Based on these political and social aspects of public transit services, it
becomes clear that it is not possible to meet the requirements under public
law and to obtain a price covering the costs of service providing at the same
time. This means that the fare revenues cannot completely cover investment
and operating costs in public transit companies since different non-economic
points of view must be included, e.g., within the establishment of tariff structures. This situation leads to a target conflict, so that governmental obligations and deregulation can be brought together with the provision of public
transit services only in a limited range.

Centralized tasks	Decentralized tasks
Line network planning	Provision of infrastructure
Timetable construction	Provision of rolling stock
Tariff design	Provision of operating staff
Central data base management	Operational planning
Marketing / Public relations	Operating / Dispatching
Monitoring and control	Decentralized monitoring
Passenger information	Fare collection
Sharing-out of revenue	
Statistics	

Table 1. Functional assignment in a 2-level-model.

A possible solution seems to be an uncompromising vertical separation into (centralized) tasks in planning and operational control on one hand and (decentralized) operational tasks on the other hand, as described in a basic version of a 2-level-model in Table 1 (cf., e.g., Daduna (1995)). Based on these functional assignments an (exclusive) governmental-owned transit service provider has to be established to fulfill the centralized tasks – being in charge of all transit licenses (cf., e.g., Recker (2000)) for a defined service area. The operations, that mainly include the decentralized tasks, are subject to bidding processes for a tied-up public transit service under participation of different (municipality or rural community owned but also private) operating companies. Here, the invitation of tender lies within the responsibility of the service provider. The necessary obligations under public law are guaranteed for such a task allocation on the level of the service providers. At the same time intermodal competition can be avoided. On the other hand, competition structures can be created in the transit markets in a certain range, that are, however, restricted to the operational level.

This compromise contains a restricted deregulation and offers the possibility to connect relevant economic and non-economic aspects in a suitable form, and it seems to represent also the basis for the most advantageous solution of a reorganization of public transit. Experiences from a similar model realized in Sweden, which, however, does not take a complete vertical separation as a basis (cf. Jeschke (1998)), show distinct cost reductions (cf., e.g., Geuckler (1994); Vierth (1995); Cox et al. (1997)). Moreover, an actual proposal to substitute the Regulation (EEC) No 1191/69, which is aimed at a concept of controlled competition, puts the competition of operators to the fore (cf. Commission of the European Communities (2000)).

Concerning the decentralized tasks, a modified structure is practicable (especially in rail transport), which shows an additional separation. In this case (cf., e.g., Nash (1996)) the provision of infrastructure, rolling stock and operating staff becomes separated from transit operators, such that these companies attaining transit service contracts have to rent the needed personal

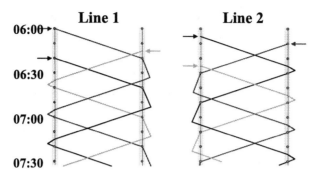

Figure 1. Line-oriented vehicle scheduling.

and technical equipment for carrying out operations. Since these structural details are of minor importance for the further discussions, they are not taken into consideration, subsequently.

4   Bidding Strategies and Vehicle Scheduling

The bidding processes, that have to be executed through the responsible service providers, constitute the formal basis in order to implement public transit market structures on the operational level. Here, different forms of terms of tenders are possible, assuming a defined line network and a detailed timetable, that is of special importance in hierarchically structured transit networks with various means of transport. An invitation to bid can be applied, e.g., to a complete transit network of a service area, to several splitted subnetworks (defined by different means of transport and/or related to space based structures) or to single lines (cf., e.g., Cox et al. (1997)). These three forms for terms of tender, which are usually discussed, show some disadvantages. In the first case, a contract, that takes on the responsibility for all operational planning tasks, is placed merely with one of the suppliers. Such a "single sourcing" decision has certain advantages, but also, as appeared in many other branches of industry, partially considerable disadvantages, mainly due to a long-term dependency of a contractor. The other two terms of tender usually represent insufficient solutions, too, because they show significant cost problems in the vehicle scheduling procedures, as demonstrated in Figures 1 and 2.

The results from the simple example in Figures 1 and 2 show very clearly the inefficiency of decentralized vehicle scheduling carried out by different operating companies. Altogether, such strategy usually leads altogether to a greater number of required busses (in this case 6 busses instead of 5). From many research projects and from practical experiences distinct advantages of network-oriented vehicle scheduling, especially based on computer-aided interlining and trip shifting approaches, are known (see, e.g., Bodin et al. (1983); Carraresi and Gallo (1984); Daduna and Mojsilovic (1988); Daduna

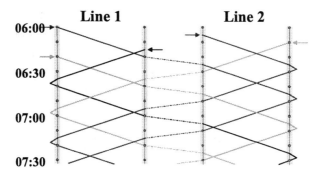

Figure 2. Network-oriented vehicle scheduling.

and Paixão (1995); Löbel (1999)). These approaches lead to savings in operating cost and, from a long-term view, also to reductions in capital expenditure. Therefore, at this point the question arises, whether it is wise to apply bidding strategies with separated tenders, which neglects cost cutting effects resulting from efficient OR-based vehicle scheduling procedures. Moreover, it is possible, that the expected cost benefits following from competition in public transit markets can not be attained, such that the aimed fundamental economic objectives of deregulation are influenced by counteracting effects.

This situation illustrates that a concept is required, that solves these conflicts between bidding strategies on the basis of subnets or single lines and the efficiency of vehicle scheduling processes and considers the above sketched cost cutting effects. A conceivable solution to avoid inefficient operations is an invitation of tender on the basis of blocks (or conceivably also duties), while these represent defined "jobs" for the bidding processes. Such a strategy determines drastic alterations of the functional responsibilities (cf. Table 1), since the vehicle scheduling (and as the occasion arises, also the duty scheduling) must be carried out by the responsible transit provider (as a centralized task), and no more by the operating companies (as a decentralized task). The fundamental responsibilities for the operational planning need to be modified as presented in Figure 3, while for the operating companies this approach inevitably leads to a loss in planning competences.

However, while making use of terms of tender that are based on blocks, it must be taken into consideration that usually during the five-year running of a contract, timetables will change to a certain extent within respective periods. The duration of contractual periods can be extended over five years, depending on a longer pay back period of contract-based investments in rolling stock and/or infrastructure (cf. Commission of the European Communities (2000)). For this reason, it may become necessary to intend contractual modifications, e.g., mileage-based calculation approaches, in order to guarantee the cost advantages of an efficient vehicle scheduling even with changed timetable data.

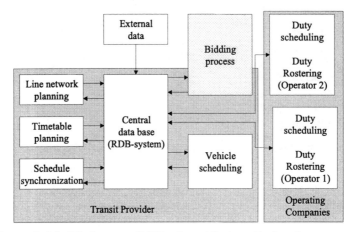

Figure 3. Modified responsibilities in public transit planning processes.

5   Information Management within Modified Organizational Structures

Besides the sketched alterations of the responsibilities for vehicle scheduling processes, the existing structures in information management have to be redesigned comprehensively. Information needs and information supply in planning and operations cannot be modeled as a unique (and mainly sequentially organized) flow structure anymore (cf., e.g., Daduna (1992)), because the realization of deregulation leads, as mentioned above, to a separation in competences and, depending on the applied tender concept, to a decentralization of carrying out operations. Therefore, the design of capable information management concepts is considerably influenced by conflicts between decentralized structures in transit operations and centralized responsibilities in determining service levels and achieving quality control.

Based on the functional structure in public transit planning and operations, the following main fields of application for computer-aided systems can be outlined (cf., e.g., Daduna and Voß (1994)):

- Centralized (basic) data management (usually based on a relational data base-system (RDB-system))
- Strategic planning (line network and timetable design, schedule synchronization)
- Operational planning (vehicle and duty scheduling, duty rostering)
- Dispatching, monitoring, and operational control (mainly based on automated vehicle monitoring-systems (AVM-systems))
- Static and dynamic passenger information
- Depot management and maintenance

Only if the invitation to bid is applied to a complete transit network, the functional responsibilities can be separated clearly. In this case data management and strategic planning are assigned to the service provider, while the

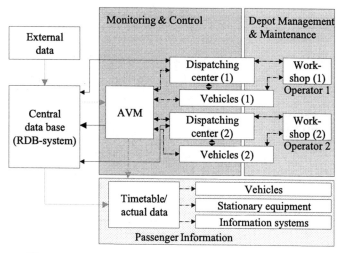

Figure 4. Functional structure and operational control.

selected operator in the bidding process will be in charge of all other fields. If the terms of tender are based on subnetworks/lines or on blocks/duties, a more differentiated structure of (partly) independent companies has to be handled and connected in a virtual network by capable interfaces (cf., e.g., Voß and Gutenschwager (2001)). The most conflicting points are the fields of dispatching, monitoring and operational control on one hand, and passenger information on the other hand. Based on the underlying functional structure, both fields show multi-directional information flows (cf. Figure 4), based on different sources (information suppliers) and sinks (information users).

Figures 3 and 4 illustrate, that information management structures are determined by two dominating components, a central RDB-system, which is mainly used in operational planning processes, and an AVM-system for (online) monitoring and control. The design of the data base system must be founded on a suitable data warehouse-concept (cf., e.g., Voß and Gutenschwager (2001)) to efficiently handle different clients. In this case two types of clients have to be taken into consideration, the internal (planning) tools used by the service provider and the external tools used by the selected contractors. In addition, an internet-based virtual structure has to be designed to connect all clients on the basis of appropriate interfaces. The AVM-system, being an important client of the RDB-system, too, is focused especially on operational activities and represents (from this point of view) the necessary dynamic component of public transit information management. Based on this framework, the responsibilities should be organized in a structure shown in Figure 5.

Essential functions of AVM-systems mainly have to support centralized tasks in carrying out transit operations (e.g., network-based operational control, inter-modal and intra-modal dynamic schedule synchronization), such that the responsibility has to be assigned definitely to the service provider.

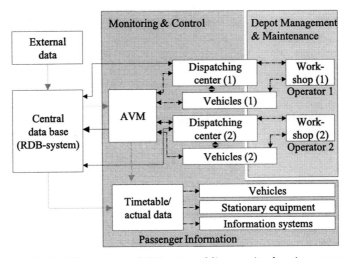

Figure 5. Modified responsibilities in public transit planning processes.

A decentralized monitoring concept cannot guarantee the necessary basis for passenger-friendly services, especially in the case of intra-modal transfers between lines operated by different contractors or in inter-modal transit. Nevertheless, the dispatching centers belonging to the operating companies need to be integrated as (local) clients in an efficient AVM-structure, because vehicle and crew (or driver) dispatching represent decentralized tasks. Moreover, passenger information must also be regarded as a centralized task, especially concerning (actual) dynamic information based on data attained from an AVM-system. Beside the use of (classical) print media, e.g., timetable booklets, wall timetables, and (online) passenger information systems for providing static information based on (planned) timetable data, actual passenger information are made available by different stationary equipment and on-board vehicle components (cf., e.g., Daduna and Voß (1996)).

Under consideration of the described changes of organizational structures of public transit, the introduced concept can be a suitable solution for structuring an effective information management. The crucial aspect is that concerning the operating companies varying combinations can be guaranteed in order to react to possible alterations, that can emerge through bidding processes. However, this concept can only be understood as a suitable basic model, that needs to be adapted to the specific (local or regional) conditions of the different service areas in each case.

6   Conclusions

The obligation for competitive tendering in public transit services, enforced by the EC commission (cf. Commission of the European Communities (2000); Heymann (2001)), must lead to considerable alterations in organizational structures and functional responsibilities in planning and operations. The necessary consequence is that the existing and at present still predominant municipality and rural community owned public (mass) transit companies will lose their monopolistic position within the next years. In order to be able to react to these basic changes, a vertical separation should be introduced into public transit, that leads to a functional partition of the concerned public transit companies. Basis for this is a modified version of the 2-level-model (described in Section 4). For urban mass transit companies, which are operating different means of transport, a horizontal separation becomes additionally necessary. An essential reason for this are the deviating cost covering rates, e.g., between bus systems and rail-based systems (light rail transit, subway transit, etc.), because these can conceivably lead with future bidding processes to relevant disadvantages in competition.

If one assumes the represented framework with comprehensive structural changes, it becomes clear, that also considerable influences arise at the design of software tools for public transit applications. A very essential aspect is that the internal information management of transit providers requires suitable interfaces in order to connect dislocated external clients, representing tools and/or technical equipment used by the different contractors. The main objective is the efficient coordination of data flows in the respective (virtual) network structures, in order to guarantee the necessary information supply of all users. To meet these requirements, the presently used tools for computer-aided planning and for monitoring and control need to be adjusted, and new tools need also be developed.

These tools must constitute a necessary basis for planning and managing public transit within subdivided intra-modal networks, especially in connection with rail-based or guided means of transport, and also in inter-modal networks. On this basis, the possibility will be given to organize and to control transit operations, that are carried out by different contractors. Furthermore, direct awards of service contracts are to be avoided, which are allowed for specific cases (e.g., guarantee of safety standards, efficient coordination) as an exception of inviting (public) tenders.

Furthermore, an efficient tendering is inseparably interconnected with the availability of capable information management tools. This relates not only to strategic and operational planning but also to monitoring and control. Suitable tools represent an essential factor in order to manage (restricted) competition in public transit markets. For this reason the design and development of computer-aided systems to be applied in public transit is a very interesting research area including commercial software development, where the pending structural changes will enforce additional impulses.

Bibliography

Bodin, L., B. Golden, A. Assad, and M.O. Ball (1983). Routing and scheduling of vehicles and crews: The state of the art. *Computers & Operations*

Research 10, 63–211.

Carraresi, P. and G. Gallo (1984). Network models for vehicle and crew scheduling. *European Journal of Operational Research 16*, 139–151.

Commission of the European Communities (2000). *Proposal for a regulation of the European Parliament and of the Council on action by Member States concerning public service requirements and the award of public servicecontracts in passenger transport by rail, road and inland waterway.* http://europa.eu.int/eur-lex/en/com/pdf/2000/en_500PC0007.pdf, 31.01.2001.

Costa, Á. (1996). The organisation of urban public transport systems in Western European metropolitan areas. *Transportation Research A 30*, 349–359.

Cox, W., J. Love, and N. Newton (1997). *Competition in public transport: International state of the art.* Paper presented to the 5th International Conference on Competition and Ownership in Passenger Transport, Leeds. http://www.publicpurpose.com/t5.htm, 23.01.2001.

Daduna, J.R. (1992). The integration of computer-aided systems for planning and operational control in public transit. In M. Desrochers and J.-M. Rousseau (Eds.), *Computer-Aided Transit Scheduling, Lecture Notes in Economics and Mathematical Systems*, 386, Springer, Berlin, 347–358.

Daduna, J.R. (1995). Organisationsstrukturen des öffentlichen Personennahverkehrs und ihre Einbindung in den kommunalen Bereich. *Zeitschrift für Verkehrswissenschaft 66*, 187–206.

Daduna, J.R. and M. Mojsilovic (1988). Computer-aided vehicle and duty scheduling using the HOT programme system. In J.R. Daduna and A. Wren (Eds.), *Computer-Aided Transit Scheduling, Lecture Notes in Economics and Mathematical Systems*, 308, Springer, Berlin, 133–146.

Daduna, J.R. and J.M.P. Paixão (1995). Vehicle scheduling for public mass transit: An overview. In J.R. Daduna, I. Branco, and J.M.P. Paixão (Eds.), *Computer-Aided Transit Scheduling, Lecture Notes in Economics and Mathematical Systems*, 430, Springer, Berlin, 76–90.

Daduna, J.R. and S. Voß (1994). Effiziente Leistungserstellung in Verkehrsbetrieben als Wettbewerbsinstrument. *Zeitschrift für Planung 5*, 227–252.

Daduna, J.R. and S. Voß (1996). Efficient technologies for passenger information systems in public mass transit. In H. Pirkul and M.J. Shaw (Eds.), *Proceedings of the 1st INFORMS Conference on Information Systems and Technology, Washington D.C.*, 386–391.

Ewers, H.-J. and G. Ilgmann (2000). Wettbewerb im öffentlichen Nahverkehr: Gefordert, gefürchtet und verteufelt. *Internationales Verkehrswesen 52*, 17–20.

Geuckler, M. (1994). Wandlungsprozesse in Schweden: Auswirkung der ÖPNV-Reform auf die Verkehrsträger. *Der Nahverkehr 9/94*, 66–74.

Gonenc, R., M. Maher, and G. Nicoletti (2000). *The implementation and the effects of regulatory reform: Past experiences and current issus.* Economics Department Working Papers 251, Organisation for Economic Co-operation and Development (OECD).

Heymann, E. (2001). *Öffentlicher Personenverkehr auf den Weg in den Wettbewerb.* Sonderbericht Deutsche Bank Research.

Ilgmann, G. and W. Petzel (1998). Szenarien des öffentlichen Verkehrs in Ballungsräumen bei Öffnung der Märkte. *Internationales Verkehrswesen 50*, 248–250.

Jeschke, C. (1998). Stand der Neuorganisation des öffentlichen Personennahverkehrs in Schweden: Erfahrungen aus Stockholm. *Internationales Verkehrswesen 50*, 128–132.

Laux, E. (1993). Deregulierung. In W. Wittmann, W. Kern, R. Köhler, H.-U. Küpper, and K. von Wysocki (Eds.), *Handwörterbuch der Betriebswirtschaft* (5 ed.)., Schäffer-Poeschel, Stuttgart. col. 743–754.

Löbel, A. (1999). Solving large-scale multiple-depot vehicle scheduling problems. In N.H.M. Wilson (Ed.), *Computer-Aided Transit Scheduling, Lecture Notes in Economics and Mathematical Systems*, 471, Springer, Berlin, 193–220.

Nash, C.A. (1996). *Separating rail infrastructure and operations: British experience.* Working paper, Institute for Transport Studies, University of Leeds.

Recker, E. (2000). Nahverkehrspläne stärken: Landkreise fordern Änderungen der Rechtsgrundlagen für den ÖPNV. *Der Nahverkehr 7-8/00*, 16–22.

Vierth, I. (1995). Regionalisierung und Deregulierung des ÖPNV. *Internationales Verkehrswesen 47*, 452–457.

Voß, S. and K. Gutenschwager (2001). *Informationsmanagement.* Springer, Berlin.

Cost-benefit-analysis of Investments into Railway Networks with Periodically Timed Schedules

Michael Kolonko and Ophelia Engelhardt-Funke

Institute for Mathematics, Technical University of Clausthal,
Erzstraße 1, D-38678 Clausthal-Zellerfeld, Germany
{maoef,kolonko}@math.tu-clausthal.de

Abstract. An efficient planning of future investments into a railway network requires a thorough analysis of possible effects. Therefore, a tool is needed for a cost-benefit-analysis at an early stage of the planning process. We present a method to obtain a cost-benefit-curve that shows the effect of investments (cost) on the quality of the network measured by the waiting time of passengers (benefit).

This curve is obtained from the solutions of a multi-criteria timetable optimization problem. Timetables are evaluated with respect to the investment they require and the benefit they bring to passengers in terms of shorter waiting times. Moreover, we show how the notion of stability of a timetable under random delays can be included into our approach. The analysis is done on a strategic level without consideration of all operational details. We use genetic algorithms to find approximate solutions to the optimization problem.

A prototype system is presently tested on a network of regional lines in Germany. We report on the first very promising results.

1 Introduction

To increase the attractiveness of public transport it is important to improve the quality of the service for the passengers and in particular to reduce the travel and waiting times in the network. An improvement usually requires a major investment. Therefore, the traffic providers have to decide how the available money should be invested into the network to obtain a maximal benefit for the passengers. This is an urgent issue, e.g., on some of the regional lines in the new states of Germany. Here, the condition of rails, crossings and switches allows only a limited speed of the trains on some of the sections. Usually there are different levels of investment possible, e.g., one could simply provide a level crossing with automatic barriers, renew a switch or rebuild the whole section. Therefore, apart from deciding which sections are to be modernized the level of modernization has to be fixed for each section.

This requires a detailed cost-benefit-analysis of possible investments taking into account the different investment scenarios; see Figure 1 for a simple example of a cost-benefit curve. From such an analysis the decision makers could expect answers to questions like:

- What is the (maximal) benefit passengers can derive from any given amount of investment? How can it be obtained, i.e., what are the detailed investment decisions? Here, the 'benefit' is the reduction of waiting time, other choices are discussed below.
- How much money would it cost to increase the quality (i.e., to reduce the waiting time) by a certain percentage?
- What is the return of investment (measured in terms of benefit), i.e., how much additional benefit could be obtained from increasing the investment over a certain level (see, e.g., levels c_1 and c_2 in Figure 1)?

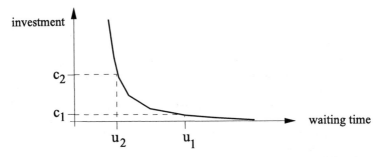

Figure 1. A simple cost-benefit-curve, where benefit is measured by the waiting time in the network.

Generally, the benefit for the passengers is the quality improvement of the timetable: e.g., higher speed of the trains shortens the travel times and also gives room to design timetables with better connections and shorter waiting times. The link between investment and passengers is provided by the timetable which, therefore, will be the main ingredient of our analysis.

The necessary investment as well as the benefit for the passengers can be formulated as properties (more formally: cost functions) of a timetable. Finding optimal timetables with respect to these multiple targets then yields a cost-benefit-analysis as is shown below. In addition, this result will also give the planner all the information on how to allocate the investment in the network and how to schedule the trains to obtain the maximal benefit.

It should be pointed out that we are only concerned with the strategic, long-term planning problem in which operational details like safety headways and capacity restrictions are not considered.

This research is performed in cooperation with the Nahverkehrsservice Sachsen-Anhalt (NASA) GmbH. NASA is a provider of regional rail traffic in the state of Sachsen-Anhalt in Germany. A prototype system for the optimization and the evaluation of the results has been developed and is presently tested on the network of NASA as explained below.

2   A Mathematical Model of Timetable Optimization

2.1   The Timetable

We consider a network with fixed *lines* $\mathfrak{L} = \{L_1,...,L_N\}$ that are served periodically each with a fixed period τ_{L_i}. A line is represented by the list of consecutive stations the train passes through. Reverse directions are modeled as separate lines. We assume that lines are strictly periodical. That means, e.g., that if some stations are skipped on a line during weak traffic hours, this has to be modeled as a separate line with a possibly large period.

Let $\mathfrak{S} = \{S_1, \ldots, S_K\}$ denote the set of all *stations* of the network and $\mathfrak{Z}$ the set of all *sections* of tracks, i.e., stretches of tracks between two neighbouring stations S, S'. Note that there may be more than one such section between two stations, in particular, if two lines have different speed on the same physical track, we model this by different sections. Let (S, S', L) denote the section from $\mathfrak{Z}$ that line L uses to travel from station S to its next station S'. To each section, there is a list of potential improvements with their costs and their effect on the running time of all lines on that section. When investing into that section one has to choose among these improvements.

Let $\mathfrak{D} = \{(L, S) \mid L \in \mathfrak{L}, S \in \mathfrak{S}; L \text{ departs from } S\}$ be the set of possible *departures*. Then a *timetable* T consists of two lists:

$$T = \Big((\pi(L, S) \mid (L, S) \in \mathfrak{D}), \quad (\delta(z) \mid z \in \mathfrak{Z}) \Big)$$

where $\pi(L, S) \in \{0, 1, \cdots, \tau_L - 1\}$ is the *departure time* (modulo the period τ_L) of line L in station S and $\delta(z)$ is the *running time* scheduled for all trains on section $z \in \mathfrak{Z}$. From this the scheduled *arrival time*

$$\gamma(L, S') := \pi(L, S) + \delta((S, S', L))$$

of L in S' can be calculated. Incorporating the running times instead of the arrival times into the timetable is more convenient for our purpose.

As was mentioned above, we only consider a strategic planning situation, in which the details of the network, like capacities of sections or stations and safety constraints (headways), are not taken into account. This greatly simplifies the problem of finding a (mathematically) feasible timetable. In fact, any list $(p(L, S) \mid (L, S) \in \mathfrak{D})$ of integers can be interpreted as a list of departure times simply by reducing them modulo the appropriate line period: $\pi(L, S) := p(L, S) \mod \tau_L$. We assume that for the running times $\delta(z)$ there is a lower bound $\underline{\delta}(z)$ that could be achieved if all improvements on that section were realized, and an upper value $\overline{\delta}(z)$, e.g., the running time of the present timetable. Then the set of feasible timetables is given as

$$\mathfrak{T} := \underset{(L,S) \in \mathfrak{D}}{\text{\Large X}} \{0, \ldots, \tau_{L-1}\} \times \underset{z \in \mathfrak{Z}}{\text{\Large X}} \{\underline{\delta}(z), \ldots, \overline{\delta}(z)\}$$

Any $T \in \mathfrak{T}$ represents a valid timetable within our framework. Note that all times are treated as integers, interpreted, e.g., as 0.1 minute.

2.2   Cost Function I: Investment

Each feasible timetable $T \in \mathfrak{T}$ requires a certain running time on each section $z \in \mathfrak{Z}$. To calculate the amount of investment necessary to enable that running time one has to be given the *local cost functions*

$$c_z : \{\underline{\delta}(z), \cdots, \overline{\delta}(z)\} \to \mathbb{N}$$

for each $z \in \mathfrak{Z}$. $c_z(\delta)$ gives the minimal amount of money needed to enable the running time δ on section z, see Figure 2 for an example of such a local cost function. As mentioned above there is a list of possible improvements on section z, e.g., building a new level crossing, modernizing a switch or rebuilding the whole track. The local cost function is built from these data by calculating the cost and the reduction in running time for all possible combinations of improvements. In particular the system checks all possible combinations for the cheapest way to achieve a certain reduction in running time. In Figure 2, a running time δ would require to install a new crossing and a new switch at the (cumulative) cost of $c_z(\delta) = c' + c$.

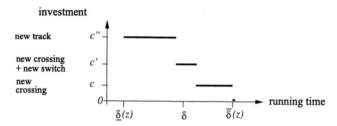

Figure 2. A simple local cost function $c_z(\cdot)$.

From these local cost functions we may calculate the *total amount of investment* required by timetable T as $C(T) := \sum_{z \in \mathfrak{Z}} c_z(\delta(z))$. Note that now the investment is just a cost function value, i.e., a property of the timetable.

2.3   Cost Function II: Waiting Time

When changing from line L to line L' at station S the waiting time in a periodic network can be given by:

$$w(L, L', S) := \pi(L', S) - (\gamma(L, S) + \alpha(L, L', S)) \mod \gcd(\tau_L, \tau_{L'})$$

Here, $\gamma(L, S)$ is the arrival time of line L at station S as defined above in § 2.1. For $L \neq L'$, $\alpha(L, L', S)$ denotes the minimal transfer time at station S from line L to L'. Then $(\gamma(L, S) + \alpha(L, S))$ denotes the earliest possible time at which a passenger changing from line L to line L' can reach the train of line L' and $\pi(L', S)$ is his or her actual departure time. For $L = L'$, $\alpha(L, L, S)$ denotes the minimal stopping time of line L in station S. Therefore,

$w(L, L, S)$ gives the amount by which the actual stopping time of line L exceeds its minimal stopping time in station S. This is the waiting time of passengers continuing their journey on line L at station S.

It is known from the literature that for any connection in a periodic network the smallest, the largest and the average waiting time occurring during a day differ only by constants from $w(L, L', S)$, see Nachtigall (1996) and the references given there. As these constants do not depend on the timetable, minimizing $w(L, L', S)$ is equivalent to minimizing any of the above target functions. The *total weighted waiting time* is now given by

$$W(T) := \sum_{S \in \mathfrak{S}} \sum_{L, L' \in \mathfrak{L}} w(L, L', S) \cdot g(L, L', S)$$

Here $g(L, L', S)$ denotes a weight, e.g., the average number of passengers changing from line L to line L' at station S. $g(L, L', S)$ will be 0 if there is no reasonable connection from L to L' at S, e.g., if L' is the reverse line of L. For $L = L'$, $g(L, L', S)$ denotes the number of passengers who continue their journey on line L as described above. There is an additional penalty factor for $L \neq L'$ to give waiting on the platform a heavier weight than waiting in the train during stops.

The weights $g(L, L', S)$ can be entered into our system if such numbers are available, e.g., from traffic counts. If such information is not available it can be estimated by our system: if there is an $OD-$matrix $M = (m(o, d))$ available giving the number of passengers $m(o, d)$ that travel from origin o to destination d then our system can calculate weights by sending the $m(o, d)$ passengers along shortest routes through the network. If even less data are available, then we can estimate the $OD-$matrix using Lill's law, see § 6.2.

In some situations it is important to consider additional waiting times like waiting for connections to other networks or waiting when entering the system at particular stations at given time points (schools, large factories). To include this kind of service into our system (e.g., arriving at 8 am at the school station) we model the corresponding events as artificial lines with fixed arrival and/or departure times. Minimizing the total waiting time $W(T)$ will then lead to small waiting times for the connections to these artificial lines (e.g., the train will arrive shortly before 8 am at the school station).

Note that we do not take into account waiting times that occur when entering the railway system from outside at some random point of time. These waiting times only depend on the periods of the lines which are considered to be fixed in our context.

2.4 Multi-criteria Optimization

The two cost functions, investment $C(T)$ and waiting time $W(T)$, reflect the quality of a timetable from different viewpoints: the traffic provider will be interested in low investments whereas the passengers will insist on short travel times and on short waiting times as the most unpleasant part of the journey.

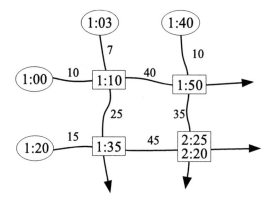

Figure 3. A simple network with four lines.

Further aspects of timetable quality may be considered by using additional cost functions like the total weighted travel time:

$$R(T) := \sum_{o,d \in \mathfrak{S}} r(o,d) \cdot g(o,d)$$

where $r(o,d)$ is the minimal scheduled travel time for a route from origin o to destination d and $g(o,d)$ is the weight for the importance of that route. We are also considering $U(T)$ as the number of vehicles necessary to run T which can be calculated from a particular periodic scheduling problem by linear optimization (see, e.g., Orlin (1982); Serafini and Ukovich (1989)). Another cost function taking into account the delays will be discussed in the next section.

Thus we are faced with an optimization problem where each timetable $T \in \mathfrak{T}$ has a multi-dimensional vector of cost function values, e.g.:

$$\Big(C(T), W(T), R(T), U(T) \Big)$$

The aim is to find good timetables under this multidimensional criterion. Obviously, these cost functions are not independent of each other.

Figure 3 shows a simple example. There are four lines starting at the oval stations. The rectangles indicate stations where lines can be changed, we assume that the stopping and transfer times are 0. The numbers on the sections denote their running times. The trains are assumed to run every 60 minutes. The schedule is determined by the departure times in the starting stations which are chosen such that there are no waiting times at the first changing stations. This necessarily leads to a conflict at the last, fourth station. Whatever departure times we prescribe at the starting stations, the waiting time in the system will be 5 min and 55 min, respectively, as the running times in the directed cycle of the changing stations sum to $5 = 40 + 35 - 45 - 25$. Only an investment that will shorten these running times, e.g., from 40 to 35 minutes will reduce the waiting time (to 0 in this case).

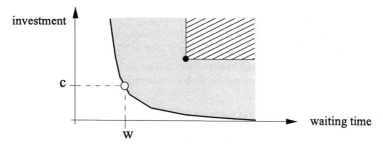

Figure 4. Pareto-optimal solutions for two cost functions.

In general one cannot expect that there is a single timetable minimizing all criteria simultaneously. Instead, one is looking for the so-called *Pareto-optimal* or undominated solutions. A timetable T is *dominated* by T' if (for the cost values considered above)

$$C(T') \leq C(T), R(T') \leq R(T), W(T') \leq W(T), \text{ and } U(T') \leq U(T),$$

and at least one of the "$\leq$" is a "$<$". In this situation T' is better than T and T should not be used. In Figure 4 the shaded area indicates the cost values of all timetables for two cost functions. The timetable belonging to the black dot dominates all timetables in the hatched quadrant. The bold line marks the Pareto-optimal solutions.

The cost function values of the Pareto-optimal timetables constitute a cost-benefit curve (or surface in case of more than two criteria). Each point (c, w) on the curve in Figure 4 gives the maximal benefit w achievable by an investment c or the least investment one has to make to obtain benefit w, respectively. From the timetable T that is represented by the point $(c, w) = (C(T), W(T))$ we can also determine how the total investment $c = C(T) = \sum_{z \in 3} c_z(\delta(z))$ should be allocated to the different sections $z \in 3$ and how the 'benefit,' i.e., the waiting time $W(T)$, spreads over the stations of the network.

An algorithm for the approximate solution of this multi-criteria optimization problem is presented below in Section 5.

3 Stability of Timetables under Delays

3.1 Scheduled vs. Actual Times

A particular focus of our research is on the stability of timetables. If we optimize timetables only with respect to their *scheduled* travel or waiting time, we shall end up with timetables that are highly synchronized but have only small time buffers at stations. These timetables may turn out to be very instable in real operation as small delays seem to be inevitable in complex networks. But then, the resulting real travel and waiting times may be much larger than the scheduled ones due to missed connections. Therefore, it is important to

take small delays into account when designing timetables (whereas untyp-
ically large delays caused, e.g., by accidents shall not be considered here).
We can incorporate this aspect into our approach by defining a suitable cost
function, e.g., $M(T) = $ mean travel time under delays.

Minimizing $M(T)$ or minimizing the scheduled travel time $R(T)$ and the
difference $M(T) - R(T)$ would result in timetables that have the additional
quality of 'stability.' It could also be of interest to examine the *variation*
of travel times that occur during a day. Typically, one would expect that
large waiting time (= large time buffers) correlates with high stability. Then
a cost-benefit-analysis including $M(T)$ would show how much stability can
be gained, e.g., by investing or by increasing the waiting time, see Goverde
(1998) for a result on time buffers for a single isolated connection. Another
target would be to examine the effect different 'waiting rules' (stating how
long a train has to wait for its delayed feeder train) have on the delays and
the waiting times.

To be able to calculate $M(T)$ one has to model the typical small operational delays on lines and at stations, their propagation through the network
by the waiting rules and their absorption by time buffers. More precisely, one
has to know the joint probability distribution of the delays in the whole network at every point in time. This is an extremely complex stochastic process
which at present cannot be handled analytically.

Therefore, simulation of the mean delays seems to be the only way, see,
e.g., Suhl and Mellouli (1999) for a slightly different context. In our system,
however, the delays are a cost function which has to be evaluated over and
over again. Exact simulation of the complete network is too time-consuming
to be included into our system at present.

Instead, we are extending the analytical model of local delays on a section
to simple tree-like (sub)networks. We intend to derive a fast approximate
macroscopic simulation of the whole network using analytical representations
of its subnets. In this program we have achieved a major step by exploring
analogies between the accumulation of delays on a single section and the
operation of a queuing system which is explained in the next section.

3.2   Modeling Delays in Simple Nets

We start with a simple model of external disturbances and possible reactions
to it. We assume that along lines, perturbations occur randomly at places
$O_i, i = 1, 2, \ldots$ and cause a sudden stop of length $Z_i, i = 1, 2, \ldots$. As long
as the train is in time, it travels at the scheduled speed of say a km/h.
As soon as it is delayed, the driver turns to the maximal speed b at which
he drives until the train is in time again or arrives at the next station. We
neglect all braking or acceleration processes. See Figure 5 for a possible place-
time-diagram between stations S and S'. Here, perturbations occur at places
O_1, O_2, O_3 causing delays of random amounts Z_1, Z_2, Z_3. If delayed, the train
runs at increased speed b, indicated by bold lines. δ indicates the scheduled
arrival time at S', ε is the arrival delay.

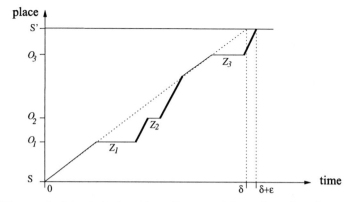

Figure 5. A typical place-time diagram with random disturbances.

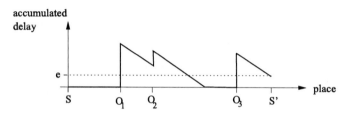

Figure 6. The delays accumulated on a trip as in Figure 5.

$c := 1/a - 1/b$ denotes the possible rate of delay reduction, i.e., $c \cdot s$ are the minutes of delay that can be made up for on a section of length s km. Hence the accumulation of delay along a section may look like Figure 6.

There is a strong analogy to queueing theory in which customers arrive at random points of time and require a random amount of service time from the server. Here the total load of service time D_t lying ahead of the server at time t (the so-called virtual waiting time) has the same profile as the accumulated delay. In fact the train can be viewed as a server that serves (reduces) all requests (perturbations) with a service rate corresponding to c. Therefore, results on virtual waiting time can be used to derive the distribution of the delay accumulated on a section.

We assume that the places $(O_i)_{i \geq 1}$ at which the perturbations occur form a Poisson process with rate λ, that the amounts $(Z_i)_{i \geq 1}$ of delay they cause are i.i.d., exponentially distributed with parameter μ and that both processes are independent. Then the accumulated delay $D_{S'}$ acquired along a section from S to S' approximately has the distribution:

$$P(D_{S'} \leq t) = \begin{cases} 1 - \frac{\lambda}{c\mu} & \text{if } t = 0 \\ \frac{\lambda}{c\mu}(1 - e^{t(\mu - \lambda/c)}) & \text{if } t > 0 \end{cases}$$

Note that $1 - \lambda/c\mu$ is the probability of arriving in time and $1 - e^{-t(\mu - \lambda/c)}$ is the conditional distribution of the delay given the train is delayed. A similar result holds under more general assumptions on the (O_i)-process, see Engelhardt-Funke and Kolonko (2000) for details.

Moreover, if the train leaves station S with a departure delay of random amount B_S then this delay can be reduced at rate c during the 'idle time' of the server which has approximate duration of $s(1 - \frac{\lambda}{c\mu})$ if s is the length of the section. Hence the arrival delay at the next station will be approximately

$$D_{S'} + [B_S - sc(1 - \frac{\lambda}{c\mu})]^+$$

Assuming that all perturbations are independent, this scheme can be iterated to give the delay distributions along a line. The analytically derived distributions have the same structure as those extracted from empirical delay data (see, e.g., Mühlhans (1990) and Herrmann (1996)). We can use this approach also to derive the propagated delays on connecting trains with delayed feeders at least in simple tree-like net structures, see also Weigand (1981).

Note that in more complex structures containing (undirected) circles, delays of consecutive trains may no longer be independent due to feedback effects. Another problem arises from the dependencies caused by the circulation of (delayed) vehicles. These problems are at present beyond our analytical model and remain to be simulated in a future version of our system.

4   The Reduced Internal Network

For an efficient solution of the multidimensional optimization problem, sketched in § 2.4, in particular when calculating the waiting times, we have to reduce the network to the data relevant for that calculation (see also Nachtigall and Voget (1997)).

We restrict ourselves to *transfer stations* $\hat{\mathfrak{S}}$ in which passengers can change between different lines and determine all *change-or-stop-relations* (L, L', S): 'change' for $L \neq L'$ and 'stop' for $L = L'$. We also aggregate the sections connecting two transfer stations into a *segment* $\hat{z}$. Note that there may be more than one segment between two transfer stations if there are different routes or if the trains have different speeds and different minimal stopping times on their way. The corresponding local cost functions $c_z(\cdot)$ are then added into one cost function $\hat{c}_{\hat{z}}(\cdot)$ for the segment. This operation requires some care as only favourable combinations of improvements for the aggregated sections should be used. $\hat{c}_{\hat{z}}(\delta)$ then gives the minimal costs to achieve a running time of δ on $\hat{z}$. The corresponding reduced timetables $\hat{T}$ only contain departure times $\hat{\pi}(L, S)$ for $S \in \hat{\mathfrak{S}}$ and running times $\hat{\delta}(\hat{z})$ for segments $\hat{z}$. Again, the resulting set of feasible timetables $\hat{\mathfrak{T}}$ has a very simple structure.

Note that the effort for reduction has to be spent only once at the beginning of the optimization. Internally, the reduced network is stored as an activity-on-arc-network, with the change-or-stop-relations as activity arcs and the segments $\hat{z}$ with their cost functions as vertices. After the optimization the reduced timetables and cost functions have to be 'inflated' again. In particular, the aggregated investments have to be decoded carefully to yield the costs and actions on the original sections.

5 A Solution with an Evolutionary Algorithm

In Nachtigall (1999) it is shown that the solution of the one-dimensional waiting time problem with fixed running times is a very complex periodic optimization problem, see also Nachtigall (1996) and Zimmermann and Lindner (2000). Here, we added more cost functions and increased the dimension of the solution space (by introducing the running times) so that the problem becomes far too complex for the methods of exact mathematical optimization. In particular, if we include cost functions like $M(T)$ that can only be simulated, exact methods are excluded.

In our prototype implementation we have successfully applied an evolutionary algorithm for the approximate solution of the multi-dimensional optimization problem. The algorithm is based on a population of timetables that are chosen randomly at the beginning, see Figure 7. The population is then enlarged by producing 'offspring' using genetic operators like crossover and mutation. This is indicated by the left arrow in Figure 7. For the crossover two timetables are chosen at random from the present population. The two lists of the timetables are then crossed either with standard operators like one-point or uniform crossover or by more sophisticated methods taking into account the regional structure of the timetable. The resulting timetable is mutated by randomly changing departure and/or running times. To make sure that the mutated timetable is still feasible the running times on each section z must be restricted to their respective ranges $[\underline{\delta}(z), \overline{\delta}(z)]$ (see § 2.1). The departure times are easy to handle as they are given as offsets to their line periods. Any result of a random mutation may, therefore, be interpreted as a departure time, possibly after modulo reduction. So the results of crossover and mutation are timetables from $\hat{\mathfrak{T}}$ again.

These crude stochastic operations are complemented by local search heuristics that may be used to improve the result of the crossover and mutation. Here, the running times on all segments of tracks are increased (within their limits) until the waiting time of the timetable becomes (locally) minimal. An alternative heuristic varies the departure times at all stations so that the

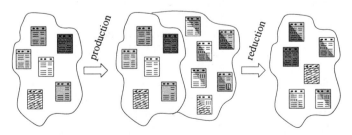

Figure 7. A production-reduction cycle of the evolutionary algorithm.

waiting time at this station becomes locally minimal. As for these heuristics all segments or all departures, respectively, have to examined, these improvements are very time-consuming compared to the genetic operators.

Invoking crossover, mutation and possibly local improvement repeatedly, a number of offspring timetables is produced enlarging the present population. This enlarged population will be reduced to its original size by selecting the 'fittest' timetables, see the step 'reduction' in Figure 7. There are different ways to take care of the multidimensional cost function during the reduction, see Ishibuchi and Murata (1996). Particularly successful is a selection procedure that adapts the type of reduction to be used to the present state of the population, see Kolonko and Voget (1998) for details on this. The reproduction-reduction cycle is repeated for a number of generations. Typically the population of timetables tends to improve quite fast (see Figure 8).

The evolution of the population can be visualized on the screen, see Figure 8 for a screenshot. The cost function values of the individuals (timetables) of each population form a cloud in the space spanned by the cost functions. Its lower envelope are the present Pareto-best solutions. They form an approximation to the Pareto-optimal set and the cost-benefit-curve. The visualization can also be used to examine the impact of the different parameter settings of the algorithm.

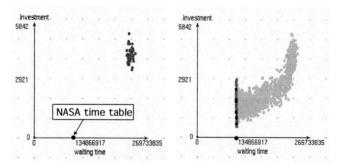

Figure 8. The evolution of the timetables.

6   Practical Results with the NASA Network

Our prototype implementation is presently applied to the network of the NASA GmbH in the state of Sachsen-Anhalt, Germany. This network consists of 467 stations and 295 lines (in the sense of our model, see § 2.1). The lines are formed of 551 different sections z and NASA decided that 190 of them could be modernized. The rest of 361 sections already have been renewed or are considered to be of less importance. After reducing the network as described in Section 4 there are 1010 different departure pairs (L, S) describing that line L departs from station S. We have 3910 change-or-stop-relations (L, L', S), so there are 3910 reasonable possibilities for passengers to change from line L to line L' in station S or – in case $L = L'$ – to stay in the train of line L during its stop at station S.

The two cost functions of interest to our partners from NASA are the necessary investment $C(T)$ and the passenger waiting time $W(T)$. Since not all necessary data were available we agreed with our partners to estimate the missing data to get a first analysis on a strategic planning level.

6.1   Local Cost Functions

On each of the 190 sections open for renovation there is only one type of general reconstruction possible that enables the train to run at an increased speed. Hence the local cost function (see § 2.2) attains only two values because no combinations of different improvements are available. As data we were given the present maximal speed and the maximal speed after a renovation of the section. We assumed that the actual average speed would increase with the same proportion. We then calculated the amount of running time that could be saved on each section from its length, its present running time and the possible increase in average speed.

Our system can cope with quite general cost functions for the improvements on the sections. However, the exact price for such an improvement is not yet known. So NASA suggested to use a linear cost estimator, meaning that reconstructing one km of the track costs one unit of money. Of course this is only a rough estimator but it was considered adequate for now by our partners. From the estimated data the investment costs $C(T)$ are calculated as described above. Note that in this situation $C(T)$ is proportional to the minimal length of tracks that have to be renewed to enable timetable T, with the restriction, that only complete sections can be reconstructed.

6.2   Waiting Time

There were no data on the number of passengers changing at stations and no $OD-$matrix available. Therefore, we used an estimation for the weights $g(L, L', S)$, used in the expression $W(T)$ for the waiting time (see § 2.3). This estimator is based on the so-called Lill's law (see Potthoff (1970)). There are five categories of cities. To each station S its category $\text{Cat}(S)$ is assigned that characterizes its size and economical importance. The average number $m(S, S')$ of passengers travelling from S to S' is then estimated by

$$m(S, S') = \frac{\text{Cat}(S) \cdot \text{Cat}(S')}{\text{distance}(S, S')^2} \cdot \text{const}$$

where $\text{distance}(S, S')$ is the length of a shortest path from S to S' in the network. Here, 'shortest' refers to the running times on the tracks from the present timetable (without any investment). The $m(S, S')$ are then collected into an OD-matrix M.

Kolonko and Engelhardt-Funke

To obtain the weights $g(L, L', S)$ for each change-or-stop-relation (L, L', S), it is counted how many passengers will pass through that relation while following their shortest connection from S to S' as described above. For simplification, we do not take into account a possible split of passenger streams if there are several shortest paths. Also, we neglect a possible feedback effect of timetables on the routes passengers choose. The weights calculated in this way present a kind of public pressure on the respective change-or-stop-relations. Timetables should respect them as they also reflect the goal to drive passengers through the network on shortest routes for economical and ecological reasons.

6.3 Optimization and Results

We included the actual NASA timetable (or rather a strictly periodic version of it) into our starting population. The cost function values of a typical starting population are shown in the screen-shots in Figure 8. In the left picture, the random solutions of the starting population form a cloud in the upper right corner, the present NASA timetable (with investment 0) is included. Naturally, it dominates all random starting solutions. During the optimization the cloud moves towards the lower left corner as the quality of the solutions increases (less costs and less waiting time). The NASA timetable is the only Pareto-optimal solution for a number of generations but it is typically reached by other solutions as shown in the right picture of Figure 8 after just a few seconds. Here, the dark grey dots represent the present population. Former solutions that have not 'survived' are drawn as light grey dots, whereas the black dots indicate the new Pareto-best solutions, which require some investment but have less waiting time than the NASA timetable.

Note that the waiting time on the x-axis has the unit 'passengers × minutes' and includes the penalty factor for waiting on the platform as mentioned in § 2.3. It attains very high values which have no direct practical meaning. They are only used as a relative measure of the quality of timetables when compared with the present one.

The diagram in Figure 9 shows the Pareto-best solutions found in several runs with different parameter settings including local improvement operators. The scale of the axis has been adjusted to give a better overview; for comparison with Figure 8, a screenshot has been added showing the same cost-benefit curve with the scaling as used in Figure 8. The NASA timetable has been included for illustration though it is dominated by other timetables. It is shown at the lower right corner of the diagram and has waiting time 94334724 units (and 0 investment). The best solution at 0 investment that our system has found (by adjusting the departure times) has waiting time 89362695, which is an improvement of 5.2 %. Figure 9 also shows that investments between 100 and 500 units are particularly efficient as the waiting time decreases drastically in this region. On the contrary, investments of more than 900 units seem not very efficient.

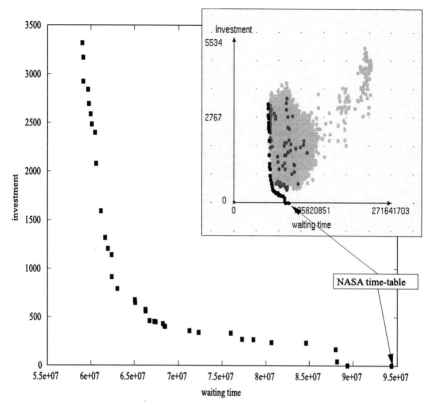

Figure 9. The lower envelope of the cloud approximates the cost-benefit curve.

One should note, however, that a direct comparison between the actual timetable of NASA and our Pareto-best results is difficult, as in contrast to the NASA timetable we do not take into account operational constraints as was mentioned above. Moreover, we had to 'rectify' the NASA timetable at some points to make it strictly periodic.

As the algorithm is highly stochastic, each run gives a different picture. The results of Figure 8 were obtained within 30 sec on a medium sized Pentium PC. With runs of about 5 minutes we are able to find timetables that dominate the present NASA timetable, i.e., they have less waiting time without any investment. The results in Figure 9 were obtained from several runs taking a total of about 5 h.

The result as shown in Figure 9 represents an approximation of the Pareto-optimal solutions. It can be examined in detail using our interactive cost-benefit-analyser. Here, the cost-benefit curve is displayed in a diagram similar to Figure 9. The map of the network is displayed in a separate window as in Figure 10. If a point (i.e., a timetable) in the diagram is activated by a mouse-click, the allocation of the investment it requires is displayed on the map by marking the corresponding sections red. In addition, there is a pop-

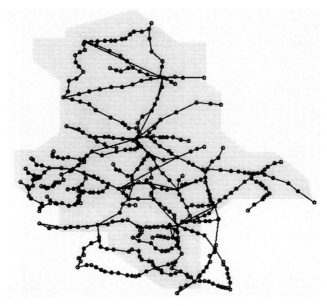

Figure 10. The interactive cost-benefit-analyzer explains the optimization results.

up list of the improvements selected for each section. In a similar fashion it can be shown how the benefit of the timetable, i.e., the reduction of waiting time spreads over the stations of the network. Of course, the timetable and the allocation of investments can be output as text.

7   Future Work

Our future work will focus on three points: First, we want to integrate the macroscopic simulation of delays into the optimization to be able to measure the robustness of a timetable under random delays. Secondly, the optimization will be improved incorporating additional local search heuristics like simulated annealing to complement the global search aspect of genetic algorithms. Using simulated annealing in a multidimensional cost environment is a particularly challenging task. Finally, we are investigating how capacity and safety constraints could be integrated into our model. As the present implementation works very fast with a network as that of NASA, we are quite optimistic about the chance to include these aspects into our system.

Bibliography

Engelhardt-Funke, O. and M. Kolonko (2000). Optimal time-tables: Modelling stochastic pertubations. In *Proceedings of the 7th International*

Workshop on Project Management and Scheduling (PMS 2000), Osnabrück, 125–127.

Goverde, R. (1998). Optimal transfertimes in railway time tables. In *Abstracts of the 6th Meeting of the EURO Working Group on Transportation*, Göteborg, Sweden, 1–5.

Herrmann, U. (1996). *Untersuchung zur Verspätungsentwicklung von Fernreisezügen auf der Datengrundlage der RZÜ Frankfurt am Main.* Dissertation, TH Darmstadt.

Ishibuchi, H. and T. Murata (1996). Multi-objective genetic local search algorithm. In *Proceedings of the IEEE International Conference on Evolutionary Computation*, IEEE, Piscataway, 119–124.

Kolonko, M. and S. Voget (1998). Multidimensional optimization using fuzzy genetic algorithms. *Journal of Heuristics 4*, 221–244.

Mühlhans, E. (1990). Berechnung der Verspätungsentwicklung bei Zugfahrten. *ETR – Eisenbahntechnische Rundschau 39*, 465–468.

Nachtigall, K. (1996). Periodic network optimization with different arc frequencies. *Discrete Applied Mathematics 69*, 1–17.

Nachtigall, K. (1999). *Periodic Network Optimization and Fixed Interval Timetables.* Habilitation thesis, University Hildesheim.

Nachtigall, K. and S. Voget (1997). Minimizing waiting times in integrated fixed interval timetables by upgrading railway tracks. *European Journal of Operational Research 103*, 610–627.

Orlin, J.B. (1982). Minimizing the number of vehicles to meet a fixed periodic schedule: An application of periodic posets. *Operations Research 30*, 760–776.

Potthoff, G. (1970). *Verkehrsströmungslehre 3: Die Verkehrsströme im Netz.* Transpress VEB Verlag für Verkehrswesen, Berlin.

Serafini, P. and W. Ukovich (1989). A mathematical model for periodic scheduling problems. *SIAM Journal on Discrete Mathematics 2*, 550–581.

Suhl, L. and T. Mellouli (1999). Requirements for, and design of, an operations control system for railways. In N.H.M. Wilson (Ed.), *Computer-Aided Transit Scheduling, Lecture Notes in Economics and Mathematical Systems*, 471, Springer, Berlin, 371–390.

Weigand, W. (1981). Verspätungsübertragung in Fernverkehrsnetzen. *ETR – Eisenbahntechnische Rundschau 30*, 915–919.

Zimmermann, U. and T. Lindner (2000). *Train Schedule Optimization in Public Transportation.* Technical report, TU Braunschweig. Available at http://www.math.tu-bs.de/mo/research/preprints.html.

Appendix 1: Referees

We greatly appreciate the helpful support and advice of the following individuals who have provided referee reports:

Andrzej Adamski, Kraków
Norbert Ascheuer, Berlin
Claus Biederbick, Paderborn
Christian Bierwirth, Bremen
Ralf Borndörfer, Berlin
Jürgen W. Böse, Braunschweig
Michael R. Bussieck, Washington D.C.
Alberto Caprara, Bologna
Gerold Carl, Saulheim
Avi Ceder, Haifa
Angel Corberán, Valencia
Joachim R. Daduna, Berlin
Thomas Emden-Weinert, Bad
    Homburg
Alexander Fay, Dossenheim
Andreas Fink, Braunschweig
Matteo Fischetti, Padova
Mark A. Fleischer, College Park
Richard Freling, Rotterdam
Luca M. Gambardella, Lugano
Michel Gendreau, Montréal
Stefan Gläser, Wolfsburg
Peter Greistorfer, Graz
Tore Grünert, Aachen
Nicolau D.F. Gualda, Sao Paulo
Kai Gutenschwager, Braunschweig
Knut Haase, Kiel
Horst W. Hamacher, Kaiserslautern
Mark Hickman, Tucson
Dennis Huisman, Rotterdam
Houyuan Jiang, Clayton
Gorazd Kandus, Ljubljana
Matthew G. Karlaftis, Athens
Robert Klein, Darmstadt
Leo Kroon, Rotterdam
Ann S.K. Kwan, Leeds

Raymond S.K. Kwan, Leeds
Klaus Ladner, Graz
Manuel Laguna, Boulder
Andreas Löbel, Berlin
Federico Malucelli, Milano
Rafael Marti, Valencia
Taïeb Mellouli, Paderborn
Pitu Mirchandani, Tucson
Gautam Mitra, Uxbridge
Karl Nachtigall, Dresden
Kai Nagel, Zurich
Maddalena Nonato, Perugia
Marta Oliveira, Lisboa
Uwe Pape, Berlin
Margaret E. Parker, Leeds
Rita Portugal, Lisboa
Warren Powell, Princeton
Edwin Romeijn, Gainesville
Jean-Marc Rousseau, Montréal
Gabriele Schneidereit, Braunschweig
Armin Scholl, Jena
Hans-Jürgen Sebastian, Aachen
Jürgen Siegmann, Berlin
Leena Suhl, Paderborn
Doris Tesch, Munich
Paolo Toth, Bologna
Marjan van den Akker, Amsterdam
Peter Värbrand, Linköping
Stefan Voß, Braunschweig
Manfred Wermuth, Braunschweig
Thomas Wiedemann, Berlin
Nigel H.M. Wilson, Cambridge
Anthony Wren, Leeds
Baichun Xiao, New York
Stephan Zelewski, Essen
Uwe T. Zimmermann, Braunschweig

Appendix 2: List of Presented Papers not Included in this Volume

Roberto Baldacci, Vittorio Maniezzo, and Aristide Mingozzi: *An Exact Algorithm for the Car Pooling Problem.*

Mohamadreza Banihashemi and Ali Haghani: *A Model for the Multiple Depot Transit Vehicle Scheduling Problem with Route Time Constraints.*

Uwe Becker, Imma Braun, Stefan Gläser, and Eckehard Schnieder: *Analysis and Modelling of a Fleet Management System.*

Bernhard Böhringer, Ralf Borndörfer, Martin Kammler, and Andreas Löbel: *Scheduling Duties by Adaptive Column Generation.*

Marco A. Boschetti and Aristide Mingozzi: *New Benchmark Results for the Multi Depot Crew Scheduling Problem.*

Angel Corberán, Elena Fernández, Manuel Laguna, and Rafael Martí: *Approximation Procedures for Multiobjective Scheduling of School Buses.*

Ken Darby-Dowman, James Little, Gautam Mitra, and Vincent Nwana: *Parallel Branch, Fix and Relax for Bus Crew Scheduling.*

Robert-Jan van Egmond: *An Algebraic Approach for Scheduling Train Movements.*

Thomas Emden-Weinert, Hans-Georg Kotas, and Ulf Speer: *DISSY – Driver Scheduling System for Public Transport.*

Andreas Ernst, Mohan Krishnamoorthy, Helen Nott, Bowie Owens, Daniel Prager, and David Sier: *SORT: Staff Optimisation Rostering Toolkit.*

Dietmar Fiehn and Paul Levi: *Connecting Distributed Routing Systems.*

Markus Friedrich and Helmut Prungel: *open.P: An open Database for Public Transport Operators and Agencies.*

Knut Haase, Guy Desaulniers, and Jacques Desrosiers: *Improved Results for the Simultaneous Vehicle and Crew Scheduling Problem in Urban Mass Transit Systems.*

Horst W. Hamacher and Anita Schöbel: *An Optimization Approach for the Delay Management Problem.*

Arnd-Dietrich zur Horst, Jürgen Böse, and Stefan Voß: *General Proceeding to Install Passenger Information Systems in the Supply Area of Public Mass Transit Companies.*

Ann S.K. Kwan, Raymond S.K. Kwan, Margaret E. Parker, and Anthony Wren: *Proving the Versatility of Automatic Driver Scheduling on Difficult Train & Bus Problems.*

Ann S.K. Kwan and Raymond S.K. Kwan: *An Improved Genetic Algorithm for Public Transport Driver Scheduling.*

Thomas Lindner and Uwe T. Zimmermann: *Cost-Oriented Train Scheduling.*

Helena R. Lourenço, José P. Paixão, and Rita Portugal: *Multiobjective Metaheuristics for the Bus-Driver Scheduling Problem.*

Marco E. Lübbecke and Uwe T. Zimmermann: *Scheduling of Switching Engines in Practice.*

Oli B.G. Madsen: *Vehicle Routing Problems with Time Windows – Some Recent Developments.*

Romualdas Mickus: *The Programme System "PIKAS'2000" to Calculate and Synchronize Timetables for all Kinds of Urban Public Transport as Means of Saving Fuel and Energy.*

Nuno A. Moreira and Rui C. Oliveira: *A Decision Support System for Operational Planning in Railway Networks.*

Martin Müller Elschner and Jürgen Roß: *Passenger Information Systems on the Internet – The European Dimension.*

Karl Nachtigall: *Air Traffic Flow Management in Europe.*

Somnuk Ngamchai and David J. Lovell: *Optimal Time Transfer in Bus Transit Route Design Using A Genetic Algorithm.*

Kurra V. Krishna Rao, S. Muralidhar, and S.L. Dhingra: *Public Transport Routing and Scheduling Using Genetic Algorithms.*

Michel Rizzi: *@llegr@© A New Transportation Service Design Software for Urban Transit Networks.*

Jean-Marc Rousseau: *Scheduling Regional Transportation with HASTUS.*

David M. Ryan and Jody N. Snowdon: *A Model and Solution Method for the Simultaneous Optimisation of Train and Driver Schedules.*

Gustavo P. Silva, Nicolau D.F. Gualda, and Raymond S.K. Kwan: *Bus Scheduling Based on an Arc Generation – Network Flow Approach.*

Alex Wardrop and Peter Pudney: *Development of Strategic Infrastructure and Train Operations Modelling.*

Robert Watson, Grahame B. Cooper, Simon N. Nicholls, Barry Grant, and Daren Wood: *The Role of Automated Timetable Generation in the Development of Improved Railway/Mass Transit Timetables and Infrastructure Configurations – A Simulated Annealing Approach Assessed from a Business Perspective.*

Appendix 3: Exhibitors and Sponsors

Berlin-Info, Berlin (Germany)

http://www.berlin-info.de

Berliner Kindl Brauerei, Berlin (Germany)

http://www.binding.de/gruppe/
beteiligungen/berliner-kindl.html

Berliner Volksbank, Berlin (Germany)

http://www.berliner-volksbank.de

Die Stern und Kreisschiffahrt, Berlin (Germany)

http://www.STERNundKREIS.de

Giro Inc., Montréal (Canada)

http://www.giro.ca/

IVU Traffic Technologies AG, Berlin (Germany)

http://www.ivu.de/

Konrad-Adenauer-Foundation, Berlin (Germany)

http://www.kas.de/

mdv Mentz Datenverarbeitung GmbH, München (Germany)

http://www.mentzdv.de/

MERAKAS Ltd., Vilnius (Lithuania)

http://www2.omnitel.net/merakas/

OPT, Lda., Porto (Portugal)

PTV AG, Karlsruhe (Germany)

http://www.ptv.de/

SYSTRA, Paris (France)

http://www.systra.com/

Trapeze Software Group, Inc., Mississauga (Canada)

http://www.trapezesoftware.com

VSS Gesellschaft für Beratung, Projektmanagement und Informationstechnologien mbH, Bremen (Germany)

http://www.vss.com/

Vol. 405: S. Komlósi, T. Rapcsák, S. Schaible (Eds.), Generalized Convexity. Proceedings, 1992. VIII, 404 pages. 1994.

Vol. 406: N. M. Hung, N. V. Quyen, Dynamic Timing Decisions Under Uncertainty. X, 194 pages. 1994.

Vol. 407: M. Ooms, Empirical Vector Autoregressive Modeling. XIII, 380 pages. 1994.

Vol. 408: K. Haase, Lotsizing and Scheduling for Production Planning. VIII, 118 pages. 1994.

Vol. 409: A. Sprecher, Resource-Constrained Project Scheduling. XII, 142 pages. 1994.

Vol. 410: R. Winkelmann, Count Data Models. XI, 213 pages. 1994.

Vol. 411: S. Dauzère-Péres, J.-B. Lasserre, An Integrated Approach in Production Planning and Scheduling. XVI, 137 pages. 1994.

Vol. 412: B. Kuon, Two-Person Bargaining Experiments with Incomplete Information. IX, 293 pages. 1994.

Vol. 413: R. Fiorito (Ed.), Inventory, Business Cycles and Monetary Transmission. VI, 287 pages. 1994.

Vol. 414: Y. Crama, A. Oerlemans, F. Spieksma, Production Planning in Automated Manufacturing. X, 210 pages. 1994.

Vol. 415: P. C. Nicola, Imperfect General Equilibrium. XI, 167 pages. 1994.

Vol. 416: H. S. J. Cesar, Control and Game Models of the Greenhouse Effect. XI, 225 pages. 1994.

Vol. 417: B. Ran, D. E. Boyce, Dynamic Urban Transportation Network Models. XV, 391 pages. 1994.

Vol. 418: P. Bogetoft, Non-Cooperative Planning Theory. XI, 309 pages. 1994.

Vol. 419: T. Maruyama, W. Takahashi (Eds.), Nonlinear and Convex Analysis in Economic Theory. VIII, 306 pages. 1995.

Vol. 420: M. Peeters, Time-To-Build. Interrelated Investment and Labour Demand Modelling. With Applications to Six OECD Countries. IX, 204 pages. 1995.

Vol. 421: C. Dang, Triangulations and Simplicial Methods. IX, 196 pages. 1995.

Vol. 422: D. S. Bridges, G. B. Mehta, Representations of Preference Orderings. X, 165 pages. 1995.

Vol. 423: K. Marti, P. Kall (Eds.), Stochastic Programming. Numerical Techniques and Engineering Applications. VIII, 351 pages. 1995.

Vol. 424: G. A. Heuer, U. Leopold-Wildburger, Silverman's Game. X, 283 pages. 1995.

Vol. 425: J. Kohlas, P.-A. Monney, A Mathematical Theory of Hints. XIII, 419 pages, 1995.

Vol. 426: B. Finkenstädt, Nonlinear Dynamics in Economics. IX, 156 pages. 1995.

Vol. 427: F. W. van Tongeren, Microsimulation Modelling of the Corporate Firm. XVII, 275 pages. 1995.

Vol. 428: A. A. Powell, Ch. W. Murphy, Inside a Modern Macroeconometric Model. XVIII, 424 pages. 1995.

Vol. 429: R. Durier, C. Michelot, Recent Developments in Optimization. VIII, 356 pages. 1995.

Vol. 430: J. R. Daduna, I. Branco, J. M. Pinto Paixão (Eds.), Computer-Aided Transit Scheduling. XIV, 374 pages. 1995.

Vol. 431: A. Aulin, Causal and Stochastic Elements in Business Cycles. XI, 116 pages. 1996.

Vol. 432: M. Tamiz (Ed.), Multi-Objective Programming and Goal Programming. VI, 359 pages. 1996.

Vol. 433: J. Menon, Exchange Rates and Prices. XIV, 313 pages. 1996.

Vol. 434: M. W. J. Blok, Dynamic Models of the Firm. VII, 193 pages. 1996.

Vol. 435: L. Chen, Interest Rate Dynamics, Derivatives Pricing, and Risk Management. XII, 149 pages. 1996.

Vol. 436: M. Klemisch-Ahlert, Bargaining in Economic and Ethical Environments. IX, 155 pages. 1996.

Vol. 437: C. Jordan, Batching and Scheduling. IX, 178 pages. 1996.

Vol. 438: A. Villar, General Equilibrium with Increasing Returns. XIII, 164 pages. 1996.

Vol. 439: M. Zenner, Learning to Become Rational. VII, 201 pages. 1996.

Vol. 440: W. Ryll, Litigation and Settlement in a Game with Incomplete Information. VIII, 174 pages. 1996.

Vol. 441: H. Dawid, Adaptive Learning by Genetic Algorithms. IX, 166 pages.1996.

Vol. 442: L. Corchón, Theories of Imperfectly Competitive Markets. XIII, 163 pages. 1996.

Vol. 443: G. Lang, On Overlapping Generations Models with Productive Capital. X, 98 pages. 1996.

Vol. 444: S. Jørgensen, G. Zaccour (Eds.), Dynamic Competitive Analysis in Marketing. X, 285 pages. 1996.

Vol. 445: A. H. Christer, S. Osaki, L. C. Thomas (Eds.), Stochastic Modelling in Innovative Manufacturing. X, 361 pages. 1997.

Vol. 446: G. Dhaene, Encompassing. X, 160 pages. 1997.

Vol. 447: A. Artale, Rings in Auctions. X, 172 pages. 1997.

Vol. 448: G. Fandel, T. Gal (Eds.), Multiple Criteria Decision Making. XII, 678 pages. 1997.

Vol. 449: F. Fang, M. Sanglier (Eds.), Complexity and Self-Organization in Social and Economic Systems. IX, 317 pages, 1997.

Vol. 450: P. M. Pardalos, D. W. Hearn, W. W. Hager, (Eds.), Network Optimization. VIII, 485 pages, 1997.

Vol. 451: M. Salge, Rational Bubbles. Theoretical Basis, Economic Relevance, and Empirical Evidence with a Special Emphasis on the German Stock Market.IX, 265 pages. 1997.

Vol. 452: P. Gritzmann, R. Horst, E. Sachs, R. Tichatschke (Eds.), Recent Advances in Optimization. VIII, 379 pages. 1997.

Vol. 453: A. S. Tangian, J. Gruber (Eds.), Constructing Scalar-Valued Objective Functions. VIII, 298 pages. 1997.

Vol. 454: H.-M. Krolzig, Markov-Switching Vector Autoregressions. XIV, 358 pages. 1997.

Vol. 455: R. Caballero, F. Ruiz, R. E. Steuer (Eds.), Advances in Multiple Objective and Goal Programming. VIII, 391 pages. 1997.

Vol. 456: R. Conte, R. Hegselmann, P. Terna (Eds.), Simulating Social Phenomena. VIII, 536 pages. 1997.

Vol. 457: C. Hsu, Volume and the Nonlinear Dynamics of Stock Returns. VIII, 133 pages. 1998.

Vol. 458: K. Marti, P. Kall (Eds.), Stochastic Programming Methods and Technical Applications. X, 437 pages. 1998.

Vol. 459: H. K. Ryu, D. J. Slottje, Measuring Trends in U.S. Income Inequality. XI, 195 pages. 1998.

Vol. 460: B. Fleischmann, J. A. E. E. van Nunen, M. G. Speranza, P. Stähly, Advances in Distribution Logistic. XI, 535 pages. 1998.

Vol. 461: U. Schmidt, Axiomatic Utility Theory under Risk. XV, 201 pages. 1998.

Vol. 462: L. von Auer, Dynamic Preferences, Choice Mechanisms, and Welfare. XII, 226 pages. 1998.

Vol. 463: G. Abraham-Frois (Ed.), Non-Linear Dynamics and Endogenous Cycles. VI, 204 pages. 1998.

Vol. 464: A. Aulin, The Impact of Science on Economic Growth and its Cycles. IX, 204 pages. 1998.

Vol. 465: T. J. Stewart, R. C. van den Honert (Eds.), Trends in Multicriteria Decision Making. X, 448 pages. 1998.

Vol. 466: A. Sadrieh, The Alternating Double Auction Market. VII, 350 pages. 1998.

Vol. 467: H. Hennig-Schmidt, Bargaining in a Video Experiment. Determinants of Boundedly Rational Behavior. XII, 221 pages. 1999.

Vol. 468: A. Ziegler, A Game Theory Analysis of Options. XIV, 145 pages. 1999.

Vol. 469: M. P. Vogel, Environmental Kuznets Curves. XIII, 197 pages. 1999.

Vol. 470: M. Ammann, Pricing Derivative Credit Risk. XII, 228 pages. 1999.

Vol. 471: N. H. M. Wilson (Ed.), Computer-Aided Transit Scheduling. XI, 444 pages. 1999.

Vol. 472: J.-R. Tyran, Money Illusion and Strategic Complementarity as Causes of Monetary Non-Neutrality. X, 228 pages. 1999.

Vol. 473: S. Helber, Performance Analysis of Flow Lines with Non-Linear Flow of Material. IX, 280 pages. 1999.

Vol. 474: U. Schwalbe, The Core of Economies with Asymmetric Information. IX, 141 pages. 1999.

Vol. 475: L. Kaas, Dynamic Macroeconomics with Imperfect Competition. XI, 155 pages. 1999.

Vol. 476: R. Demel, Fiscal Policy, Public Debt and the Term Structure of Interest Rates. X, 279 pages. 1999.

Vol. 477: M. Théra, R. Tichatschke (Eds.), Ill-posed Variational Problems and Regularization Techniques. VIII, 274 pages. 1999.

Vol. 478: S. Hartmann, Project Scheduling under Limited Resources. XII, 221 pages. 1999.

Vol. 479: L. v. Thadden, Money, Inflation, and Capital Formation. IX, 192 pages. 1999.

Vol. 480: M. Grazia Speranza, P. Stähly (Eds.), New Trends in Distribution Logistics. X, 336 pages. 1999.

Vol. 481: V. H. Nguyen, J. J. Strodiot, P. Tossings (Eds.). Optimation. IX, 498 pages. 2000.

Vol. 482: W. B. Zhang, A Theory of International Trade. XI, 192 pages. 2000.

Vol. 483: M. Königstein, Equity, Efficiency and Evolutionary Stability in Bargaining Games with Joint Production. XII, 197 pages. 2000.

Vol. 484: D. D. Gatti, M. Gallegati, A. Kirman, Interaction and Market Structure. VI, 298 pages. 2000.

Vol. 485: A. Garnaev, Search Games and Other Applications of Game Theory. VIII, 145 pages. 2000.

Vol. 486: M. Neugart, Nonlinear Labor Market Dynamics. X, 175 pages. 2000.

Vol. 487: Y. Y. Haimes, R. E. Steuer (Eds.), Research and Practice in Multiple Criteria Decision Making. XVII, 553 pages. 2000.

Vol. 488: B. Schmolck, Ommitted Variable Tests and Dynamic Specification. X, 144 pages. 2000.

Vol. 489: T. Steger, Transitional Dynamics and Economic Growth in Developing Countries. VIII, 151 pages. 2000.

Vol. 490: S. Minner, Strategic Safety Stocks in Supply Chains. XI, 214 pages. 2000.

Vol. 491: M. Ehrgott, Multicriteria Optimization. VIII, 242 pages. 2000.

Vol. 492: T. Phan Huy, Constraint Propagation in Flexible Manufacturing. IX, 258 pages. 2000.

Vol. 493: J. Zhu, Modular Pricing of Options. X, 170 pages. 2000.

Vol. 494: D. Franzen, Design of Master Agreements for OTC Derivatives. VIII, 175 pages. 2001.

Vol. 495: I Konnov, Combined Relaxation Methods for Variational Inequalities. XI, 181 pages. 2001.

Vol. 496: P. Weiß, Unemployment in Open Economies. XII, 226 pages. 2001.

Vol. 497: J. Inkmann, Conditional Moment Estimation of Nonlinear Equation Systems. VIII, 214 pages. 2001.

Vol. 498: M. Reutter, A Macroeconomic Model of West German Unemployment. X, 125 pages. 2001.

Vol. 499: A. Casajus, Focal Points in Framed Games. XI, 131 pages. 2001.

Vol. 500: F. Nardini, Technical Progress and Economic Growth. XVII, 191 pages. 2001.

Vol. 501: M. Fleischmann, Quantitative Models for Reverse Logistics. XI, 181 pages. 2001.

Vol. 502: N. Hadjisavvas, J. E. Martínez-Legaz, J.-P. Penot (Eds.), Generalized Convexity and Generalized Monotonicity. IX, 410 pages. 2001.

Vol. 503: A. Kirman, J.-B. Zimmermann (Eds.), Economics with Heterogenous Interacting Agents. VII, 343 pages. 2001.

Vol. 504: P.-Y. Moix (Ed.),The Measurement of Market Risk. XI, 272 pages. 2001.

Vol. 505: S. Voß, J. R. Daduna (Eds.), Computer-Aided Scheduling of Public Transport. XI, 466 pages. 2001.